Get the eBook FREE!

(PDF, ePub, Kindle, and liveBook all included)

We believe that once you buy a book from us, you should be able to read it in any format we have available. To get electronic versions of this book at no additional cost to you, purchase and then register this book at the Manning website.

Go to https://www.manning.com/freebook and follow the instructions to complete your pBook registration.

That's it!
Thanks from Manning!

Kafka Streams in Action

SECOND EDITION
EVENT-DRIVEN APPLICATIONS AND MICROSERVICES

BILL BEJECK
FOREWORD BY JUN RAO

MANNING
SHELTER ISLAND

For online information and ordering of this and other Manning books, please visit www.manning.com. The publisher offers discounts on this book when ordered in quantity. For more information, please contact

Special Sales Department
Manning Publications Co.
20 Baldwin Road
PO Box 761
Shelter Island, NY 11964
Email: orders@manning.com

♾ Recognizing the importance of preserving what has been written, it is Manning's policy to have the books we publish printed on acid-free paper, and we exert our best efforts to that end. Recognizing also our responsibility to conserve the resources of our planet, Manning books are printed on paper that is at least 15 percent recycled and processed without the use of elemental chlorine.

The authors and publisher have made every effort to ensure that the information in this book was correct at press time. The authors and publisher do not assume and hereby disclaim any liability to any party for any loss, damage, or disruption caused by errors or omissions, whether such errors or omissions result from negligence, accident, or any other cause, or from any usage of the information herein.

Manning Publications Co.		
20 Baldwin Road	Development editor:	Frances Lefkowitz
PO Box 761	Technical development editor:	John Guthrie
Shelter Island, NY 11964	Review editors:	Aleksandar Dragosavljević and Dunja Nikitović
	Production editor:	Keri Hales
	Copy editor:	Alisa Larson
	Proofreader:	Jason Everett
	Technical proofreader:	Karsten Strøbæk
	Typesetter:	Dennis Dalinnik
	Cover designer:	Marija Tudor

ISBN: 9781617298684
Printed in the United States of America

brief contents

contents

        10.4   Integrating the Processor API and the Kafka
                Streams API   319

11   ksqlDB   321
        11.1   Understanding ksqlDB   322
        11.2   More about streaming queries   325
        11.3   Persistent vs. push vs. pull queries   333
        11.4   Creating Streams and Tables   338
        11.5   Schema Registry integration   341
        11.6   ksqlDB advanced features   345

12   Spring Kafka   352
        12.1   Introducing Spring   352
        12.2   Using Spring to build Kafka-enabled applications   355
                Spring Kafka application components   358 ▪ Enhanced
                application requirements   362
        12.3   Spring Kafka Streams   367

13   Kafka Streams Interactive Queries   374
        13.1   Kafka Streams and information sharing   375
        13.2   Learning about Interactive Queries   376
                Building an Interactive Queries app with Spring Boot   379

14   Testing   389
        14.1   Understanding the difference between unit and
                integration testing   390
                Testing Kafka producers and consumers   391 ▪ Creating tests
                for Kafka Streams operators   395 ▪ Writing tests for a Kafka
                Streams topology   398 ▪ Testing more complex Kafka Streams
                applications   401 ▪ Developing effective integration tests   405

        appendix A   Schema compatibility workshop   412
        appendix B   Confluent resources   422
        appendix C   Working with Avro, Protobuf, and JSON Schema   424
        appendix D   Understanding Kafka Streams architecture   446

                      index   463

foreword

When a business event occurs, traditional data-at-rest systems record it but leave the use of the data to a much later time. In contrast, Apache Kafka is a data streaming platform designed to react to business events in real time. Over the past decade, Apache Kafka has become the standard for data streaming. Hundreds of thousands of organizations, including most of the largest enterprises in the world, are using Kafka to take action on what's happening to their business. Those actions allow them to enhance customer experience, gain new business insights, improve efficiency, reduce risks, and so on, all within a few short seconds.

Applications built on top of Kafka are event driven. They typically take one or more data streams as the input and continuously transform these data into a new stream. The transformation often includes streaming operations such as filtering, projection, joining, and aggregation. Expressing those operations in low-level Java code is inefficient and error prone. Kafka Streams provides a handful of high-level abstractions for developers to express those common streaming operations concisely and is a very powerful tool for building event-driven applications in Java.

Bill has been a long-time contributor to Kafka and is an expert in Kafka Streams. What's unique about Bill is that not only does he understand the technology behind Kafka Streams, but he also knows how to use Kafka Streams to solve real problems. In this book, you will hear the key concepts in Kafka Streams directly from Bill. You will also see many hands-on examples of how to build end-to-end event-driven applications using Kafka Streams, together with other Kafka APIs and Schema Registry. If you are a developer wanting to learn how to build the next-gen event-driven applications on Kafka, you'll find this book invaluable. Enjoy the book and the power of data streaming!

—Jun Rao, co-founder, Confluent and Apache Kafka

preface

After completing the first edition of *Kafka Streams in Action*, I thought I had accomplished everything I set out to do. But as time went on, my understanding of the Kafka ecosystem and my appreciation for Kafka Streams grew. I saw that Kafka Streams was more powerful than I had initially thought. Additionally, I noticed other important pieces in building event-streaming applications; Kafka Streams is still a key player but not the only requirement. I realized that Apache Kafka could be considered the central nervous system for an organization's data. If Kafka is the central nervous system, then Kafka Streams is a vital organ performing some necessary operations.

But Kafka Streams relies on other components to bring events into Kafka or export them to the outside world where its results and calculations can be put to good use. I'm talking about the producer and consumer clients and Kafka Connect. As I put the pieces together, I realized you need these other components to complete the event-streaming picture. Couple all this with some significant improvements to Kafka Streams since 2018, and I knew I wanted to write a second edition.

But I didn't just want to add cosmetic touches to the previous edition; I wanted to express my improved understanding and add complete coverage of the entire Kafka ecosystem. This meant expanding the scope of some subjects from sections of chapters to whole chapters (like the producer and consumer clients), or adding entirely new chapters (such as the new chapters on Connect and Schema Registry). For the existing Kafka Streams chapters, writing a second edition meant updating and improving the existing material to clarify and communicate my deeper understanding.

   Taking on the second edition with this new focus during the Covid-19 pandemic wasn't easy, and not without some serious personal challenges along the way. But in the end, it was worth every minute of revision, and if I went back in time, I'd make the same decision.

   I hope that new readers of *Kafka Streams in Action* will find the book an essential resource and that readers from the first edition will enjoy and apply the improvements as well.

acknowledgments

First, I want to thank my wife, Beth, for supporting my signing up for a second edition. Writing the first edition of a book is very time-consuming, so you'd think the second edition would be more straightforward, just making adjustments for things like API changes. But in this case, I wanted to expand on my previous work and decided to do an entire rewrite. Beth never questioned my decision and fully supported my new direction, and as before, I couldn't have completed this without her support. Beth, you are fantastic, and I'm very grateful to have you as my wife. I'd also like to thank my three children for having great attitudes and supporting me in doing a second edition.

Next, I thank my editor at Manning, Frances Lefkowitz, whose continued expert guidance and patience made the writing process fun this time. I also thank John Guthrie for his excellent, precise technical feedback and Karsten Strøbæk, the technical proofreader, for superb work reviewing the code. Many hands at Manning contributed to the production of this edition; thanks to all of them.

I'd also like to thank the Kafka Streams developers and community for being so engaging and brilliant in making Kafka Streams the best stream processing library available. I want to acknowledge all the Kafka developers for building such high-quality software, especially Jay Kreps, Neha Narkhede, and Jun Rao, not only for starting Kafka in the first place but also for creating such a great place to work in Confluent. In addition, another special thanks to Jun for writing the foreword of this edition.

Last but certainly not least, I thank the reviewers for their hard work and invaluable feedback in making the quality of this book better for all readers: Alain Couniot, Allen Gooch, Andres Sacco, Balakrishnan Balasubramanian, Christian Thoudahl,

Daniela Zapata, David Ong, Ezra Simeloff, Giampiero Granatella, John Roesler, Jose San Leandro Armendáriz, Joseph Pachod, Kent Spillner, Manzur Mukhitdinov, Michael Heil, Milorad Imbra, Miloš Milivojević, Najeeb Arif, Nathan B. Crocker, Robin Coe, Rui Liu, Sambasiva Andaluri, Scott Huang, Simon Hewitt, Simon Tschöke, Stanford S. Guillory, Tan Wee, Thomas Peklak, and Zorodzayi Mukuya.

about this book

I wrote the second edition of *Kafka Streams in Action* to teach you how to build event streaming applications in Kafka Streams and include other components of the Kafka ecosystem, Producer and Consumer clients, Connect, and Schema Registry. I took this approach because for your event-streaming application to be as effective as possible, you'll need not just Kafka Streams but other essential tools. My approach to writing this book is a pair-programming perspective; I imagine myself sitting next to you as you write the code and learn the API. You'll learn about the Kafka broker and how the producer and consumer clients work. Then, you'll see how to manage schemas, their role with Schema Registry, and how Kafka Connect bridges external components and Kafka. From there, you'll dive into Kafka Streams, first building a simple application, then adding more complexity as you dig deeper into Kafka Streams API. You'll also learn about ksqlDB, testing, and, finally, integrating Kafka with the popular Spring framework.

Who should read this book

Kafka Streams in Action is for any developer wishing to get into stream processing. While not strictly required, knowledge of distributed programming will help understand Kafka and Kafka Streams. Knowledge of Kafka is beneficial but not required; I'll teach you what you need to know. Experienced Kafka developers and those new to Kafka will learn how to develop compelling stream-processing applications with Kafka Streams. Intermediate-to-advanced Java developers familiar with topics like serialization will learn how to use their skills to build a Kafka Streams application. The book's source code is written in Java 17 and extensively uses Java lambda syntax, so experience with lambdas (even from another language) will be helpful.

How this book is organized: A road map

This book has three parts spread over 14 chapters. While the book's title is "Kafka Streams in Action," it covers the entire Kafka event-streaming platform. As a result, the first five chapters cover the different components: Kafka brokers, consumer and producer clients, Schema Registry, and Kafka Connect. This approach makes sense, especially considering that Kafka Streams is an abstraction over the consumer and producer clients. So, if you're already familiar with Kafka, Connect, and Schema Registry or if you're excited to get going with Kafka Streams, then by all means, skip directly to Part 3.

Part 1 introduces event streaming and describes the different parts of the Kafka ecosystem to show you the big-picture view of how it all works and fits together. These chapters also provide the basics of the Kafka broker for those who need them or want a review:

- Chapter 1 provides some context on what an event and event-streaming are and why they are vital for working with real-time data. It also presents the mental model of the different components we'll cover: the broker, clients, Kafka Connect, Schema Registry, and, of course, Kafka Streams. I don't go over any code but describe how they all work.
- Chapter 2 is a primer for developers who are new to Kafka, and it covers the role of the broker, topics, partitions, and some monitoring. Those with more experience with Kafka can skip this chapter.

Part 2 moves on and covers getting data into and out of Kafka and managing schemas:

- Chapter 3 covers using Schema Registry to help you manage the evolution of your data's schemas. Spoiler alert: you're always using a schema—if not explicitly, it's implicitly there.
- Chapter 4 discusses the Kafka producer and consumer clients. The clients are how you get data into and out of Kafka and provide the building blocks for Kafka Connect and Kafka Streams.
- Chapter 5 is about Kafka Connect. Kafka Connect provides the ability to get data into Kafka via source connectors and export it to external systems with sink connectors.

Part 3 gets to the book's heart and covers developing Kafka Streams applications. In this section, you'll also learn about ksqlDB and testing your event-streaming application, and it concludes with integrating Kafka with the Spring Framework:

- Chapter 6 is your introduction to Kafka Streams, where you'll build a Hello World application and, from there, build a more realistic application for a fictional retailer. Along the way, you'll learn about the Kafka Streams DSL.

- Chapter 7 continues your Kafka Streams learning path, where we discuss application state and why it's required for streaming applications. In this chapter, some of the things you'll learn about are aggregating data and joins.
- Chapter 8 teaches you about the KTable API. Whereas a KStream is a stream of events, a KTable is a stream of related events or an update stream.
- Chapter 9 covers windowed operations and timestamps. Windowing an aggregation allows you to bucket results by time, and the timestamps on the records drive the action.
- Chapter 10 dives into the Kafka Streams Processor API. Up to this point, you've been working with the high-level DSL, but here, you'll learn how to use the Processor API when you need more control.
- Chapter 11 takes you further into the development stack, where you'll learn about ksqlDB. ksqlDB allows you to write event-streaming applications without any code but using SQL.
- Chapter 12 discusses using the Spring Framework with Kafka clients and Kafka Streams. Spring allows you to write more modular and testable code by providing a dependency injection framework for wiring up your applications.
- Chapter 13 introduces you to Kafka Streams Interactive Queries or IQ. IQ is the ability to directly query the state store of a state operation in Kafka Streams. You'll use what you learned in Chapter 12 to build a Spring-enabled IQ web application.
- Chapter 14 covers the all-important topic of testing. You'll learn how to test client applications with a Kafka Streams topology, the difference between unit testing and integration testing, and when to apply them.

Finally, there are four appendices for further explanations:

- Appendix A contains a workshop on Schema Registry to get hands-on experience with the different schema compatibility modes.
- Appendix B presents information on using Confluent Cloud to help develop your event streaming applications.
- Appendix C is a survey of working with the different schema types Avro, Protobuf, and JSON Schema.
- Appendix D covers the architecture and internals of Kafka Streams.

About the code

This book contains many examples of source code both in numbered listings and inline with normal text. In both cases, source code is formatted in a `fixed-width font like this` to separate it from ordinary text.

In many cases, the original source code has been reformatted; we've added line breaks and reworked indentation to accommodate the available page space in the book. In rare cases, even this was not enough, and listings include line-continuation-markers (➡).

Additionally, comments in the source code have often been removed from the listings when the code is described in the text. Code annotations accompany many of the listings, highlighting important concepts.

Finally, it's important to note that many of the code examples aren't meant to stand on their own: they're excerpts containing only the most relevant parts of what's currently under discussion. You'll find all the examples from the book in the accompanying source code in their complete form.

Source code for the book's examples is available from GitHub at https://github.com/bbejeck/KafkaStreamsInAction2ndEdition and the publisher's website at www.manning.com/books/kafka-streams-in-action-second-edition. The source code for the book is an all-encompassing project using the build tool Gradle (https://gradle.org). You can import the project into either IntelliJ or Eclipse using the appropriate commands. Full instructions for using and navigating the source code can be found in the accompanying README.md file.

liveBook discussion forum

Purchase of *Kafka Streams in Action, Second Edition,* includes free access to liveBook, Manning's online reading platform. Using liveBook's exclusive discussion features, you can attach comments to the book globally or to specific sections or paragraphs. It's a snap to make notes for yourself, ask and answer technical questions, and receive help from the author and other users. To access the forum, go to https://livebook.manning.com/book/kafka-streams-in-action-second-edition/discussion. You can also learn more about Manning's forums and the rules of conduct at https://livebook.manning.com/discussion.

Manning's commitment to our readers is to provide a venue where a meaningful dialogue between individual readers and between readers and the author can take place. It is not a commitment to any specific amount of participation on the part of the author, whose contribution to the forum remains voluntary (and unpaid). We suggest you try asking the author some challenging questions lest his interest stray! The forum and the archives of previous discussions will be accessible from the publisher's website as long as the book is in print.

Other online resources

- Apache Kafka documentation: https://kafka.apache.org
- Confluent documentation: https://docs.confluent.io/current
- Kafka Streams documentation: https://kafka.apache.org/documentation/streams/
- ksqlDB documentation: https://ksqldb.io/
- Spring Framework: https://spring.io/

about the author

BILL BEJECK, Apache Kafka® committer and PMC member, has been a software engineer for over 20 years. Currently, he is working at Confluent as a Staff DevX Engineer. Previously, Bill was an engineer on the Kafka Streams team for three-plus years. Before Confluent, he worked on various ingest applications as a U.S. Government contractor using distributed software such as Apache Kafka, Spark, and Hadoop. Bill maintains a blog, "Random Thoughts on Coding" (http://codingjunkie.net).

about the cover illustration

The figure on the cover of *Kafka Streams in Action, Second Edition*, "Habit of a Turkish Gentleman in 1700," is taken from a book by Thomas Jefferys, published between 1757 and 1772.

In those days, it was easy to identify where people lived and what their trade or station in life was just by their dress. Manning celebrates the inventiveness and initiative of the computer business with book covers based on the rich diversity of regional culture centuries ago, brought back to life by pictures from collections such as this one.

Part 1

In part 1, you'll learn about events and event streaming in general. Event streaming is a software development approach that considers events as an application's primary input and output. But to develop an effective event streaming application, you'll first need to learn what an event is (spoiler alert: it's everything!). Then you'll read about what use cases are good candidates for event-streaming applications and which are not.

First, you'll discover what a Kafka broker is, how it's at the heart of the Kafka ecosystem, and the various jobs it performs. Then you'll learn what Schema Registry, producer and consumer clients, Connect, and Kafka Streams are and their different roles. You'll also learn about the Apache Kafka event streaming platform. Although this book focuses on Kafka Streams, it's part of a larger whole that allows you to develop event-streaming applications. If this first part leaves you with more questions than answers, don't fret; I'll explain them all in subsequent chapters.

Welcome to the Kafka
event streaming platform

This chapter covers

- Defining event streaming and events
- Introducing the Kafka event streaming platform
- Applying the platform to a concrete example

While the constant influx of data creates more entertainment and opportunities for the consumer, increasingly, the users of this information are software systems using other software systems. Think, for example, of the fundamental interaction of watching a movie from your favorite movie streaming application. You log into the application, search for and select a film, and then watch it, and afterward, you may provide a rating or some indication of how you enjoyed the movie. Just this simple interaction generates several events captured by the movie streaming service. But this information needs analysis if it's to be of use to the business. That's where all the other software comes into play.

1.1 Event streaming

Software systems consume and store all the information obtained from your interaction and the interactions of other subscribers. Then, additional software systems use that information to make recommendations to you and to provide the streaming service with insight on what programming to provide in the future. Now, consider that this process occurs hundreds of thousands or even millions of times per day, and you can see the massive amount of information that businesses need to harness and that their software needs to make sense of to meet customer demands and expectations and stay competitive.

Another way to think of this process is that everything modern-day consumers do, from streaming a movie online to purchasing a pair of shoes at a brick-and-mortar store, generates an *event*. For an organization to survive and excel in our digital economy, it must have an efficient way of capturing and acting on these events. In other words, businesses must find ways to keep up with the demand of this endless flow of events if they want to satisfy customers and maintain a robust bottom line. Developers call this constant flow an *event stream*. And, increasingly, they are meeting the demands of this endless digital activity with an *event-streaming platform*, which utilizes a series of event-streaming applications.

An event-streaming platform is analogous to our central nervous system, which processes millions of events (nerve signals) and, in response, sends out messages to the appropriate parts of the body. Our conscious thoughts and actions generate some of these responses. When we are hungry and open the refrigerator, the central nervous system gets the message and sends out another one, telling the arm to reach for a nice red apple on the first shelf. Other actions, such as your heart rate increasing in anticipation of exciting news, are handled unconsciously.

An event-streaming platform captures events generated from mobile devices, customer interaction with websites, online activity, shipment tracking, and other business transactions. But the platform, like the nervous system, does more than capture events. It also needs a mechanism to reliably transfer and store the information from those events in the order in which they occurred. Then, other applications can process or analyze the events to extract different bits of that information.

Processing the event stream in real time is essential for making time-sensitive decisions. For example, does this purchase from customer X seem suspicious? Are the signals from this temperature sensor indicating something has gone wrong in a manufacturing process? Has the routing information been sent to the appropriate department of a business?

But the value of an event-streaming platform goes beyond gaining immediate information. Providing durable storage allows us to go back and look at event-stream data in its raw form, perform some manipulation of the data for more insight, or replay a sequence of events to try to understand what led to a particular outcome. For example, an e-commerce site offers a fantastic deal on several products on the weekend after a big holiday. The response to the sale is so strong that it crashes a few servers and brings the business down for a few minutes. By replaying all customer events, engineers can better understand what caused the breakdown and how to fix the system so it can handle a large, sudden influx of activity.

So, where do you need event-streaming applications? Since everything in life can be considered an event, any problem domain will benefit from processing event streams. But there are some areas where it's more important to do so. Here are some typical examples:

- *Credit card fraud*—A credit card owner may be unaware of unauthorized use. By reviewing purchases as they happen against established patterns (location, general spending habits), you may be able to detect a stolen credit card and alert the owner.
- *Intrusion detection*—The ability to monitor aberrant behavior in real time is critical for the protection of sensitive data and the well-being of an organization.
- *The Internet of Things*—With IoT, sensors are located in all kinds of places and send back data frequently. The ability to quickly capture and process this data meaningfully is essential; anything less diminishes the effect of deploying these sensors.
- *The financial industry*—The ability to track market prices and direction in real time is essential for brokers and consumers to make effective decisions about when to sell or buy.
- *Sharing data in real-time*—Large organizations, like corporations or conglomerates, that have many applications need to share data in a standard, accurate, and real-time way.

Bottom line: if the event stream provides essential and actionable information, businesses and organizations need event-driven applications to capitalize on the information provided.

But streaming applications are only a fit for some situations. Event-streaming applications become necessary when you have data in different places or a large volume of events requiring distributed data stores. So, if you can manage with a single database instance, streaming is unnecessary. For example, a small e-commerce business or a local government website with primarily static data aren't good candidates for building an event-streaming solution.

In this book, you'll learn about event-stream development, when and why it's essential, and how to use the Kafka event-streaming platform to build robust and responsive applications. You'll learn how to use the Kafka streaming platform's various components to capture events and make them available for other applications. We'll cover using the platform's components for simple actions such as writing (producing) or reading (consuming) events to advanced stateful applications requiring complex transformations so you can solve the appropriate business challenges with an event-streaming approach. This book is suitable for any developer looking to get into building event-streaming applications.

Although the title, *Kafka Streams in Action*, focuses on Kafka Streams, this book teaches the entire Kafka event-streaming platform, end to end. That platform includes crucial components, such as producers, consumers, and schemas, that you must work with before building your streaming apps, which you'll learn in part 1. As a result, we won't get into the subject of Kafka Streams itself until later in the book, in chapter 6.

But the enhanced coverage is worth it. Kafka Streams is an abstraction built on top of components of the Kafka event streaming platform, so understanding these components gives you a better grasp of how you can use Kafka Streams.

1.2   *What is an event?*

So we've defined an event stream, but what is an event? We'll define an event simply as "something that happens" (https://www.merriam-webster.com/dictionary/event). While the term *event* probably brings to mind something *notable* happening, like the birth of a child, a wedding, or a sporting event, we're going to focus on smaller, more constant events like a customer making a purchase (online or in-person) or clicking a link on a web page or a sensor transmitting data. Either people or machines can generate events. It's the sequence of events and the constant flow of them that make up an event stream.

Events conceptually contain three main components:

- *Key*—An identifier for the event
- *Value*—The event itself
- *Timestamp*—When the event occurred

Let's discuss each of these parts of an event in more detail. The key could be an identifier for the event, and as we'll learn in later chapters, it plays a role in routing and grouping events. Think of an online purchase: using the customer ID is an excellent example of the key. The value is the event payload itself. The event value could be a trigger, such as activating a sensor when someone opens a door, or a result of some action like the item purchased in the online sale. Finally, the timestamp is the date-time when recording when the event occurred. As we go through the various chapters in this book, we'll encounter all three components of this "event trinity" regularly.

I've used a lot of different terms in this introduction, so let's wrap this section up with a table of definitions (table 1.1).

Table 1.1   Definitions

Event	Something that occurs and attributes about it recorded
Event Stream	A series of events captured in real-time from sources such as mobile or IoT devices
Event Streaming Platform	Software to handle event streams—capable of producing, consuming, processing, and storage of event streams
Apache Kafka	The premier event streaming platform—provides all the components of an event streaming platform in one battle-tested solution
Kafka Streams	The native event stream processing library for Kafka

1.3   An event stream example

Let's say you've purchased a flux capacitor and are excited to receive your new purchase. Let's walk through the events leading up to the time you get your brand new flux capacitor, using the illustration in figure 1.1 as your guide.

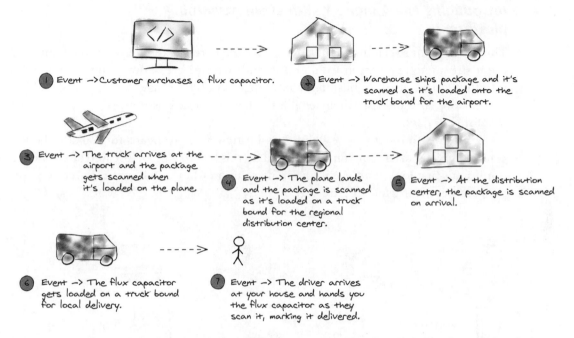

Figure 1.1   A sequence of events comprising an event stream starting with the online purchase of the flux capacitor

Consider the steps toward receiving your flux capacitor: ❶ You complete the purchase on the retailer's website, and the site provides a tracking number. ❷ The retailer's warehouse receives the purchase event information and puts the flux capacitor on a shipping truck, recording the date and time your purchase left the warehouse. ❸ The truck arrives at the airport, and the driver loads the flux capacitor on a plane and scans a barcode that records the date and time. ❹ The plane lands, and the package is loaded on a truck again headed for the regional distribution center. The delivery service records the date and time they loaded your flux capacitor. ❺ The truck from the airport arrives at the regional distribution center. A delivery service employee unloads the flux capacitor, scanning the date and time of the arrival at the distribution center. ❻ Another employee takes your flux capacitor, scans the package, saves the date and time, and loads it on a truck bound for delivery to you. ❼ The driver arrives at your house, scans the package one last time, and hands it to you. You can start building your time-traveling car!

From our example here, you can see how everyday actions create events, hence an event stream. The individual events are the initial purchase, each time the package changes custody, and the final delivery. This scenario represents events generated by just one purchase. But if you think of the event streams generated by purchases from Amazon and the various shippers of the products, the number of events could easily number in the billions or trillions.

1.4 *Introducing the Apache Kafka event streaming platform*

The Kafka event streaming platform provides the core capabilities to implement your event streaming application from end to end. We can break down these capabilities into three main areas: publishing/consuming, durable storage, and processing. This move, store, and process trilogy enables Kafka to operate as the central nervous system for your data.

Before we go on, it will be helpful to illustrate what it means for Kafka to be the central nervous system for your data. We'll do this by showing before and after illustrations. Let's first look at an event-streaming solution where each input source requires a separate infrastructure (figure 1.2).

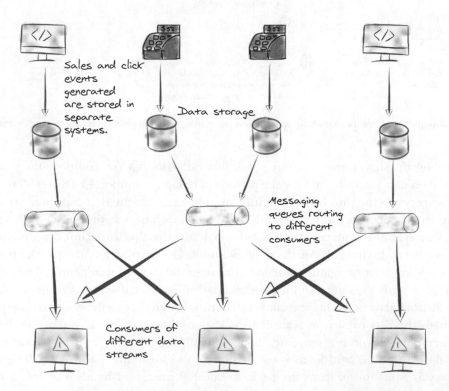

Figure 1.2 Initial event-streaming architecture leads to complexity as the different departments and data stream sources need to be aware of the other sources of events.

In the illustration, individual departments create separate infrastructures to meet their requirements. However, other departments may be interested in consuming the same data, which leads to a more complicated architecture to connect the various input streams. Let's look at figure 1.3, which shows how the Kafka event streaming platform can change things.

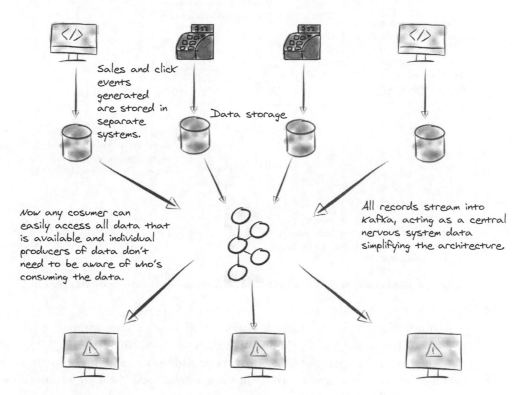

Figure 1.3   Using the Kafka event streaming platform with a simplified architecture

As you can see from this updated illustration, adding the Kafka event streaming platform simplifies the architecture dramatically. All components now send their records to Kafka. Additionally, consumers read data from Kafka with no awareness of the producers.

At a high level, Kafka is a distributed system of servers and clients. The servers are called brokers; the clients are record producers sending records to the brokers, and the consumer clients read records for the processing of events.

1.4.1   *Kafka brokers*

Kafka brokers durably *store* your records in contrast with traditional messaging systems (RabbitMQ or ActiveMQ), where the messages are ephemeral. The brokers store the data agnosticically as the key-value pairs (and some other metadata fields) in byte format and are somewhat of a black box to the broker.

Preserving events has more profound implications concerning the difference between messages and events. You can think of messages as "tactical" communication between two machines, while events represent business-critical data you don't want to throw away (figure 1.4).

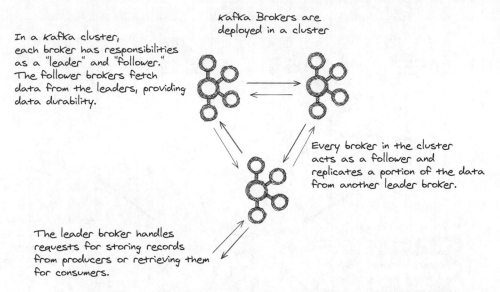

Figure 1.4   You deploy brokers in a cluster, and brokers replicate data for durable storage.

This illustration shows that Kafka brokers are the storage layer within the Kafka architecture and sit in the "storage" portion of the event-streaming trilogy. But in addition to acting as the storage layer, the brokers provide other essential functions such as serving client requests and coordinating with consumers. We'll go into details of broker functionality in chapter 2.

1.4.2   Schema Registry

Data governance is vital to begin with, and its importance only increases as the size and diversity of an organization grows. Schema Registry stores schemas of the event records (figure 1.5). Schemas enforce a contract for data between producers and consumers. Schema Registry also provides serializers and deserializers, supporting different tools that are Schema Registry aware. Providing (de)serializers means you don't have to write your serialization code. We'll cover Schema Registry in chapter 3. There's also a workshop on migrating Schema Registry schemas in appendix A.

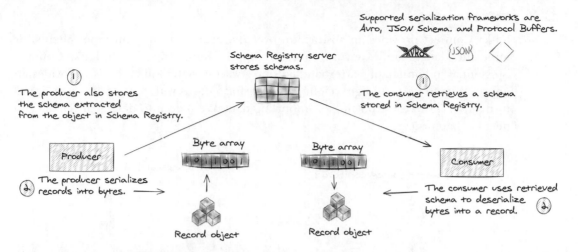

Figure 1.5   Schema registry enforces data modeling across the platform.

1.4.3   Producer and consumer clients

The producer client is responsible for sending records into Kafka, and the consumer is responsible for reading records from Kafka (figure 1.6). These two clients form the basic building blocks for creating an event-driven application and are agnostic to each other, allowing for greater scalability. The producer and consumer client also form the foundation for any higher-level abstraction working with Apache Kafka. We cover clients in chapter 4.

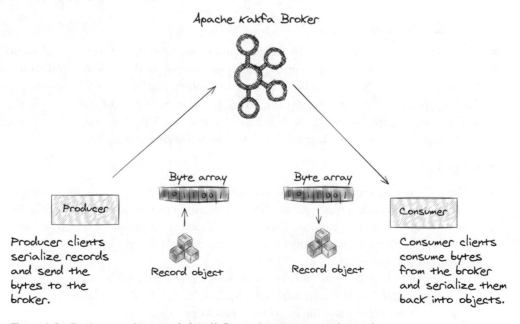

Figure 1.6   Producers write records into Kafka, and consumers read records.

1.4.4   *Kafka Connect*

Kafka Connect provides an abstraction over the producer and consumer clients for importing data to and exporting data from Apache Kafka (figure 1.7). Kafka Connect is essential in connecting external data stores with Apache Kafka. It also provides an opportunity to perform lightweight data transformations with Simple Message Transforms (SMTs) when exporting or importing data. We'll go into details of Kafka Connect in chapter 5.

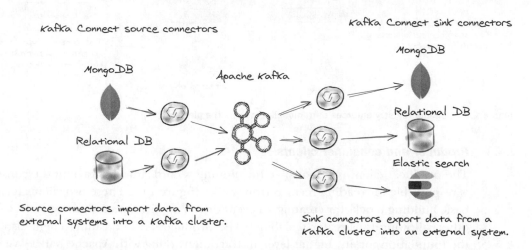

Figure 1.7   Kafka Connect bridges the gap between external systems and Apache Kafka.

1.4.5   *Kafka Streams*

Kafka Streams is Kafka's native stream processing library (figure 1.8). Kafka Streams is written in the Java programming language and is used by client applications at the perimeter of a Kafka cluster; it is not run inside a Kafka broker. It supports performing operations on event data, including transformations and stateful operations like joins and aggregations. Kafka Streams is where you'll do the heart of your work when dealing with events—chapters 6 to 10 cover Kafka Streams in detail.

1.4.6   *ksqlDB*

ksqlDB is an event streaming database (figure 1.9). It does this by applying an SQL interface for event stream processing. Under the covers, ksqlDB uses Kafka Streams to perform its event streaming tasks. A key advantage of ksqlDB is that it allows you to specify your event streaming operation in SQL; no code is required. We'll discuss ksqlDB in chapter 11.

Now that we've gone over how the Kafka event streaming platform works, including the individual components, let's apply a concrete example of a retail operation demonstrating how the Kafka event streaming platform works.

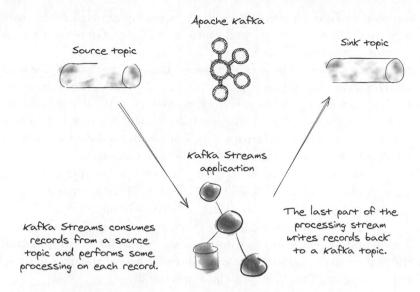

Apache Kafka

Source topic

Sink topic

Kafka Streams
application

Kafka Streams consumes
records from a source
topic and performs some
processing on each record.

The last part of the
processing stream
writes records back
to a Kafka topic.

Figure 1.8   Kafka Streams is the stream processing API for Kafka.

```
CREATE TABLE activePromotions AS
    SELECT rideId,
           qualifyPromotion(KmToDst) AS promotion
    FROM locations
    GROUP BY rideId
    EMIT CHANGES;


    SELECT rideId, promotion
    FROM activePromotions
    WHERE ROWKEY = '6fdOfcdb';
```

Figure 1.9   ksqlDB provides streaming database capabilities.

1.5   A concrete example of applying the Kafka event streaming platform

Let's say there is a consumer named Jane Doe, and she checks her email. There's one email from ZMart with a link to a page on the ZMart website containing a coupon for 15 percent off the total purchase price. Once on the web page, Jane clicks another link to activate and print the coupon. While this whole sequence is just another online purchase for Jane, it represents clickstream events for ZMart.

Let's pause our scenario to discuss the relationship between these simple events and how they interact with the Kafka event streaming platform. The data generated by the initial clicks to navigate to and print the coupon creates clickstream information captured and produced directly into Kafka with a producer microservice. The marketing department started a new campaign and wants to measure its effectiveness, so the clickstream events available here are valuable.

The first sign of a successful project is that users click on the email links to retrieve the coupon. The data science group is also interested in the prepurchase of clickstream data. The data science team can track customers' actions and attribute purchases to those initial clicks and marketing campaigns. The amount of data from this single activity may seem minor. You have a significant amount of data when you factor in a large customer base and several different marketing campaigns.

Now, let's resume our shopping example. It's late summer, and Jane has meant to go shopping to get her children back-to-school supplies. Since tonight is a rare night with no family activities, she stops off at ZMart on her way home.

Walking through the store after grabbing everything she needs, Jane walks by the footwear section and notices some new designer shoes that would go great with her new suit. She realizes that's not what she came in for, but what the heck? Life is short (ZMart thrives on impulse purchases!), so Jane gets the shoes.

As Jane reaches the self-checkout aisle, she scans her ZMart member card. After scanning all the items, she scans the coupon, which reduces the purchase by 15 percent. Then Jane pays for the transaction with her debit card, takes the receipt, and walks out of the store. A little later that evening, Jane checked her email, and there was a message from ZMart thanking her for her patronage with coupons for discounts on a new line of designer clothes.

Let's dissect the purchase transaction and see whether this event triggers a sequence of operations performed by the Kafka event streaming platform. So now ZMart's sales data streams into Kafka. In this case, ZMart uses Kafka Connect to create a source connector to capture the sales as they occur and send them to Kafka. The sale transaction brings us to the first requirement: the protection of customer data. In this case, ZMart uses an SMT to mask the credit card data as it goes into Kafka (figure 1.10).

As Connect writes records into Kafka, different organizations within ZMart immediately consume them. The department in charge of promotions created an application for consuming sales data to assign purchase rewards if the customer is a loyalty club member. If the customer reaches a threshold for earning a bonus, an email with a coupon goes out to them (figure 1.11).

It's important to note that ZMart processes sales records immediately after the sale. So, customers get timely emails with their rewards within a few minutes of completing their purchases. Acting on the purchase events as they happen allows ZMart a quick response time to offer customer bonuses.

The Data Science group within ZMart uses the sales data topic as well. It uses a Kafka Streams application to process the sales data, building up purchase patterns of what customers in different locations are purchasing the most. The Kafka Streams

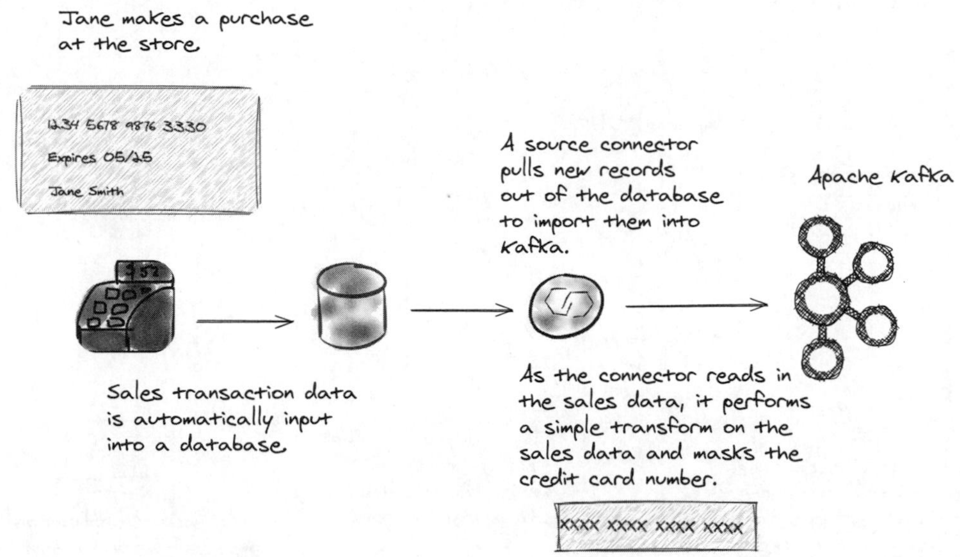

Jane makes a purchase
at the store.

4434 5678 9876 3330

Expires 05/25

Jane Smith

A source connector
pulls new records
out of the database
to import them into
Kafka.

Apache Kafka

Sales transaction data
is automatically input
into a database

As the connector reads in
the sales data, it performs
a simple transform on the
sales data and masks the
credit card number.

XXXX XXXX XXXX XXXX

Figure 1.10   Sending all of the sales data directly into Kafka with Connect masking the credit card numbers as part of the process

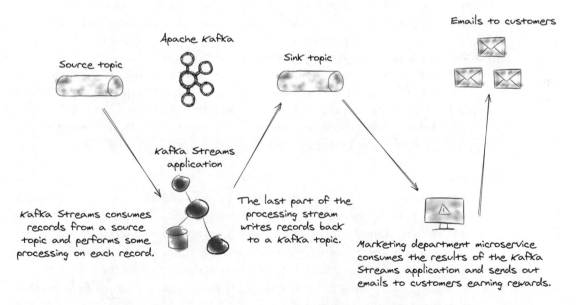

Emails to customers

Apache Kafka

Source topic

Sink topic

Kafka Streams
application

Kafka Streams consumes
records from a source
topic and performs some
processing on each record.

The last part of the
processing stream
writes records back
to a Kafka topic.

Marketing department microservice
consumes the results of the Kafka
Streams application and sends out
emails to customers earning rewards.

Figure 1.11   Marketing department application for processing customer points and sending out earned emails

application crunches the data in real time and sends the results to a sales-trends topic (figure 1.12).

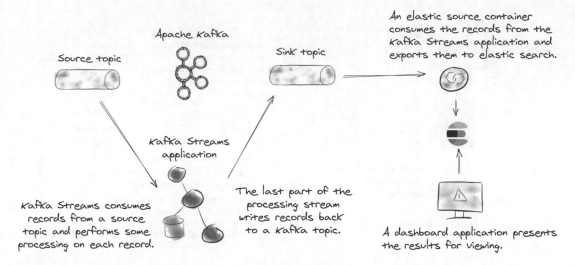

Figure 1.12   **Kafka Streams application crunching sales data and Kafka Connect exporting the data for a dashboard application**

ZMart uses another Kafka connector to export the sales trends to an external application that publishes the results in a dashboard. Another group also consumes from the sales topic to keep track of inventory and order new items if they drop below a given threshold, signaling the need to order more of those products.

At this point, you can see how ZMart uses the Kafka platform. It is important to remember that with an event streaming approach, ZMart responds to data as it arrives, allowing it to make quick and efficient decisions. Also, note that you write into Kafka once, yet multiple groups consume it at different times, independently, so that one group's activity doesn't impede another's.

Summary

- Event streaming captures events generated from different sources like mobile devices, customer interaction with websites, online activity, shipment tracking, and business transactions. Event streaming is analogous to our nervous system.
- An event is "something that happens," and the ability to react immediately and review later is an essential concept of an event streaming platform.
- Kafka acts as a central nervous system for your data and simplifies your event stream processing architecture.
- The Kafka event streaming platform provides the core capabilities for you to implement your event streaming application from end to end by delivering the three main components of publish/consume, durable storage, and processing.

- Kafka brokers are the storage layer and service requests from clients for writing and reading records. The brokers store records as bytes and do not touch or alter the contents.
- Schema Registry provides a way to ensure compatibility of records between producers and consumers.
- Producer clients write (produce) records to the broker. Consumer clients consume records from the broker. The producer and consumer clients are agnostic of each other. Additionally, the Kafka broker doesn't know who the individual clients are; they only process the requests.
- Kafka Connect provides a mechanism for integrating existing systems, such as external storage for getting data into and out of Kafka.
- Kafka Streams is the native stream processing library for Kafka. It runs at the perimeter of a Kafka cluster, not inside the brokers, and provides support for transforming data, including joins and stateful transformations.
- ksqlDB is an event streaming database for Kafka. It allows you to build robust real-time systems with just a few lines of SQL.

Kafka brokers

2

This chapter covers

- Explaining how the Kafka broker is the storage layer in the Kafka event streaming platform
- Describing how Kafka brokers handle requests from clients for writing and reading records
- Understanding topics and partitions
- Using JMX metrics to check for a healthy broker

In chapter 1, I provided an overall view of the Kafka event streaming platform and the different components that make up the platform. This chapter will focus on the system's heart, the Kafka broker. The Kafka broker is the server in the Kafka architecture and serves as the storage layer.

In describing the broker behavior in this chapter, we'll get into some lower-level details. It's essential to cover them to give you an understanding of how the broker operates. Additionally, some of the things we'll cover, such as topics and partitions, are essential concepts you'll need to understand when we get into the client chapter. But as a developer, you won't have to handle these topics daily.

2.1   *Introducing Kafka brokers*

As the storage layer, the broker manages data, including retention and replication. Retention is how long the brokers store records. Replication is how brokers make copies of the data for durable storage, meaning you won't lose data if you lose a machine.

But the broker also handles requests from clients. Figure 2.1 shows the client applications and the brokers.

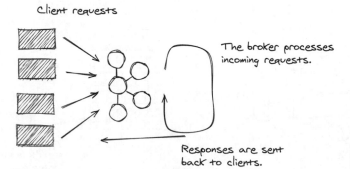

Figure 2.1   Clients communicating with brokers

To give you a quick mental model of the broker's role, let's summarize the illustration: clients send requests to the broker, and the broker then processes those requests and sends a response. While I'm glossing over several details of the interaction, that is the gist of the operation.

> **NOTE** Kafka is a deep subject, so I won't cover every aspect. I'll review enough information to get you started working with the Kafka event streaming platform. For in-depth coverage, look at *Kafka in Action* by Dylan Scott, Viktor Gamov, and Dave Klein (Manning, 2022).

You can deploy Kafka brokers on commodity hardware, containers, virtual machines, or cloud environments. In this book, you'll use Kafka in a docker container, so you won't need to install it directly.

While you're learning about the Kafka broker, I'll need to talk about the producer and consumer clients. But since this chapter is about the broker, I'll focus more on the broker's responsibilities. So, I'll leave out some of the client details. But don't worry; we'll get to those details in chapter 4.

So, let's get started with some walk-throughs of how a broker handles client requests, starting with producing.

2.2   *Produce requests*

When a client wants to send records to the broker, it does so with a produce request. Clients send records to the broker for storage so that consuming clients can later read those records.

Figure 2.2 is an illustration of a producer sending records to a broker. It's important to note these illustrations aren't drawn to scale. Typically, you'll have many clients communicating with several brokers in a cluster. A single client will work with more than one broker. But it's easier to get a mental picture of what's happening if I keep the illustrations simple. Also, note that I'm simplifying the interaction, but we'll cover more details when discussing clients in chapter 4.

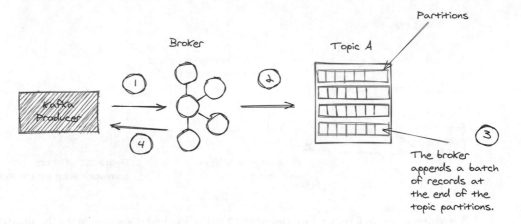

Figure 2.2   Brokers handling produce records request

Let's walk through the steps in the illustration:

1 The producer sends a batch of records to the broker. Whether a producer or consumer, the client APIs always work with a collection of records to encourage batching.
2 The broker takes the produce request out of the request queue.
3 The broker stores the records in a topic. Inside the topic are partitions. A single batch of records belongs to a specific partition within a topic, and the records are always appended at the end.
4 Once the broker stores the records, it responds to the producer. We'll talk more about what makes up a successful write later in this chapter and again in chapter 4.

Now that we've walked through an example produce request, let's walk through another request type, fetch, which is the logical opposite of producing records: consuming records.

2.3   Fetch requests

Now, let's look at the other side of the coin, from a produce request to a fetch request. Consumer clients issue requests to a broker to read (or consume) records from a topic with a fetch request. A critical point to understand is that consuming records does not affect data retention or records availability to other consuming clients. Kafka brokers can handle hundreds of consumer requests for records from the same topic, and each request has no effect on the others. We'll get into data retention later, but the broker handles it separately from consumers.

It's also important to note that producers and consumers are unaware of each other. The broker handles produce and consume requests separately; one has nothing to do with the other. The example in figure 2.3 is simplified to emphasize the overall action from the broker's point of view.

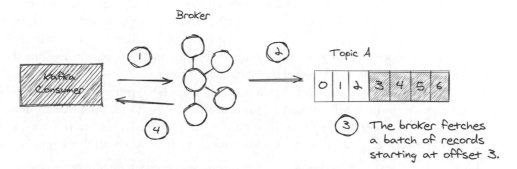

Figure 2.3 Brokers handling requests from a consumer

So, let's go through the steps of the illustrated consumer request:

1 The consumer sends a fetch request specifying the offset from which it wants to start reading records. We'll discuss offsets in more detail later in the chapter.
2 The broker takes the fetch request out of the request queue.
3 Based on the offset and the topic partition in the request, the broker fetches a batch of records.
4 The broker sends the fetched batch of records in the response to the consumer.

Now that we've completed a walk-through of two common request types, produce and fetch, I'm sure you noticed a few terms I still need to describe: topics, partitions, and offsets. Topics, partitions, and offsets are fundamental, essential concepts in Kafka, so let's take some time now to explore what they mean.

2.4 Topics and partitions

In chapter 1, we discussed that Kafka provides storage for data. Kafka durably stores your data as an unbounded series of key-value pair messages for as long as you want (messages contain other fields, such as a timestamp, but we'll get to those details later). Kafka replicates data across multiple brokers, so losing a disk or an entire broker means no data is lost.

Specifically, Kafka brokers use the filesystem for storage by appending the incoming records to the end of a file in a topic. A topic represents the directory's name containing the file to which the Kafka broker appends the records.

> **NOTE** Kafka receives the key-value pair messages as raw bytes, stores them that way, and serves the read requests in the same format. The Kafka broker is unaware of the type of record that it handles. By merely working with raw bytes, the brokers don't spend time deserializing or serializing the data, allowing for higher performance. We'll see how you can ensure that topics contain the expected byte format when we cover Schema Registry in chapter 3.

Topics have partitions, which is a way of further organizing the topic data into slots or buckets. A partition is an integer starting at 0. So, if a topic has three partitions, the partition's numbers are 0, 1, and 2. Kafka appends the partition number to the end of the topic name, creating the same number of directories as partitions with the form `topic-N` where the `N` represents the partition number.

Kafka brokers have a configuration, `log.dirs`, where you place the top-level directory's name, which will contain all topic-partition directories. Let's take a look at an example in listing 2.1. I will assume you've configured `log.dirs` with the value `/var/kafka/topic-data`, and you have a topic named `purchases` with three partitions.

Listing 2.1 Topic directory structure example

```
root@broker:/#  tree /var/kafka/topic-data/purchases*

/var/kafka/topic-data/purchases-0
├── 00000000000000000000.index
├── 00000000000000000000.log
├── 00000000000000000000.timeindex
└── leader-epoch-checkpoint
/var/kafka/topic-data/purchases-1
├── 00000000000000000000.index
├── 00000000000000000000.log
├── 00000000000000000000.timeindex
└── leader-epoch-checkpoint
/var/kafka/topic-data/purchases-2
├── 00000000000000000000.index
├── 00000000000000000000.log
├── 00000000000000000000.timeindex
└── leader-epoch-checkpoint
```

As you can see here, the topic `purchases` with three partitions ends up as three directories, `purchases-0`, `purchases-1`, and `purchases-2` on the filesystem. The topic name is more of a logical grouping, while the partition is the storage unit.

> **TIP** The directory structure shown here was generated using the `tree` command, a small command-line tool used to display all contents of a directory.

While we'll want to discuss those directories' contents, we still have some details about topic partitions to cover.

Topic partitions are the unit of parallelism in Kafka. For the most part, the higher the number of partitions, the higher your throughput. As the primary storage mechanism, topic partitions allow for the spreading of messages across several machines. The given topic's capacity isn't limited to the available disk space on a single broker. Also, as mentioned before, replicating data across several brokers ensures you won't lose data should a broker lose disks or die.

Later in this chapter, we'll discuss load distribution more when discussing replication, leaders, and followers. We'll also cover a new feature, tiered storage, where data is seamlessly moved to external storage, providing virtually limitless capacity later in the chapter.

So, how does Kafka map records to partitions? The producer client determines the topic and partition for the record before sending it to the broker. Once the broker processes the record, it appends it to a file in the corresponding topic-partition directory.

There are three possible ways of setting the partition for a record:

1 Kafka works with records in key-value pairs. Suppose the key is non-null (keys are optional). In that case, the producer maps the record to a partition using the deterministic formula of taking the hash of the key modulo the number of partitions. This approach means that records with identical keys always land on the same partition.
2 When building the `ProducerRecord` in your application, you can explicitly set the partition for that record, which the producer then uses before sending it.
3 If the message has no key or partition specified, partitions are alternated per batch. I'll detail how Kafka handles records without keys and partition assignments in chapter 4.

Now that we've covered how topic partitions work, let's revisit that Kafka always appends records to the end of the file. I'm sure you noticed the files in the directory example with an extension of `.log` (we'll talk about how Kafka names this file in section 2.6.3). But these `log` files aren't the type developers think of, where an application prints its status or execution steps. The term *log* here is a transaction log, storing a sequence of events in the order of occurrence. So, each topic partition directory contains its transaction log. At this point, asking a question about log file growth would be fair. We'll discuss log file size and management when we cover segments later in this chapter.

2.4.1   *Offsets*

As the broker appends each record, it assigns it an ID called an offset. An offset is a number (starting at 0) the broker increments by 1 for each record. In addition to being a unique ID, it represents the logical position in the file. The term *logical position* means it's the *n*th record in the file, but its physical location is determined by the size in bytes of the preceding records. In section 2.6.3, we'll talk about how brokers use an offset to find the physical position of a record. The illustration in figure 2.4 demonstrates the concept of offsets for incoming records.

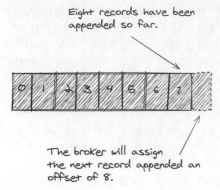

Eight records have been appended so far.

The broker will assign the next record appended an offset of 8.

Figure 2.4 Assigning the offset to incoming records

Since new records always go at the end of the file, they are in order by offset. Kafka guarantees that records are in order within a partition but not *across* partitions. Since records are in order by offset, we could also be tempted to think they are in order by time, but that's not necessarily the case. The records are in order by their *arrival* time at the broker, but not necessarily by *event time*. We'll get more into time semantics in chapter 4 when we discuss timestamps. We'll also cover event-time processing in depth when we get to chapter 9 on Kafka Streams.

Consumers use offsets to track the position of records they've already consumed. That way, the broker fetches records starting with an offset one higher than the last one read by a consumer. Let's look at figure 2.5 to explain how offsets work.

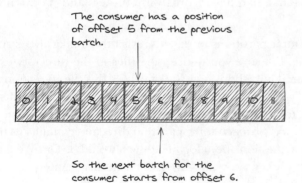

The consumer has a position of offset 5 from the previous batch.

So the next batch for the consumer starts from offset 6.

Figure 2.5 Offsets indicate where a consumer has left off reading records.

In the illustration, if a consumer reads records with offsets 0–5, the broker only fetches records starting at offset 6 in the following consumer request. The offsets used are unique for each consumer and stored in an internal topic named __consumer_ offsets. We'll go into more detail about consumers and offsets in chapter 4.

Now that we've covered topics, partitions, and offsets, let's quickly discuss some tradeoffs regarding the number of partitions to use.

2.4.2 Determining the correct number of partitions

Choosing the number of partitions to use when creating a topic is part art and part science. One of the critical considerations is the amount of data flowing into a given topic. More data implies more partitions for higher throughput. But as with anything in life, there are tradeoffs.

Increasing the number of partitions increases the number of TCP connections and open file handles. How long it takes to process an incoming record in a consumer will also determine throughput. If you have heavyweight processing in your consumer, adding more partitions may help, but the slower processing will ultimately hinder performance (Jun Rao, "How to Choose the Number of Topics/Partitions in a Kafka Cluster?" http://mng.bz/4C03).

Here are some things to consider when setting the number of partitions. You want to choose a high enough number to cover high-throughput situations, but not so high that you hit limits for the number of partitions a broker can handle as you create more and more topics. A good starting point could be the number 30, which is evenly divisible by several numbers, which results in a more even distribution of keys in the processing layer (Michael Noll, "Streams and Tables in Apache Kafka: Topics, Partitions, and Storage Fundamentals," http://mng.bz/K9qg). We'll talk more about the importance of key distribution in chapter 4 on clients and chapter 7 on Kafka Streams.

At this point, you've learned that the broker handles client requests and is the storage layer for the Kafka event streaming platform. You've also learned about topics, partitions, and their role in the storage layer.

Your next step is to get your hands dirty, producing and consuming records to see these concepts in action.

> **NOTE** We'll cover the producer and consumer clients in chapter 4. Console clients are helpful for learning, quick prototypes, and debugging. But in practice, you'll use the clients in your code.

2.5 Sending your first messages

You'll need to run a Kafka broker to run the following examples. In the previous edition of this book, the instructions were to download a binary version of the Kafka tar file and extract it locally. In this edition, I've opted to run Kafka via Docker instead. Specifically, we'll use Docker Compose, making running a multicontainer docker application easy. If you are running macOS or Windows, you can install Docker Desktop, which includes Docker Compose. For more information on installing Docker, see the installation instructions on the Docker site: https://docs.docker.com/get-docker/. Note that you can also use Confluent Cloud (https://www.confluent.io/confluent-cloud/) for a broker for running your Kafka applications. See appendix B for details on resources from Confluent.

Let's start working with a Kafka broker by producing and consuming some records.

2.5.1   *Creating a topic*

Your first step for producing or consuming records is to create a topic. But you'll need a running Kafka broker to do that, so let's take care of that now. I assume you've already installed Docker at this point. To start Kafka, download the `docker-compose .yml` file from the source code repo here https://github.com/bbejeck/KafkaStreamsIn Action2ndEdition. After downloading the file, open a new terminal window and CD to the directory with the `docker-compose.yml` file, and run this command `docker-compose up -d`.

> **TIP**   Starting docker-compose with the `-d` flag runs the docker services in the background. While it's OK to start Docker Compose without the `-d` flag, the containers print their output to the terminal, so you need to open a new terminal window to do any further operations.

Wait a few seconds and then run the following command to open a shell on the docker broker container: `docker-compose exec broker bash`.

Using the Docker broker container shell you just opened up, run this command to create a topic:

```
kafka-topics --create --topic first-topic\          The host:port to connect
  --bootstrap-server localhost:9092\                to the broker
  --replication-factor 1\
  --partitions 1                  Specifies the
                                  replication factor
      The number of partitions
```

> **NOTE**   Although you're using Kafka in a Docker container, the commands to create topics and run the console producer and consumer are the same.

Since you're running a local broker for testing, you don't need a replication factor greater than 1. The same thing goes for the number of partitions; at this point, you only need one partition for this local development.

Now you have a topic, let's write some records to it.

2.5.2   *Producing records on the command line*

Now, from the same window you ran the `--create --topic` command, run the following command to start a console producer:

```
kafka-console-producer --topic first-topic\          The topic you created
  --bootstrap-server localhost:9092\                 in the previous step
  --property parse.key=true\
  --property key.separator=":"            The host:port for the
                                          producer client to connect
      Specifies the separator of          to the broker
        the key and value
                                     Specifies that you'll provide a key
```

When using the console producer, you need to specify whether you will provide keys. Although Kafka works with key-value pairs, the key is optional and can be null. Since the key and value go on the same line, you must specify how Kafka can parse the key and value by providing a delimiter.

After you enter the previous command and press Enter, you should see a prompt waiting for your input. Enter some text like the following:

```
key:my first message
 key:is something
 key:very simple
```

You type in each line and then press Enter to produce the records. Congratulations, you have sent your first messages to a Kafka topic! Now, let's consume the records you just wrote to the topic. Keep the console producer running, as you'll use it again in a few minutes.

2.5.3   *Consuming records from the command line*

Now, it's time to consume the records you just produced. Open a new terminal window and run the `docker-compose exec broker bash` command to get a shell on the broker container. Then run the following command to start the console consumer:

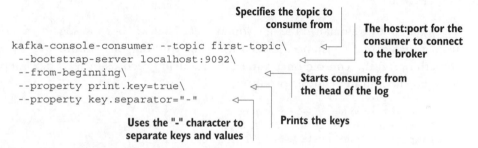

You should see the following output on your console:

```
key-my first message
 key-is something
 key-very simple
```

I should briefly discuss why you used the `--from-beginning` flag. You produced values before starting the consumer. As a result, you wouldn't have seen those messages as the console consumer reads from the end of the topic. So, the `--from-beginning` parameter sets the consumer to read from the beginning of the topic. Now, return to the producer window and enter a new key-value pair. The console window with your consumer will update by adding the latest record at the end of the current output.

This completes your first example, but let's go through one more example so you can see how partitions come into play.

2.5.4   *Partitions in action*

In the previous exercise, you produced and consumed some key-value records, but the topic only has one partition, so you didn't see the effect of partitioning. Let's do one more example, but this time, we'll create a new topic with two partitions, produce records with different keys, and see the differences.

You should still have a console producer and consumer running now. Go ahead and shut both down by entering a CTRL+C command on the keyboard.

Now, let's create a new topic with partitions. Execute the following command from one of the terminal windows you used to either produce or consume records:

```
kafka-topics --create --topic second-topic\
 --bootstrap-server localhost:9092\
 --replication-factor 1\
 --partitions 2
```

For your next step, let's start a console consumer:

```
kafka-console-consumer --topic second-topic\
 --bootstrap-server broker:9092 \
 --property print.key=true \
 --property key.separator="-" \
 --partition 0
```

Specifies the partition we'll consume from

This command is similar to the one you executed before, but you specify the partition from which you'll consume the records. After running this command, you will see something on the console once you start producing records in your next step. Now, let's start up another console producer:

```
kafka-console-producer --topic second-topic\
  --bootstrap-server localhost:9092\
  --property parse.key=true\
  --property key.separator=":"
```

After you've started the console producer, enter these key-value pairs:

```
key1:The lazy
key2:brown fox
key1:jumped over
key2:the lazy dog
```

You should only see the following records from the console consumer you have running:

```
key1:The lazy
key1:jumped over
```

You don't see the other records here because the producer assigned them to partition 1. You can test this for yourself by executing a CTRL+C in the terminal window of the current console consumer and then running the following:

```
kafka-console-consumer --topic second-topic\
  --bootstrap-server broker:9092\
  --property print.key=true\
  --property key.separator="-"\
  --partition 1\
  --from-beginning
```

You should see the following results:

```
key2:brown fox
key2:the lazy dog
```

You will see all the records produced for the topic if you rerun the previous consumer without specifying a partition. We'll go into more detail about consumers and topic partitions in chapter 4.

We're done with the examples at this point, so you can shut down the producer and the consumer by entering a CTRL+C command. Then you can stop all the Docker containers now by running docker-compose down.

To quickly recap this exercise, you've just worked with the core Kafka functionality. You produced some records to a topic; then, in another process, you consumed them. While in practice, you'll use topics with higher partition counts, a much higher volume of messages, and something more sophisticated than the console tools, the concepts are the same.

We've also covered the basic unit of storage the broker uses, partitions. We discussed how Kafka assigns each incoming record a unique, per-partition ID, the offset, and always appends records at the end of the topic partition log. But as more data flows into Kafka, do these files continue to grow indefinitely? The answer to this question is no, and we'll cover how the brokers manage data in the next section.

2.6 Segments

So far, you've learned that brokers append incoming records to a topic partition file. But they don't continue to append to the same one, which would create substantial monolithic files. Instead, brokers break up the files into discrete parts called segments. Using segments to enforce the data retention settings and retrieve records by offset for consumers is much easier.

Earlier in the chapter, I stated the broker writes to a partition; it appends the record to a file. But a more accurate statement is the broker appends the record to the *active segment*. The broker creates a new segment when a log file reaches a specific size (1 MB by default). The broker still uses previous segments to serve read (consume) consumer requests. Let's look at an illustration of this process in figure 2.6.

Following along in the illustration here, the broker appends incoming records to the currently active segment. Once it reaches the configured size, the broker creates a new segment that is considered the active segment. This process is repeated indefinitely.

The configuration controlling the size of a segment is log.segment.bytes, which again has a default value of 1 MB. Additionally, the broker will create new segments

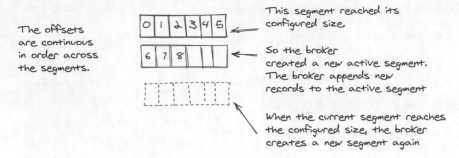

The offsets are continuous in order across the segments.

This segment reached its configured size

So the broker created a new active segment. The broker appends new records to the active segment

When the current segment reaches the configured size, the broker creates a new segment again

Figure 2.6 Creating new segments

over time. The `log.roll.ms` or `log.roll.hours` governs the maximum time before the broker creates a new segment. The `log.roll.ms` is the primary configuration but has no default value. The default value of the `log.roll.hours` is 168 hours (7 days). When a broker creates a new segment based on time, a new record has a timestamp greater than the earliest timestamp in the currently active segment plus the `log.roll.ms` or `log.roll.hours` configuration. It's not based on wall clock time or when the file was last modified.

> **NOTE** The number of records in a segment won't necessarily be uniform, as figure 2.6 might suggest here. In practice, they could vary in the total number of records. Remember, the total size or age of the segment triggers the broker to create a new one.

Now that we have covered how brokers create segments, we can discuss their data retention role.

2.6.1 *Data retention*

As records continue to come into the brokers, the brokers will need to remove older records to free up space on the filesystem over time. Brokers use a two-tiered approach to deleting data, time, and size. For time-based deletion, Kafka deletes records older than a configured retention time based on the record's timestamp. If the broker placed all records in one big file, it would have to scan the file to find all those records eligible for deletion. But with the records stored in segments, the broker can remove segments where the latest timestamp exceeds the configured retention time. There are three time-based configurations for data deletion presented here in order of priority:

- `log.retention.ms`—How long to keep a log file in milliseconds
- `log.retention.minutes`—How long to keep a log file in minutes
- `log.retention.hours`—How long to keep a log file in hours

By default, only the `log.retention.hours` configuration has a default value of 168 (7 days). Kafka has the `log.retention.bytes` configuration for size-based retention. By default, it's set to -1. If you configure size- and time-based retention, brokers delete segments whenever they meet either condition.

So far, we've focused our discussion on data retention based on the elimination of entire segments. If you remember, Kafka records are in key-value pairs. What if you wanted to retain the latest record per key? That would mean not removing entire segments but only removing the oldest records for each key. Kafka provides just such a mechanism called compacted topics.

2.6.2 Compacted topics

Consider the case where you have keyed data and receive updates for that data over time, meaning a new record with the same key will update the previous value. For example, a stock ticker symbol could be the key, and the price per share would be the regularly updated value. Imagine you're using that information to display stock values, and you have a crash or restart. You need to be able to start back up with the latest data for each key (see the Kafka documentation, "Log Compaction," http://kafka .apache.org/documentation/#compaction).

If you use the deletion policy, a broker could remove a segment between the last update and the application's crash or restart. You wouldn't have all the records on startup. Retaining the final known value for a given key would be better than treating the next record with the same key as an update to a database table.

Updating records by key is the behavior that compacted topics (logs) deliver. Instead of taking a coarse-grained approach and deleting entire segments based on time or size, compaction is more fine-grained and deletes old records *per key* in a log. At a high level, the log cleaner (a pool of threads) runs in the background, recopying log-segment files and removing records if there's an occurrence later in the log with the same key. Figure 2.7 illustrates how log compaction retains the most recent message for each key.

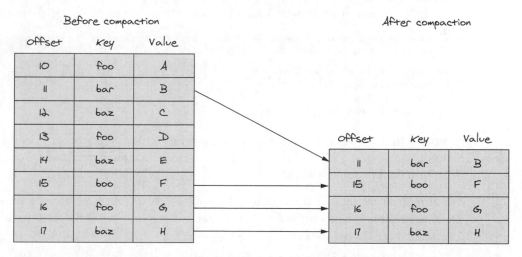

Figure 2.7 On the left is a log before compaction. You'll notice duplicate keys with different values. These duplicates are updates. On the right is after compaction, which retains the latest value for each key, but it's smaller in size.

This approach guarantees that the last record for a given key is in the log. You can specify log retention per topic, so it's entirely possible to use time-based retention and for other topics, use compaction.

By default, the log cleaner is enabled. To use compaction for a topic, you must set the `log.cleanup.policy=compact` property when creating it.

Compaction is used in Kafka Streams when using state stores, but you won't create those logs/topics yourself—the framework handles that task. Nevertheless, it's essential to understand how compaction works. Log compaction is a broad subject, and we've only touched on it here. For more information, see the Kafka documentation: http://kafka.apache.org/documentation/#compaction.

> **NOTE** With a `cleanup.policy` of `compact`, you might wonder how to remove a record from the log. You delete with compaction by using a `null` value for the given key, creating a tombstone marker. Tombstones ensure that compaction removes prior records with the same key. Kafka removes the tombstone marker later to free up space.

The key takeaway from this section is that if you have independent, standalone events or messages, use log deletion. If you have updates to events or messages, you'll want to use log compaction.

Now that we've covered how Kafka brokers manage data using segments, it would be an excellent time to reconsider and discuss the topic partition directory contents.

2.6.3   *Topic partition directory contents*

Earlier in this chapter, we discussed that a topic is a logical grouping for records, and the partition is the actual physical storage unit. Kafka brokers append each incoming record to a file in a directory corresponding to the topic and partition specified in the record. For review, the following listing provides the contents of a topic-partition.

Listing 2.2   Contents of topic partition directory

```
/var/kafka/topic-data/purchases-0
├── 00000000000000000000.index
├── 00000000000000000000.log
├── 00000000000000000000.timeindex
```

> **NOTE** In practice, you'll most likely not interact with a Kafka broker on this level. We're going into this level of detail to provide a deeper understanding of how broker storage works.

We already know the `log` file contains the Kafka records, but what are the `index` and `timeindex` files? When a broker appends a record, it stores other fields along with the key and value. Three fields are

- The offset (which we've already covered)
- The size
- The record's physical position in the segment

The `index` is a memory-mapped file that maps offset to position. The `timeindex` is also a memory-mapped file containing a mapping of the timestamp to offset. Let's look at the `index` files first in figure 2.8.

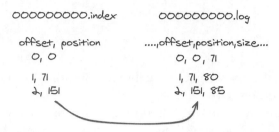

Figure 2.8 **Searching for a start point based on offset 2**

Brokers use the index files to find the starting point for retrieving records based on the given offset. The brokers do a binary search in the `index` file, looking for an index–position pair with the largest offset that is less than or equal to the target offset. The offset stored in the `index` file is relative to the base offset. That means if the base offset is 100, offset 101 is stored as 1; offset 102 is stored as 2, etc. Using the relative offset, the `index` file can use two 4-byte entries, one for the offset and the other for the position. The base offset is the number used to name the file, which we'll cover soon.

The `timeindex` is a memory-mapped file that maintains a mapping of timestamp to offset (figure 2.9).

OOOOOOOO.timeindex
  timestamp, offset
   122456789, O
      ....

123985789, 100   Figure 2.9   `timeindex` **file**

NOTE A memory-mapped file is a particular file in Java that stores a portion of the file in memory, allowing for faster reads from the file. For a more detailed description, read the excellent entry "What Is Memory-Mapped File in Java" from GeeksForGeeks site (http://mng.bz/wj47).

The file's physical layout is an 8-byte timestamp and a 4-byte entry for the "relative" offset. The brokers search for records by looking at the timestamp of the earliest segment. If the timestamp is smaller than the target timestamp, the broker does a binary search on the `timeindex` file looking for the closest entry.

So what about the names then? The broker names these files based on the first offset in the `log` file. A segment in Kafka comprises the `log`, `index`, and `timeindex` files. So, in our previous example directory listing, there is one active segment. Once the broker creates a new segment, the directory will look something like the following listing.

Listing 2.3   Contents of the directory after creating a new segment

```
/var/kafka/topic-data/purchases-0
├── 00000000000000000000.index
├── 00000000000000000000.log
├── 00000000000000000000.timeindex
├── 00000000000000037348.index
├── 00000000000000037348.log
├── 00000000000000037348.timeindex
```

Based on this directory structure, the first segment contains records with offset 0–37347, and in the second segment, the offsets start at 37348.

The files in the topic partition directory are stored in a binary format and aren't suitable for viewing. As I mentioned, you won't interact with the files on the broker, but sometimes, you may need to view the files' contents when looking into an issue.

> **WARNING**   You should *never modify or directly access* the files stored in the topic-partition directory. Only use the tools provided by Kafka to view the contents.

2.7   *Tiered storage*

We've discussed that brokers are the storage layer in the Kafka architecture. We've also covered how brokers store data in immutable, append-only files and manage data growth by deleting segments when the data reaches an age exceeding the configured retention time. But as Kafka can be used for your data's central nervous system, meaning all data flows into Kafka, the disk space requirements will continue to grow. Figure 2.10 depicts this situation.

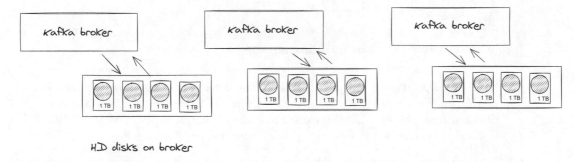

Figure 2.10   Kafka brokers store all data on a local disk, which will continue to grow over time.

As you can see, if you want to store data longer, you'll have to add more disks to make space for newly arriving records. This situation means that Kafka users wanting to keep data longer than the required retention period must offload data from the cluster to more scalable, long-term storage. One could use Kafka Connect (which we'll cover in chapter 5) to move the data, but long-term storage requires building different applications to access that data.

A new feature in Kafka called Tiered Storage separates the compute and storage layers. I'll only give a brief description here, but for more details, you can read KIP-405 (http://mng.bz/qjZK). At a high level, the proposal is for the Kafka brokers to have a concept of local and remote storage. Local storage is the same as the brokers use today, but remote storage would be something more scalable, say S3, for example, but the Kafka brokers still manage it. Figure 2.11 is another illustration demonstrating how brokers can handle data retention with tiered storage.

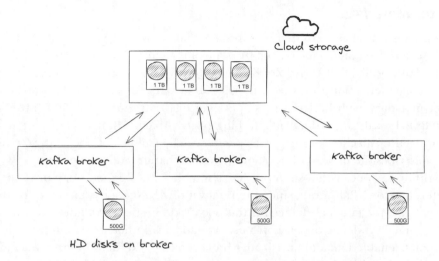

Figure 2.11   Kafka brokers only keep "hot" data locally; all warm and cold records are migrated to cloud storage.

The concept is that the brokers migrate older data to remote storage over time. This tiered storage approach is essential for two reasons:

- The Kafka brokers handle the data migration as part of normal operations. Setting up a separate process to move older data is unnecessary.
- The older data is still accessible via the Kafka brokers, so no additional applications are required to process older data. The use of tiered storage will be seamless to client applications. They won't know or need to know if the records consumed are local or from the tiered storage.

Using the tiered storage approach effectively gives Kafka brokers the ability to have infinite storage capabilities. Another benefit of tiered storage, which might not be evident at first blush, is the improvement in elasticity. Before tiered storage, when adding a new broker, entire partitions needed to get moved across the network. Remember from our previous conversation that Kafka distributes topic partitions among the brokers. So, adding a new broker means calculating new assignments and moving the data accordingly. But with tiered storage, most segments will be in the storage tier beyond the active ones. This means there is much less data that needs to get moved around, so changing the number of brokers will be much faster.

As of the writing of this book (October 2023) with version 3.6.0, tiered storage for Apache Kafka is available as an early access feature. Still, at this time, it's not recommended for production. Again, for the reader interested in the details involved in the tiered storage feature, I encourage you to read the details found in KIP-405 (http://mng.bz/qjZK).

2.8   *Cluster metadata*

Kafka is a distributed system that requires metadata to manage all activity and state in the cluster. Having metadata to keep the state of the cluster is integral to Kafka's architecture. Historically, Kafka used ZooKeeper (https://zookeeper.apache.org/) for metadata management. But now, with KIP-500, Kafka can use Kafka brokers to store the cluster metadata, with Kafka servers operating in KRaft mode. KIP-500 (http://mng.bz/7vPx) describes the details. The blog post, "Apache Kafka Needs No Keeper: Removing the Apache ZooKeeper Dependency" by Colin McCabe (http://mng.bz/mjrn) describes the process of how and when the changes to Kafka occur.

Right now, you can choose to run Kafka in either ZooKeeper mode or KRaft, with a preference for KRaft mode since the removal of ZooKeeper is targeted for the 4.0 release. Since the target audience for this book is developers and not cluster administrators, some knowledge of *how* Kafka uses metadata is sufficient. The storage and use of metadata enable Kafka to have leader brokers and to do such things as tracking the replication of topics.

The use of metadata in a cluster is involved in the following aspects of Kafka operations:

- *Cluster membership*—Joining and maintaining membership in a cluster. If a broker becomes unavailable, ZooKeeper removes the broker from cluster membership.
- *Topic configuration*—Keeping track of the topics in a cluster, which broker is the leader for a topic, how many partitions there are, and any specific configuration overrides for a topic.
- *Access control*—Identifying which users (a person or other software) can read from and write to particular topics.

This has been a quick overview of how Kafka manages metadata. I don't want to go into too much detail about metadata management as my approach to this book is more from the developer's point of view and not someone who will manage a Kafka cluster. Now that we've briefly discussed Kafka's need for metadata and how it's used, let's resume our discussion on leaders and followers and their role in replication.

2.9   *Leaders and followers*

So far, we've discussed topics' role in Kafka and how and why topics have partitions. You've seen that partitions aren't all located on one machine but are spread out on brokers throughout the cluster. Now, it's time to look at how Kafka provides data availability in the face of machine failures.

In the Kafka cluster for each topic partition, one broker is the *leader*, and the rest are followers (figure 2.12).

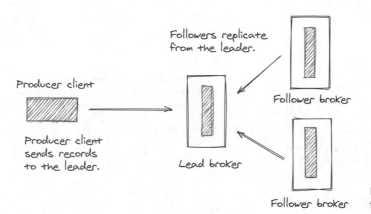

Figure 2.12   Leader and follower example

Figure 2.12 has a simplified view of the leader and follower concept. The lead broker for a topic partition handles all of the produce and consume requests (although it is possible to have consumers work with followers, and we'll cover that in chapter 4 on clients). The follower brokers replicate records from the leader for a given topic partition. Kafka uses this leader-and-follower relationship for data integrity. Remembering that the leadership for the topic is essential, partitions are spread around the cluster. No single broker is the leader for all partitions of a given topic.

But before we discuss how leaders, followers, and replication work, we must consider what Kafka does to enable replication with leaders and followers.

2.9.1   *Replication*

In the previous section on leaders and followers, I mentioned that topic partitions have a leader broker and one or more followers. Figure 2.12 shows this concept. Once the leader adds records to the log, the followers read from the leader.

Kafka replicates records among brokers to ensure data availability should a broker in the cluster fail. Figure 2.13 demonstrates the replication flow between brokers. A user configuration determines the replication level, but using a setting of three is recommended. With a replication factor of 3, the lead broker is considered replica 1, and two followers are replicas 2 and 3.

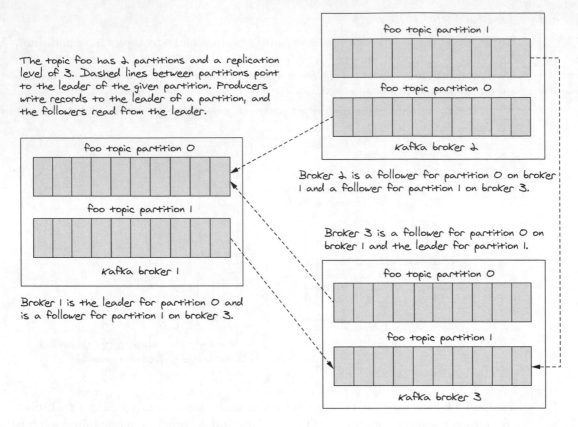

The topic foo has 2 partitions and a replication level of 3. Dashed lines between partitions point to the leader of the given partition. Producers write records to the leader of a partition, and the followers read from the leader.

foo topic partition 0

foo topic partition 1

Kafka broker 1

Broker 1 is the leader for partition 0 and is a follower for partition 1 on broker 3.

foo topic partition 1

foo topic partition 0

Kafka broker 2

Broker 2 is a follower for partition 0 on broker 1 and a follower for partition 1 on broker 3.

Broker 3 is a follower for partition 0 on broker 1 and the leader for partition 1.

foo topic partition 0

foo topic partition 1

Kafka broker 3

Figure 2.13   The Kafka replication process

The Kafka replication process is straightforward. Brokers following a topic partition consume messages from the topic-partition leader. After the leader appends new records to the log, followers consume from the leader and append them to their logs. After the followers add the records, their logs replicate the leader's log with the same data and offsets. When fully caught up to the leader, these following brokers are considered an in-sync replica (ISR).

When a producer sends a batch of records, the leader must append them before the followers can replicate them. There is a small window of time where the leader will be ahead of the followers. Figure 2.14 demonstrates this concept.

In practical terms, this slight lag of replication records is no issue. But, we must ensure that it does not fall too far behind, as this could indicate a problem with the follower. So, how do we determine what's not too far back? Kafka brokers have the configuration replica.lag.time.max.ms (figure 2.15).

The replica lag time configuration sets an upper bound on how long followers must issue a fetch request or be entirely caught up for the leader's log. Followers fail-

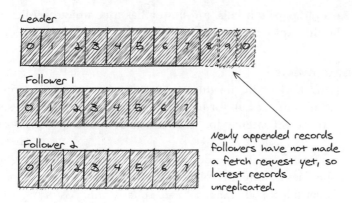

Newly appended records
followers have not made
a fetch request yet, so
latest records
unreplicated.

Figure 2.14 The leader may have a few unreplicated messages in its topic partition.

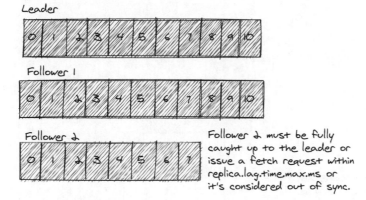

Follower 2 must be fully
caught up to the leader or
issue a fetch request within
replica.lag.time.max.ms or
it's considered out of sync.

Figure 2.15 Followers must issue a fetch request or be caught up within the lag time configuration.

ing to do so within the configured time are considered too far behind and removed from the ISR list.

As I previously stated, follower brokers caught up with their leader broker are considered an ISR. ISR brokers can be elected leaders should the current leader fail or become unavailable (see the Kafka documentation, "Replication," http://kafka.apache.org/documentation/#replication).

In Kafka, consumers never see records that all ISRs haven't written. The offset of the latest record stored by all replicas is known as the high-water mark, and it represents the highest offset accessible to consumers. This property of Kafka means that consumers don't worry about recently read records disappearing. As an example, consider the situation in figure 2.15. Since offsets 8–10 have yet to be written to all the replicas, 7 is the highest offset available to consumers of that topic.

Should the lead broker become unavailable or die before records 8–10 persist, an acknowledgment isn't sent to the producer, and it will retry sending the records. There's a little more to this scenario, and we'll talk about it more in chapter 4 on clients.

If the leader for a topic partition fails, a follower has a complete replica of the leader's log. But we should explore the relationship between leaders, followers, and replicas.

REPLICATION AND ACKNOWLEDGMENTS

When writing records to Kafka, the producer can wait for acknowledgment of record persistence of none, some, or all for in-sync replicas. These different settings allow the producer to trade off latency for data durability. But there is a crucial point to consider.

The leader of a topic-partition is considered a replica itself. The configuration `min.insync.replicas` specifies how many replicas must be in sync to consider a record committed. The default setting for `min.insync.replicas` is one. Assuming a broker cluster size of three and a replication factor of 3 with a setting of `acks=all`, only the leader must acknowledge the record. Figure 2.16 demonstrates this scenario.

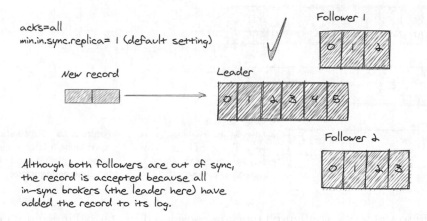

Figure 2.16   acks set to all with default in-sync replicas

How can something like this happen? Imagine that the two followers temporarily lag enough for the controller to remove them from the ISR. This means that even with setting `acks=all` on the producer, there is potential data loss should the leader fail before the followers can recover and become in sync again.

To prevent such a scenario, you need to set the `min.insync.replicas=2`. Setting the `min` in-sync replicas configuration to 2 means the leader checks the number of in-sync replicas before appending a new record to its log. The leader only processes the produce request if the required number of in-sync replicas is met. Instead, the leader throws a `NotEnoughReplicasException`, and the producer will retry the request.

Let's look at another illustration in figure 2.17 to help get a clear idea of what is going on.

As you can see in figure 2.17, a batch of records arrives. But the leader won't append them because there aren't enough in-sync replicas. Doing so increases your data dura-

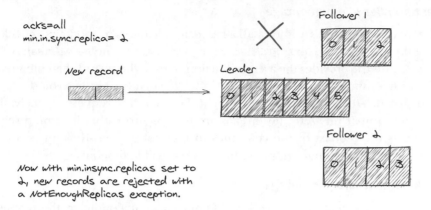

Figure 2.17 Setting the `min` ISR to a value greater than 1 increases data durability.

bility as the produce request will only succeed once enough in-sync replicas exist. This discussion of message acknowledgments and in-sync replicas is broker-centric. In chapter 4, when we discuss clients, we'll revisit this idea from the producer client's perspective to discuss the performance tradeoffs.

2.10 *Checking for a healthy broker*

At the beginning of the chapter, we covered how a Kafka broker handles requests from clients and processes them in the order of their arrival. Kafka brokers handle several types of requests, including, for example

- *Produce*—A request to append records to the log
- *Fetch*—A request to consume records from a given offset
- *Metadata*—A request for the cluster's current state—broker leaders for topic partitions, topic partitions available, etc.

These are a small subset of all possible requests made to the broker. The broker processes requests in first-in,-first-out processing order, passing them off to the appropriate handler based on the request type.

Simply put, a client makes a request, and the broker responds. If requests come in faster than the broker can reply, they queue up. Internally, Kafka has a thread pool dedicated to handling incoming requests. This process leads us to the first line of checking for issues should your Kafka cluster performance suffer.

With a distributed system, you must embrace failure as a way of life. However, this doesn't mean the system should shut down at the first sign of an issue. Network partitions are expected in a distributed system, frequently resolving quickly. It makes sense to have a notion of retryable errors versus fatal errors. If you are experiencing issues with your Kafka installation or timeouts for producing or consuming records, for example, where's the first place to look?

2.10.1 *Request handler idle percentage*

When experiencing issues with a Kafka-based application, a good first check is to examine the `RequestHandlerAvgIdlePercent` JMX metric. The `RequestHandlerAvgIdlePercent` metric provides the average fraction of time the threads handling requests are idle, with a number between 0 and 1. Under normal conditions, you'd expect an idle ratio of 0.7–0.9, indicating that the broker handles requests quickly. If the request-idle number hits zero, no threads are left for processing incoming requests, which means the request queue continues to increase. A massive request queue is problematic, meaning longer response times and possible timeouts.

2.10.2 *Network handler idle percentage*

The `NetworkProcessorAvgIdlePercent` JMX metric is analogous to the request-idle metric. The network-idle metric measures the average time the network processors are busy. In the best scenarios, you want to see the number above 0.5. If it's consistently below 0.5, that indicates a problem.

2.10.3 *Underreplicated partitions*

The `UnderReplicatedPartitions` JMX metric represents the number of partitions belonging to a broker removed from the ISR. We discussed ISR and replication in section 2.9.1. A value higher than zero means a Kafka broker is not keeping up with replicating for assigned follower topic partitions. Causes of a nonzero `UnderReplicatedPartitions` metric could indicate network issues or that the broker is overloaded and can't keep up. Note that you always want to see the `UnderReplicatedPartitions` metric at 0.

Summary

- The Kafka broker is the storage layer and handles client requests for producing (writing) and consuming (reading) records.
- Kafka brokers receive records as bytes, store them in the same format, and send them out for consume requests in byte format.
- Kafka brokers durably store records in topics.
- Topics represent a directory on the filesystem and are partitioned, meaning the records in a topic are placed in different buckets.
- Kafka uses partitions for throughput and for distributing the load as it spreads them out on different brokers.
- Kafka brokers replicate data from each other for durable storage.

Part 2

In part 1, you were introduced to the Apache Kafka event streaming platform. You learned, at a high level, the various components that make up the Kafka platform. You went on from there to learn about how the Kafka broker operates and the various functions it performs in acting as a central nervous system for data. In this part, you're going to dive in and learn in detail about getting data into Kafka.

First up is Schema Registry. Schema Registry helps enforce the implied contract between Kafka producers and consumers (If you don't know what I mean by an implied contract, don't worry; I'll explain it). But you might say, "I don't use schemas." Well, here's the rub: you are always using one; it just depends on whether it's an explicit or implicit schema. By the end of chapter 3, you'll fully understand what I mean by that statement and how Schema Registry solves the problem.

Next, you'll move on to the workhorses of the Kafka platform, producer and consumer clients. You'll see how producers get data into Kafka and how consumers get it out. Learning about the Kafka clients is essential because there are important tools in Kafka that are abstractions over producer and consumer client, so understanding how they work is essential.

Finally, you'll discover Kafka Connect. Built on top of Kafka producer and consumer clients, Connect is a bridge between the outside world and Kafka. Source connectors pull data into Kafka from relational databases or just about any external data store or system such as ElasticSearch, Snowflake, or S3. Sink connectors do the opposite, exporting data from Kafka into external systems.

Schema Registry 3

In chapter 2, you learned about the heart of the Kafka streaming platform, the Kafka broker. In particular, you learned how the broker is the storage layer appending incoming messages to a topic, serving as an immutable, distributed log of events. A topic represents the directory containing the log file(s).

Since the producers send messages over the network, they must be serialized first into binary format, an array of bytes. The Kafka broker does not change the messages in any way; it stores them in the same format. It's the same when the broker responds to fetch requests from consumers; it retrieves the already serialized messages and sends them over the network.

By only working with messages as arrays of bytes, the broker is entirely agnostic to the data type the messages represent and utterly independent of the applications producing and consuming the messages and the programming languages those applications use. Decoupling the broker from the data format permits any client using the Kafka protocol to produce or consume messages.

3.1 *Objects*

While bytes are essential for storage and transport over the network, developers are far more efficient working at a higher level of abstraction: the object. So, where does this transformation occur, then? At the client level, the producers and consumers of messages (figure 3.1).

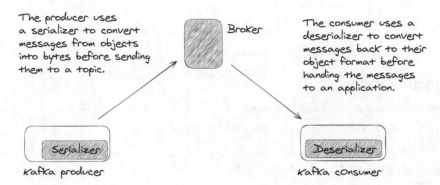

Figure 3.1   The conversion of objects to bytes and bytes to objects happens at the client level.

Looking at this illustration, the message producer uses an instance of a `Serializer` to convert the message object into bytes before sending it to the topic on the broker. The message consumer does the opposite process: it receives bytes from the topic and uses an instance of a `Deserializer` to convert the bytes back into the same object format. The producer and consumer are decoupled from the (de)serializers; they call either the `serialize` or `deserialize` method (figure 3.2).

As depicted in figure 3.2, the producer expects to use an instance of the `Serializer` interface and calls the `Serializer.serialize` method, passing in an object of a given type and getting back bytes. The consumer works with the `Deserializer` interface. The consumer provides an array of bytes to the `Deserializer.deserialize` method and receives an object of a given type in return.

The producer and consumer get the (de)serializers via configuration parameters. We'll see examples later in the chapter.

NOTE I'm mentioning producers and consumers here and throughout the chapter, but we'll only go into enough detail to understand the context

Kafka Producers execute -> Serializer.serialize(T message)

Kafka Consumers execute -> Deserializer.deserialize(byte[] bytes)

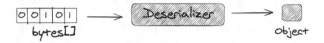

Figure 3.2 The serializer and deserializer are agnostic of the producer and consumer and perform the expected action when executing the `serialize` and `deserialize` methods.

required for this chapter. We'll cover producer and consumer client details in the next chapter.

The point I'm emphasizing here is that the object type the producer serializes for a given topic is expected to be the same object type that a consumer deserializes. Since producers and consumers are completely agnostic of each other, these messages or event domain objects represent an implicit contract between the producers and consumers.

So, now the question is, does something exist that developers of producers and consumers can use that informs them of the proper structure of messages? The answer to that question is, yes, the schema.

3.2 What is a schema, and why do you need one?

When you mention the word *schema* to developers, their first thought is database schemas. A database schema describes its structure, including the names and startups of the columns in database tables and the relationship between tables. But the schema I'm referring to here, while similar in purpose, is quite different.

For our purposes, I'm referring to a language-agnostic description of an object, including the name, the fields on the object, and the type of each field. The following listing is an example of a potential schema in JSON format.

Listing 3.1 Basic example of a schema in JSON format

```
{
"name":"Person",              ◁── The name of the object
  "fields": [                 ◁── Defines the fields on the object
    {"name": "name", "type":"string"},   ◁── The names of the fields and their types
    {"name": "age", "type": "int"},
```

```
      {"name": "email", "type":"string"}
   ]
}
```

Here, our fictional schema describes an object named `Person` with fields we'd expect to find on such an object. Now, we have a structured description of an object that producers and consumers can use as an agreement or contract on what the object should look like before and after serialization. In section 3.2.9, I'll cover details on how you use schemas in message construction and (de)serialization.

But for now, I'd like to review some key points we've established so far:

- The Kafka broker only works with messages in binary format (byte arrays).
- Kafka producers and consumers are responsible for messages' (de)serialization. Additionally, since these two are unaware of each other, the records form a contract between them.

We also learned that we can make the contract between producers and consumers explicit by using a schema. So we have our *why* for using a schema, but what we've defined so far is a bit abstract, and we need to answer these questions for the *how*:

- How do you put schemas to use in your application development lifecycle?
- Given that serialization and deserialization are decoupled from the Kafka producers and consumers, how can they use serialization that ensures messages are in the correct format?
- How do you enforce the correct version of a schema to use? After all, changes are inevitable.

The answer to these *how* questions is Schema Registry.

3.2.1   *What is Schema Registry?*

Schema Registry provides a centralized application for storing schemas, schema validation, and sane schema evolution (message structure changes) procedures. Perhaps more importantly, it serves as the source of truth of schemas that producer and consumer clients can quickly discover. Schema Registry provides serializers and deserializers that you can use to configure Kafka producers and Kafka consumers, easing the development of applications working with Kafka.

The Schema Registry serializing code supports schemas in the serialization frameworks Avro (https://avro.apache.org/docs/current/) and Protocol Buffers (https://developers.google.com/protocol-buffers). I'll refer to Protocol Buffers as "Protobuf" going forward. Additionally, Schema Registry supports schemas written using the JSON Schema (https://json-schema.org/), but this is more of a specification versus a framework. I'll get into working with Avro and Protobuf JSON Schema as we progress through the chapter, but for now, let's take a high-level view of how Schema Registry works, as shown in figure 3.3.

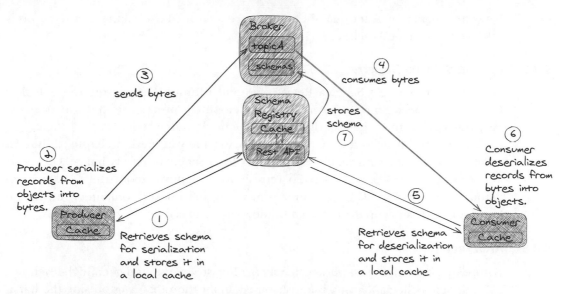

Figure 3.3 Schema Registry ensures consistent data format between producers and consumers.

Let's quickly walk through how Schema Registry works based on this illustration:

1 As a producer calls the `serialize` method, a Schema Registry–aware serializer retrieves the schema (via HTTP) and stores it in its local cache.
2 The serializer embedded in the producer serializes the record.
3 The producer sends the serialized message (bytes) to Kafka.
4 A consumer reads the bytes.
5 The Schema Registry–aware deserializer in the consumer retrieves the schema and stores it in its local cache.
6 The consumer deserializes the bytes based on the schema.
7 The Schema Registry servers produce a message with the schema storing it in the _schemas topic.

TIP While I'm presenting Schema Registry as an essential part of the Kafka event streaming platform, it's not required. Remember, Kafka producers and consumers are decoupled from the serializers and deserializers they use. Providing a class that implements the appropriate interface will work fine with the producer or consumer. But you will lose the validation checks that come from using Schema Registry. I'll cover serializing without Schema Registry at the end of this chapter.

While the previous illustration gave you a good idea of how Schema Registry works, there's an important detail I'd like to point out here. While it's true that the serializer or deserializer will reach out to Schema Registry to retrieve a schema for a given record type, it only does so *once*—the first time it encounters a record type it doesn't have the schema for. After that, the schema needed for (de)serialization operations is retrieved from a local cache.

3.2.2   *Getting Schema Registry*

Our first step is to get Schema Registry up and running. Again, you'll use Docker Compose to speed up your learning and development process, so grab the `docker-compose.yml` file from the root directory in the book source code.

This file is similar to the `docker-compose.yml` file you used in chapter 2. But, in addition to the Kafka image, there is an entry for a Schema Registry image. Go ahead and run `docker-compose up -d`. To refresh your memory about the Docker commands, the `-d` is for "detached" mode, meaning the docker containers run in the background, freeing up the terminal window you've executed the command.

3.2.3   *Architecture*

Before we go into how you work with Schema Registry, we should get a high-level view of its design. Schema Registry is a distributed application that lives outside the Kafka brokers. Clients communicate with Schema Registry via a REST API. A client could be a serializer (producer), deserializer (consumer), a build tool plugin, or a command-line request using `curl`. I'll cover using build tool plugins—Gradle, in this case—in section 3.2.6.

Schema Registry uses Kafka as storage (write-ahead log) for all its schemas in _schemas, a single-partitioned, compacted topic. It has a primary architecture, meaning there is one leader node in the deployment, and the other nodes are secondary.

> **NOTE**   The double underscore characters (__)are a Kafka topic naming convention denoting internal topics not meant for public consumption. From this point forward, we'll refer to this topic simply as `schemas`.

This primary architecture means that only the primary node in the deployment writes to the schemas topic. Any node in the deployment will accept a request to store or update a schema, but secondary nodes forward the request to the primary node. Let's look at figure 3.4 to demonstrate.

Anytime a client registers or updates a schema, the primary node produces a record to the _schemas topic. Schema Registry uses a Kafka producer to write, and all the nodes use a consumer to read updates. So, you can see that Schema Registry backs up its local state in a Kafka topic, making schemas very durable.

> **NOTE**   When working with Schema Registry throughout all the examples in the book, you'll only use a single node deployment suitable for local development.

But all Schema Registry nodes serve read requests from clients. If any secondary nodes receive a registration or update request, they forward it to the primary node. Then, the secondary node returns the response from the primary node. Let's take a

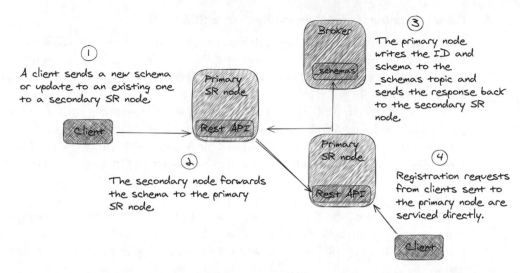

Figure 3.4 Schema Registry is a distributed application where only the primary node communicates with Kafka.

look at an illustration of this architecture in figure 3.5 to solidify your mental model of how this works.

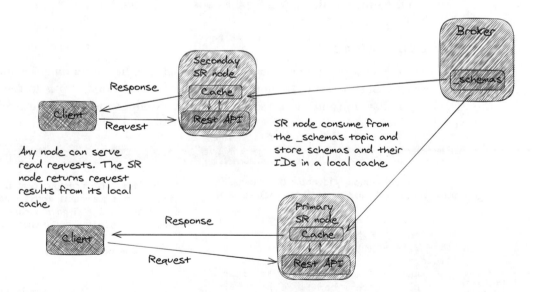

Figure 3.5 All Schema Registry nodes can serve read requests.

Now that we've given an overview of the architecture, let's get to work by issuing a few basic commands using Schema Registry REST API.

3.2.4 *Communication: Using Schema Registry's REST API*

So far, we've covered how Schema Registry works. Now, it's time to see it in action by uploading a schema and running some additional commands to get more information about your uploaded schema. For the initial commands, you'll use `curl` and `jq` in a terminal window.

`curl` (https://curl.se/) is a command-line utility for working with data via URLs. `jq` (https://stedolan.github.io/jq/) is a command-line JSON processor. To install `jq` for your platform, visit the `jq` download site: https://stedolan.github.io/jq/download/. For `curl`, it should come with Windows 10+ and macOS. On Linux, you can install it via a package manager. If you use macOS, you can install both using homebrew (https://brew.sh/).

In later sections, you'll use a Gradle plugin to interact with Schema Registry. After you get an idea of how the different REST API calls work, you'll use the Gradle plugins and some simple producer and consumer examples to see the serialization in action.

Typically, you'll use the build tool plugins to perform Schema Registry actions. First, they make the development process much faster than running the API calls from the command line, and second, they will automatically generate source code from schemas. We'll cover using build tool plugins in section 3.2.6.

> **NOTE** There are Maven and Gradle plugins for working with Schema Registry, but the source code project for the book uses Gradle, so that's the plugin you'll use.

3.2.5 *Registering a schema*

Before we start, ensure you've run `docker-compose up -d` to have a running Schema Registry instance. But there will be nothing registered, so your first step is to register a schema. Let's have fun and create a schema for Marvel Comic superheroes, the Avengers. You'll use Avro for your first schema, and let's take a second now to discuss the format.

Listing 3.2 Avro schema for Avengers

The namespace uniquely identifies the schema. For generated Java code, the namespace is the package name.

The type is record, a complex type. Other complex types are enums, arrays, maps, unions, and fixed. We'll go into more detail about Avro types later in this chapter.

```
{"namespace": "bbejeck.chapter_3",
 "type": "record",
 "name": "Avenger",
 "fields": [
     {"name": "name", "type": "string"},
     {"name": "real_name", "type": "string"},
     {"name": "movies", "type":
                     {"type": "array", "items": "string"},
      "default": []
     }
  ]
}
```

The name of the record

Declares the fields of the record

Describes the individual fields. Fields in Avro are either simple or complex.

Provides a default value. Avro uses the default value when deserializing if the serialized bytes don't contain this field.

You defined Avro schemas in JSON format. You'll use this same schema file in section 3.2.6 when we discuss the Gradle plugin for code generation and interactions with Schema Registry. Since Schema Registry supports Protobuf and JSON Schema formats, let's take a look at the same type in those schema formats here as well.

Listing 3.3 Protobuf schema for Avengers

```
syntax = "proto3";                          ◄───  Defines the version of Protobuf;
                                                  we're using version 3 in this book.

package bbejeck.chapter_3.proto;            ◄───  Declares the package name
option java_multiple_files = true;          ◄───

message Avenger {                           ◄───  Instructs Protobuf to generate
    string name = 1;                        ◄───  separate files for the message
    string real_name = 2;
    repeated string movies = 3;             ◄───  Defines the message

}                              A repeated field;        Unique field number
                               corresponds to a list
```

The Protobuf schema looks closer to regular code as the format is not JSON. Protobuf uses the numbers you see assigned to the fields to identify those fields in the message binary format. While Avro specification allows for setting default values, in Protobuf (version 3), every field is considered optional, but you don't provide a default value. Instead, Protobuf uses the type of the field to determine the default. For example, the default for a numerical field is 0; for strings, it's an empty string, and repeated fields are an empty list.

> **NOTE** Protobuf is a deep subject, and since this book is about the Kafka event streaming pattern, I'll only cover enough of the Protobuf specification for you to get started and feel comfortable using it. For full details, you can read the language guide found here: http://mng.bz/5oB1.

Now let's take a look at the JSON Schema version.

Listing 3.4 JSON Schema schema for Avengers

```
{
  "$schema": "http://json-schema.org/draft-07/schema#",    ◄───  References the
  "title": "Avenger",                                            specific schema spec
  "description": "A JSON schema of Avenger object",
  "type": "object",                                        ◄───  Specifies the
  "javaType": "bbejeck.chapter_3.json.SimpleAvengerJson",  ◄───  type is an object
  "properties": {                              ◄───
    "name": {                                        Lists the fields   The javaType
      "type": "string"                               of the object      used when
    },                                                                  deserializing
    "realName": {
      "type": "string"
    },
```

```
      "movies": {
        "type": "array",
        "items": {
          "type": "string"
        },
        "default": []          ◁──┤ Specifies a
      }                            default value
    },
    "required": [
      "name",
      "realName"
    ]
}
```

The JSON Schema schema resembles the Avro version as both use JSON for the schema file. The most significant difference between the two is that in the JSON Schema, you list the object fields under a `properties` element versus a `fields` array, and in the fields themselves, you simply declare the name versus having a `name` element.

> **NOTE**   Please note there is a difference between a schema written in JSON format and one that follows the JSON Schema format. JSON Schema is "JSON Schema is the vocabulary that enables JSON data consistency, validity, and interoperability at scale" (https://json-schema.org/). As with Avro and Protobuf, I will focus on enough for you to use it in your projects, but for in-depth coverage, you should visit https://json-schema.org/ for more information.

I've shown the different schema formats here for comparison. But in the rest of the chapter, I'll usually only show one version of a schema in an example to save space.

Now that we've reviewed the schemas, let's go ahead and register one. The command to register a schema with REST API on the command line looks like the following listing.

Listing 3.5   Registering a schema on the command line

Uses the jq tojson function to format the avenger.avsc file (new lines aren't valid JSON) for uploading and then pipes the result to the curl command

The POST URL for adding the schema; the -s flag suppresses the progress info output from curl.

```
jq '. | {schema: tojson}' src/main/avro/avenger.avsc | \
curl -s -X
    POST http://localhost:8081/subjects/avro-avengers-value/versions\   ◁──
    -H "Content-Type: application/vnd.schemaregistry.v1+json" \
    -d @-  \                ◁──
    | jq       ◁──
```

The content header

The -d flag specifies the data, and @- means read from STDIN (i.e., the data provided by the jq command preceding the curl command).

Pipes the JSON response through jq to get a nicely formatted response

The result from running this command should look like the following listing.

Listing 3.6 Expected response from uploading a schema

```
{
  "id": 1
}
```

The response from the POST request is the ID that Schema Registry assigned to the new schema. Schema Registry gives a unique ID (a monotonically increasing number) to each newly added schema. Clients use this ID to store schemas in their local cache.

Before we move on to another command, I want to call your attention to listing 3.5, specifically `subjects/avro-avengers-value/`, which specifies the subject name for the schema. Schema Registry uses the subject name to manage the scope of any changes made to a schema. In this case, it's `avro-avengers-value`, which means that values (in the key-value pairs) going into the `avro-avengers` topic need to be in the format of the registered schema. We'll cover subject names and their role in making changes in section 3.3.

Next, let's look at some available commands to retrieve information from Schema Registry. Imagine you are working on building a new application to work with Kafka. You've heard about Schema Registry, and you'd like to look at a particular schema one of your co-workers developed, but you can't remember the name, and it's the weekend, and you don't want to bother anyone. What you can do is list all the subjects of registered schemas with the command in the following listing.

Listing 3.7 Listing the subjects of registered schemas

```
curl -s "http://localhost:8081/subjects" | jq
```

The response from this command is a JSON array of all the subjects. Since we've only registered once schema so far, the results should look like this:

```
[
  "avro-avengers-value"
]
```

Great, you find here what you are looking for—the schema registered for the `avro-avengers` topic.

Now let's consider there have been some changes to the latest schema, and you'd like to see what the previous version was. The problem is you don't know the version history. The next listing shows you all of the versions for a given schema.

Listing 3.8 Getting all versions for a given schema

```
curl -s "http://localhost:8081/subjects/avro-avengers-value/versions" | jq
```

This command returns a JSON array of the versions of the given schema. In our case here, the results should look like this:

```
[
  1
]
```

Now that you have the version number you need, you can run another command to retrieve the schema at a specific version.

Listing 3.9   Retrieving a specific version of a schema

```
curl -s "http://localhost:8081/subjects/avro-avengers-value/versions/1"\
 | jq '.'
```

After running this command, you should see something resembling this:

```
{
  "subject": "avro-avengers-value",
  "version": 1,
  "id": 1,
  "schema": "{\"type\":\"record\",\"name\":\"AvengerAvro\",
    \"namespace\":\"bbejeck.chapter_3.avro\",\"fields\"
    :[{\"name\":\"name\",\"type\":\"string\"},{\"name\"
     :\"real_name\",\"type\":\"string\"},{\"name\"
       :\"movies\",\"type\":{\"type\":\"array\"
         ,\"items\":\"string\"},\"default\":[]}]}"
}
```

The value for the `schema` field is formatted as a string, so the quotes are escaped, and all new-line characters are removed. With a couple of quick commands from a console window, you've been able to find a schema, determine the version history, and view the schema of a particular version.

As a side note, if you don't care about previous versions of a schema and you only want the latest one, you don't need to know the actual latest version number. You can use the REST API call in the following listing to retrieve the latest schema.

Listing 3.10   Getting the latest version of a schema

```
curl -s "http://localhost:8081/subjects/avro-avengers-value/
  versions/latest" | jq '.'
```

I won't show the results of this command here, as it is identical to the previous command.

That has been a quick tour of some of the commands available in the REST API for Schema Registry. This is just a small subset of the available commands. For a full reference, go to http://mng.bz/OZEo.

Next, we'll move on to using Gradle plugins to work with Schema Registry and Avro, Protobuf, and JSON Schema schemas.

3.2.6   *Plugins and serialization platform tools*

So far, you've learned that the event objects written by producers and read by consumers represent the contract between the producer and consumer clients. You've also learned that this implicit contract can be a concrete one in the form of a schema. Additionally, you've seen how you can use Schema Registry to store the schemas and make them available to the producer and consumer clients when they need to serialize and deserialize records.

You'll see even more functionality with Schema Registry in the upcoming sections. I'm referring to testing schemas for compatibility, different compatibility modes, and how changing or evolving a schema can be a relatively painless process for the involved producer and consumer clients.

But so far, you've only worked with a schema file, and that's still a bit abstract. As mentioned earlier in the chapter, developers work with objects when building applications. Our next step is to see how we can convert these schema files into concrete objects you can use in an application.

Schema Registry supports schemas in Avro, Protobuf, and JSON Schema formats. Avro and Protobuf are serialization platforms that provide tooling for working with schemas in their respective formats. One of the most essential tools is the ability to generate objects from the schemas.

Since JSON Schema is a standard and not a library or platform, you'll need to use an open source tool for code generation. For this book, we're using the https://github.com/eirnym/js2p-gradle project. For (de)serialization without Schema Registry, I recommend using `ObjectMapper` from the https://github.com/FasterXML/jackson-databind project.

Generating code from the schema makes your life as a developer more manageable, as it automates the repetitive boilerplate process of creating domain objects. Additionally, since you maintain the schemas in source control (Git, in our case), the chance for error, such as making a field a string type when it should be a `long` when creating the domain objects, is significantly reduced. Also, when making a change to a schema, you commit the change, and other developers pull the update and regenerate the code, and everyone is updated quickly.

We'll use the Gradle build tool (https://gradle.org/) to manage the book's source code in this book. Fortunately, there are Gradle plugins we can use for working with Schema Registry, Avro, Protobuf, and JSON Schema. Specifically, we'll use the following plugins:

- *https://github.com/ImFlog/schema-registry-plugin*—Used for interacting with Schema Registry (i.e., testing schema compatibility, registering schemas, and configuring schema compatibility)
- *https://github.com/davidmc24/gradle-avro-plugin*—Used for Java code generation from Avro schema (`.avsc`) files
- *https://github.com/google/protobuf-gradle-plugin*—Used for Java code generation from Protobuf schema (`.proto`) files

- *https://github.com/eirnym/js2p-gradle*—Used for Java code generation for schemas using the JSON Schema specification

> **NOTE**   It's important to note the distinction between schema files written in JSON, such as Avro schemas, and those using JSON Schema format (https://json-schema.org/). In the case of Avro files, they are written as JSON but follow the Avro specification. The JSON Schema files follow the official specification for JSON Schemas.

Using the Gradle plugins for Avro, Protobuf, and JSON Schema, you don't need to learn how to use the individual tools for each component; the plugins handle all the work. We'll also use a Gradle plugin to handle most interactions with Schema Registry. Let's start by uploading a schema using a Gradle command instead of a REST API command in the console.

3.2.7   *Uploading a schema file*

The first thing we'll do is use Gradle to register a schema. We'll use the same Avro schema from the REST API commands section. To upload the schema, make sure to change your current directory (CD) to the base directory of project and run this Gradle command:

```
./gradlew streams:registerSchemasTask
```

After running this command, you should see something like BUILD SUCCESSFUL in the console. Notice that all you need to enter on the command line is the name of the Gradle task (from the schema registry plugin), and the task registers all the schema inside the register { } block in the streams/build.gradle file.

Now, let's take a look at the configuration of the Schema Registry plugin in the streams/build.gradle file.

Listing 3.11   Configuration for Schema Registry plugin in streams/build.gradle

```
schemaRegistry {
    url = 'http://localhost:8081'

register {
        subject('avro-avengers-value',
                'src/main/avro/avenger.avsc',
                'AVRO')

    //other entries left out for clarity
    }

  // other configurations left out for clarity
}
```

Start of the Schema Registry configuration block in the build.gradle file

Specifies the URL to connect to Schema Registry

Registers a schema by subject name

Specifies the Avro schema file to register

The type of the schema you are registering

In the `register` block, you provide the same information, just in the format of a method call versus a URL in a REST call. Under the covers, the plugin code is still using the Schema Registry REST API via a `SchemaRegistryClient`. As a side note, you'll notice several entries in the `register` block in the source code. You'll use all of them when you go through the examples in the source code.

We'll cover using more Gradle Schema Registry tasks soon, but let's move on to generating code from a schema.

3.2.8   Generating code from schemas

As I said earlier, one of the best advantages of using the Avro and Protobuf platforms is the code generation tools. Using the Gradle plugin for these tools takes the convenience a bit further by abstracting away the details of using the individual tools. To generate the objects represented by the schemas, all you need to do is run the Gradle task in the following listing.

Listing 3.12   Generating the model objects

```
./gradlew clean build
```

Running this Gradle command generates Java code for all the types of schemas in the project—Avro, Protobuf, and JSON Schema. Now we should talk about where you place the schemas in the project. The default locations for the Avro and Protobuf schemas are the `src/main/avro` and `src/main/proto` directories, respectively. The location for the JSON Schema schemas is the `src/main/json` directory, but you need to explicitly configure this in the `build.gradle` file.

Listing 3.13   Configuring the location of JSON Schema schema files

```
jsonSchema2Pojo {

  source = files("${project.projectDir}/src/main/json")
  targetDirectory = file("${project.buildDir}/generated-main-json-java")
  // other configurations left out for clarity
}
```

The source configuration specifies where the generation tools can locate the schemas

The targetDirectory is where the tool writes the generated Java objects

> **NOTE**   All examples here refer to the schemas found in the `streams` subdirectory unless otherwise specified.

Here you can see the configuration of the input and output directories for the `js2p-gradle` plugin. The Avro plugin, by default, places the generated files in a subdirectory under the `build` directory named `generated-main-avro-java`. For Protobuf, we configure the output directory to match the pattern of JSON Schema and Avro in the `Protobuf` block of the `build.gradle` file, as shown in the following listing.

Listing 3.14   Configuring Protobuf output

```
protobuf {

    protoc {
        artifact = 'com.google.protobuf:protoc:3.25.0'    ⟵
    }
}
```
Specifies the location of the protoc compiler

I'd like to take a quick second to discuss the annotation in listing 3.14. To use Proto-buf, you need to have the compiler `protoc` installed. By default the plugin searches for a `protoc` executable. But we can use a pre-compiled version of `protoc` from Maven Central, which means you don't have to explicitly install it. But if you prefer to use your local install, you can specify the path inside the `protoc` block with `path = path/to/protoc/compiler`.

So we've wrapped up generating code from the schemas. Now it's time to run an end-to-end example.

3.2.9   *End-to-end example*

We will take everything you've learned and run a simple end-to-end example. So far, you have registered the schemas and generated the required Java files. So your next steps are to

1 Create some domain objects from the generated Java files.
2 Produce your created objects to a Kafka topic.
3 Consume the objects you just sent from the same Kafka topic.

While steps 2 and 3 have more to do with clients than Schema Registry, I want to think about it from this perspective. You're creating instances of Java objects created from the schema files, so pay attention to fields and notice how the objects conform to the structure of the schema. Also, focus on the Schema Registry–related configuration items, serializer or deserializer, and the URL for communicating with it.

> **NOTE** In this example, you will use a Kafka producer and a Kafka consumer, but I won't cover any details of working with them. If you're still getting famil-iar with the producer and consumer clients, that's fine. I'll go into detail about producers and consumers in the next chapter. But for now, go through the examples as is.

If you still need to register the schema files and generate the Java code, let's do so now. I'll repeat the steps and confirm you have run `docker-compose up -d` to ensure your Kafka broker and Schema Registry are running.

Listing 3.15   Registering schemas and generating Java files

```
./gradlew streams:registerSchemasTask    ⟵  Registers the schema files

./gradlew clean build    ⟵  Builds the Java objects
                             from schemas
```

Now let's focus on the Schema Registry–specific configurations. Go to the source code and look at the `bbejeck.chapter_3.producer.BaseProducer` class. For now, we only want to look at the following two configurations; we'll cover more configurations for the producer in the next chapter:

Specifies the serializer to use

```
producerProps.put(ProducerConfig.VALUE_SERIALIZER_CLASS_CONFIG,
    keySerializer);                                                    ◄──────
producerProps.put(AbstractKafkaSchemaSerDeConfig.SCHEMA_REGISTRY_URL_CONFIG,
    "http://localhost:8081");          ◄────
```
Sets the location of Schema registry

The first configuration sets the `Serializer` the producer will use. Remember, the `Kafka-Producer` is decoupled from the type of the `Serializer`; it simply calls the `serialize` method and gets back an array of bytes to send. So you are responsible for providing the correct `Serializer` class.

In this case, we will work with objects generated from an Avro schema, so you use `KafkaAvroSerializer`. If you look at the `bbejeck.chapter_3.producer.avro.Avro-Producer` class (which extends the `BaseProducer`), you see it pass `KafkaAvroSerializer.class` to the parent object constructor. The second configuration specifies the HTTP endpoint that the `Serializer` uses to communicate with Schema Registry. These configurations enable the interactions described in figure 3.3.

Next, let's take a quick look at creating an object.

Listing 3.16 Instantiating an object from the generated code

```
var blackWidow = AvengerAvro.newBuilder()
                .setName("Black Widow")
                .setRealName("Natasha Romanova")
                .setMovies(List.of("Avengers", "Infinity Wars",
                  "End Game")).build();
```

OK, now you're thinking, "This code creates an object; what's the big deal?" While it could be a minor point, it's more what you can't do here that I'm trying to drive home. You can only populate the expected fields with the correct types, enforcing the contract of producing records in the desired format. You could update the schema and regenerate the code.

But by making changes, you have to register the new schema, and the changes have to match the current compatibility format for the subject name. So now you can see how Schema Registry enforces the "contract" between producers and consumers. We'll cover compatibility modes and the allowed changes in section 3.4.

Now, let's run the following Gradle command to produce the objects to the `avro-avengers` topic.

Listing 3.17 Running the AvroProducer

```
./gradlew streams:runAvroProducer
```

After running this command, you'll see some output similar to this:

```
DEBUG [main] bbejeck.chapter_3.producer.BaseProducer - Producing records
 [{"name": "Black Widow", "real_name": "Natasha Romanova", "movies":
["Avengers", "Infinity Wars", "End Game"]},
{"name": "Hulk", "real_name": "Dr. Bruce Banner", "movies":
["Avengers", "Ragnarok", "Infinity Wars"]},
{"name": "Thor", "real_name": "Thor", "movies":
["Dark Universe", "Ragnarok", "Avengers"]}]
```

After the application produces these few records, it shuts itself down.

> **NOTE**   It's essential to run this command exactly as shown here, including the preceding : character. We have three different Gradle modules for our Schema Registry exercises. We need to ensure the commands we run are for the specific module. In this case, the : executes the main module only; otherwise, it will run the producer for all modules, and the example will fail.

Running this command doesn't do anything exciting, but it demonstrates the ease of serializing using a Schema Registry. The producer retrieves the schema, stores it locally, and sends the records to Kafka in the correct serialized format—all without you having to write any serialization or domain model code. Congratulations, you have sent serialized records to Kafka!

> **TIP**   It could be instructive to look at the log file generated from running this command. It's in the book source code's `streams/logs/` directory. The `log4j` configuration overwrites the log file with each run, so inspect it before running the next step.

Now, let's run a consumer that will deserialize the records. But as we did with the producer, we're going to focus on the configuration required for deserialization and working with Schema Registry.

Listing 3.18   Consumer configuration for using Avro

Configured to use a SpecificAvroReader   **Uses Avro deserialization**

```
consumerProps.put(ConsumerConfig.VALUE_DESERIALIZER_CLASS_CONFIG,
    KafkaAvroDeserializer.class);
consumerProps.put(KafkaAvroDeserializerConfig.SPECIFIC_AVRO_READER_CONFIG,
    true);
consumerProps.put(AbstractKafkaSchemaSerDeConfig.SCHEMA_REGISTRY_URL_CONFIG,
    "http://localhost:8081");
```
                                              The host:port for Schema Registry

You'll notice that you set the `SPECIFIC_AVRO_READER_CONFIG` to `true`. What does `SPECIFIC_AVRO_READER_CONFIG` do? To answer that question, let's briefly discuss working with Avro, Protobuf, and JSON Schema serialized objects.

When deserializing one of the Avro, Protobuf, or JSON Schema objects, the specific object type or a nonspecific container object is deserialized. For example, with

SPECIFIC_AVRO_READER_CONFIG set to true, the deserializer inside the consumer will return an object of type AvroAvenger, the *specific* object type. However, if you set SPECIFIC_AVRO_READER_CONFIG to false, the deserializer returns an object of type GenericRecord. GenericRecrod still follows the same schema and has the same content, but the object itself is devoid of any type awareness. As the name implies, it's simply a generic container of fields. The example in the following listing should make clear what I'm saying here.

Listing 3.19 Specific Avro records vs. GenericRecord

```
AvroAvenger avenger = // returned from consumer with
  //SPECIFIC_AVRO_READER_CONFIG=true
avenger.getName();
avenger.getRealName();          ◁──┤ Accesses fields on
avenger.getMovies();                │ the specific object

GenericRecord genericRecord = // returned from consumer with
  //SPECIFIC_AVRO_READER_CONFIG=false
if (genericRecord.hasField("name")) {
   genericRecord.get("name");
}

if (genericRecord.hasField("real_name")) {    ◁──┤ Accesses fields on
   genericRecord.get("real_name");                │ the generic object
}

if (GenericRecord.hasField("movies")) {
   genericRecord.get("movies");
}
```

This simple code example shows the differences between the specific returned type and the generic. With the AvroAvenger object, we can access the available properties directly, as the object is "aware" of its structure and provides methods for accessing those fields. But with the GenericRecord object, you need to query whether it contains a specific field before attempting to access it.

> **NOTE** The specific version of the Avro schema is not just a POJO (plain old Java object) but extends the SpecificRecordBase class.

Notice that with GenericRecord, you need to access the field precisely as specified in the schema, while the specific version uses the more familiar camel case notation. The difference between the two is that with the specific type, you know the structure, but with the generic type, since it could represent any arbitrary type, you need to query for different fields to determine its structure. You will work with a GenericRecord much like a HashMap.

However, you don't have to operate entirely in the dark. You can get a list of fields from a GenericRecord by calling GenericRecord.getSchema().getFields(). Then, you could iterate over the list of Field objects and get the names by calling the

`Fields.name()`. Additionally, you could get the name of the schema with `Generic-Record.getSchema().getFullName()`; presumably, at that point, you would know which fields the record contained. Updating a field, you'd follow a similar approach.

> ### Listing 3.20   Updating or setting fields on specific and generic records

```
avenger.setRealName("updated name")
genericRecord.put("real_name", "updated name")
```

So, this small example shows that the specific object gives you the familiar setter functionality. But in the generic version, you must explicitly declare the field you are updating. Again, you'll notice the `HashMap` behavior updating or setting a field with the generic version.

Protobuf provides a similar functionality for working with specific or arbitrary types. To work with an arbitrary type in Protobuf, you'd use a `DynamicMessage`. As with the Avro `GenericRecord`, the `DynamicMessage` offers functions to discover the type and the fields. With JSON Schema, the specific types are just the objects generated from the Gradle plugin; there's no framework code associated with it like Avro or Protobuf. The generic version is a type of `JsonNode` since the deserializer uses the jackson-databind (https://github.com/FasterXML/jackson-databind) API for serialization and deserialization.

> **NOTE**   The source code for this chapter contains examples of working with the specific and generic types of Avro, Protobuf, and JSON Schema.

So, when do you use the specific type versus the generic? You'll use the specific version if you only have one type of record in a Kafka topic. However, if you have multiple event types in a topic, you'll want to use the generic version, as each consumed record could be a different type. We'll discuss multiple event types in a single topic later in this chapter and again in chapter 4 when covering Kafka clients.

The final thing to remember is to use the specific record type: set the `kafka-AvroDeserializerConfig.SPECIFIC_AVRO_READER_CONFIG` to `true`. The default for the `SPECIFIC_AVRO_READER_CONFIG` is `false`, so the consumer returns the `Generic-Record` type if the configuration is not set.

With the discussion about different record types completed, let's resume walking through your first end-to-end example using Schema Registry. You've already produced some records using the schema you uploaded previously. Now, you need to start a consumer to demonstrate deserializing those records with the schema. Again, looking at the log files should be instructive. You'll see the embedded deserializer downloading the schema for the first record only as it gets cached after the initial retrieval.

I should also note that the following example using `bbejeck.chapter_3.consumer .avro.AvroConsumer` uses both the specific class type and the `GenericRecord` type. As the example runs, the code prints out the type of the consumed record.

NOTE   The source code has similar examples for Protobuf and JSON Schema.

So let's run the consumer example now by executing the command in the following listing from the root of the book source code project.

> **Listing 3.21   Running the AvroConsumer**

```
./gradlew streams:runAvroConsumer
```

NOTE   Again, the same caveat here about running the command with the preceding : character; otherwise, it will run the consumer for all modules, and the example will not work.

The `AvroConsumer` prints out the consumed records and shuts down. Congratulations, you've just serialized and deserialized records using Schema Registry!

So far, we've covered the types of serialization frameworks supported by Schema Registry, how to write and add a schema file, and a basic example using a schema. During the portion of the chapter where you uploaded a schema, I mentioned the term `subject` and how it defines the scope of schema evolution. The following section will teach you how to use the different subject name strategies.

3.3   Subject name strategies

Schema Registry uses the concept of a subject to control the scope of schema evolution. Another way to think of the subject is a namespace for a particular schema. In other words, as your business requirements evolve, you'll need to change your schema files to make the appropriate changes to your domain objects. For example, with our AvroAvenger domain object, you want to remove the hero's real (civilian) name and add a list of their powers.

Schema Registry uses the subject to look up the existing schema and compare the changes with the new schema. It performs this check to ensure the changes are compatible with the current compatibility mode set. We'll talk about compatibility modes in section 3.4. The subject name strategy determines the scope of where Schema Registry makes its compatibility checks.

There are three types of subject name strategies: `TopicNameStrategy`, `RecordNameStrategy`, and `TopicRecordNameStrategy`. You can infer the scope of the namespacing implied by the strategy names, but it's worth reviewing the details. Let's dive in and discuss these different strategies now.

NOTE   By default, all serializers will attempt to register a schema when serializing if they don't find the corresponding ID in their local cache. Autoregistration is a great development feature, but you may need to turn it off in a production setting with a producer configuration setting of `auto.register .schemas=false`. Another example of not wanting autoregistration is when you are using an Avro union schema with references. We'll cover this topic in more detail later in the chapter.

3.3.1   *TopicNameStrategy*

The `TopicNameStrategy` is the default subject in Schema Registry. The subject name comes from the name of the topic. You saw the `TopicNameStrategy` in action earlier in the chapter when you registered a schema with the Gradle plugin. To be more precise, the subject name is `topic-name-key` or `topic-name-value` as you can have different types for the key and value requiring different schemas.

`TopicNameStrategy` ensures only one data type on a topic since you can't register a schema for a different kind with the same topic name. Having a single type per topic makes sense in a lot of cases—for example, if you name your topics based on the event type they store. It follows that they will contain only one record type.

Another advantage of `TopicNameStrategy` is that with the schema enforcement limited to a single topic, you can have another topic using the same record type but a different schema (figure 3.6). Consider the situation where two departments employ the same record type but use other topic names. With `TopicNameStrategy`, these departments can register completely different schemas for the same record type since the scope of the schema is limited to a particular topic.

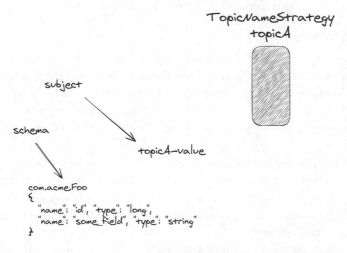

Here, the registered schema is <topic-name>-value. This restricts the type contained in the topic to that of the registered schema for the value type.

Figure 3.6   `TopicNameStrategy` enforces having the same type of domain object represented by the registered schema for the value and or the key.

Since `TopicNameStrategy` is the default, you don't need to specify any additional configurations. When you register schemas, you'll use the format of `<topic>-value` as the subject for value schemas and `<topic>-key` as the subject for key schemas. In both cases, you substitute the topic's name for the `<topic>` token.

But there could be cases where you have closely related events and want to produce those records into one topic. In that case, you'll want to choose a strategy that allows different types and schemas in a topic.

3.3.2 *RecordNameStrategy*

`RecordNameStrategy` uses the fully qualified class name (of the Java object representation of the schema) as the subject name (figure 3.7). By using the record name strategy, you can now have multiple types of records in the same topic. But the critical point is that these records have a logical relationship; just their physical layouts are different.

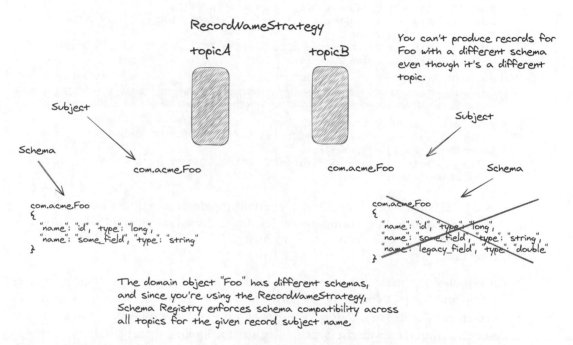

Figure 3.7 `RecordNameStrategy` **enforces having the same schema for a domain object across different topics.**

When would you choose `RecordNameStrategy`? Imagine you have different Internet of Things (IoT) sensors deployed. Some sensors measure different events so that the record structure will be dissimilar. But you still want to have them co-located on the same topic.

Since there can be different types, compatibility checks occur between schemas with the same record name. Additionally, the compatibility check extends to all topics using a subject with the same record name.

To use `RecordNameStrategy`, you use a fully qualified class name for the subject when registering a schema for a given record type. For the `AvengerAvro` object we've used in our examples, you would configure the schema registration as shown in the following listing.

Listing 3.22   Schema Registry Gradle plugin configuration for `RecordNameStrategy`

```
subject('bbejeck.chapter_3.avro.AvengerAvro','src/main/avro/avenger.avsc', 'AVRO')
```

Then, you must configure the producer and consumer with the appropriate subject name strategy.

Listing 3.23   Producer configuration for `RecordNameStrategy`

```
Map<String, Object> producerConfig = new HashMap<>();
 producerConfig.put(KafkaAvroSerializerConfig.VALUE_SUBJECT_NAME_STRATEGY,
  RecordNameStrategy.class);
 producerConfig.put(KafkaAvroSerializerConfig.KEY_SUBJECT_NAME_STRATEGY,
  RecordNameStrategy.class);
```

Listing 3.24   Consumer configuration for RecordNameStrategy

```
Map<String, Object> consumerConfig = new HashMap<>();
 config.put(KafkaAvroDeserializerConfig.KEY_SUBJECT_NAME_STRATEGY,
  RecordNameStrategy.class);
 config.put(KafkaAvroDeserializerConfig.VALUE_SUBJECT_NAME_STRATEGY,
  RecordNameStrategy.class);
```

NOTE   If you are only using Avro for serializing/deserializing the values, you don't need to add the configuration for the key. Also, the key and value subject name strategies do not need to match; I've only presented them that way here.

For Protobuf, use `KafkaProtobufSerializerConfig` and `KafkaProtobufDeserializerConfig`, and for JSON Schema, use `KafkaJsonSchemaSerializerConfig` and `KafkaJsonSchemaDeserializerConfig`. These configurations only affect how the serializer/deserializer interacts with the Schema Registry for looking up schemas. Again, serialization is decoupled from the production and consumption processes.

One thing to consider is that by using only the record name, all topics must use the same schema. If you want to use different records in a topic but want only to consider the schemas for that particular topic, you'll need to use another strategy.

3.3.3   *TopicRecordNameStrategy*

As you can infer from the name, this strategy also allows multiple record types within a topic. However, the registered schemas for a given record are only considered within the scope of the current topic. Let's look at figure 3.8 to understand better what this means.

As you can see in figure 3.8, `topic-A` can have a different schema for the record type `Foo` from `topic-B`. This strategy allows you to have multiple logically related types

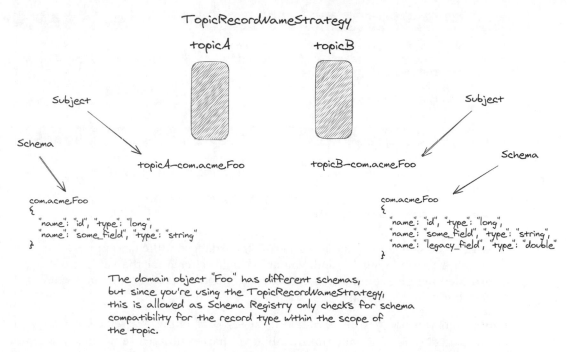

Figure 3.8   `TopicRecordNameStrategy` **allows different schemas for the same domain object across different topics.**

on one topic, but it's isolated from other topics where you have the same type but use different schemas.

Why would you use the `TopicRecordNameStrategy`? For example, consider this situation. You have one version of the `CustomerPurchaseEvent` event object in the `interactions` topic that groups all customer event types (`CustomerSearchEvent`, `CustomerLoginEvent`, etc.). But you have an older topic, `purchases`, that also contains `CustomerPurchaseEvent` objects. Still, it's for a legacy system, so the schema is older and contains different fields from the newer one. `TopicRecordNameStrategy` allows these two topics to contain the same *type* but with different schema versions. Similar to the `RecordNameStrategy` you'll need to configure the strategy, as shown in the following listing.

> **Listing 3.25   Schema Registry Gradle plugin configuration for**
> **`TopicRecordNameStrategy`**

```
subject('avro-avengers-bbejeck.chapter_3.avro.AvengerAvro',
  'src/main/avro/avenger.avsc', 'AVRO')
```

Then you must configure the producer and consumer with the appropriate subject name strategy. See, for example, the following listings.

Listing 3.26   Producer configuration for `TopicRecordNameStrategy`

```
Map<String, Object> producerConfig = new HashMap<>();
 producerConfig.put(KafkaAvroSerializerConfig.VALUE_SUBJECT_NAME_STRATEGY,
  TopicRecordNameStrategy.class);
 producerConfig.put(KafkaAvroSerializerConfig.KEY_SUBJECT_NAME_STRATEGY,
  TopicRecordNameStrategy.class);
```

Listing 3.27   Consumer configuration for TopicRecordNameStrategy

```
Map<String, Object> consumerConfig = new HashMap<>();
 config.put(KafkaAvroDeserializerConfig.KEY_SUBJECT_NAME_STRATEGY,
  TopicRecordNameStrategy.class);
 config.put(KafkaAvroDeserializerConfig.VALUE_SUBJECT_NAME_STRATEGY,
  TopicRecordNameStrategy.class);
```

NOTE   The same caveat about registering the strategy for the key applies here as well. You would only do so if you are using a schema for the key; it's only provided here for completeness. Also, the key and value subject name strategies don't need to match.

Why would you use `TopicRecordNameStrategy` over either `TopicNameStrategy` or `RecordNameStrategy`? If you want the ability to have multiple event types in a topic, you need the flexibility to have different schema versions for a given type across your topics.

But when considering multiple types in a topic, neither `TopicRecordNameStrategy` nor `RecordNameStrategy` can constrain a topic to a fixed set of types. Using either of those subject name strategies opens up the topic to an unbounded number of different types. In section 3.5, we'll cover how to improve this situation when we cover schema references.

Here's a quick summary of the different subject name strategies (Table 3.1). Think of the subject name strategy as a function that accepts the topic name and record schema as arguments, returning a subject name. `TopicNameStrategy` only uses the topic name and ignores the record schema. `RecordNameStrategy` does the opposite: it ignores the topic name and only uses the record schema. But `TopicRecordName-Strategy` uses both of them for the subject name.

Table 3.1   Schema strategies summary

Strategy	Multiple types in a topic	Different versions of objects across topics
`TopicNameStrategy`	Maybe	Yes
`RecordNameStrategy`	Yes	No
`TopicRecordNameStrategy`	Yes	Yes

So far, we've covered the subject naming strategies and how Schema Registry uses subjects for name-spacing schemas. But schema management has another dimension: evolving changes within the schema itself. How do you handle changes like the removal or addition of a field? Do you want your clients to have forward or backward compatibility? The following section will cover exactly how you handle schema compatibility.

3.4 Schema compatibility

When there are schema changes, you need to consider the compatibility with the existing schema and the producer and consumer clients. If you make a change by removing a field, how does this affect the producer serializing the records or the consumer deserializing this new format?

To handle these compatibility concerns, Schema Registry provides four base compatibility modes: BACKWARD, FORWARD, FULL, and NONE. There are also three additional compatibility modes, BACKWARD_TRANSITIVE, FORWARD_TRANSITIVE, and FULL_TRANSITIVE, that extend on the base compatibility mode with the same name. The base compatibility modes only guarantee that a new schema is compatible with the immediate previous version. Transitive compatibility specifies that the new schema is compatible with all earlier versions of a given schema applying the compatibility mode. You can specify a global compatibility level or a compatibility level per subject.

What follows is a description of the valid changes for a given compatibility mode and an illustration demonstrating the sequence of changes you'd need to make to the producers and consumers. See appendix C for a hands-on tutorial on changing a schema.

3.4.1 Backward compatibility

Backward compatibility is the default migration setting. With backward compatibility, you update the consumer code first to support the new schema (figure 3.9). The updated consumers can read records serialized with the new or immediate previous schema.

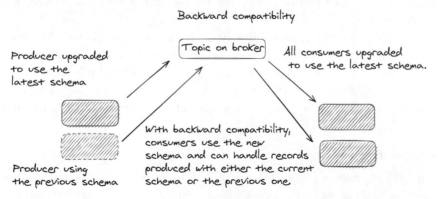

Figure 3.9 Backward compatibility updates consumers first to use the new schema. Then, they can handle records from producers using either the new schema or the previous one.

As shown in figure 3.9, the consumer can work with the previous and the new schemas. The allowed changes with backward compatibility are deleting fields or adding optional fields. A field is optional when the schema provides a default value. If the serialized bytes don't contain the optional field, the deserializer uses the specified default value when deserializing the bytes back into an object.

3.4.2   *Forward compatibility*

Forward compatibility is a mirror image of backward compatibility regarding field changes. With forward compatibility, you can add fields and delete optional ones (figure 3.10).

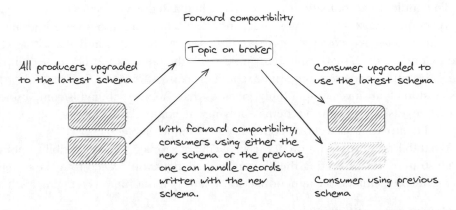

Figure 3.10   Forward compatibility updates producers first to use the new schema, and consumers can handle the records either the new schema or the previous one.

Upgrading the producer code first ensures the new fields are correctly populated and only records in the new format are available. Consumers that need updating can still work with the new schema as it will ignore the new fields, and the deleted fields have default values.

Now, you've seen two compatibility types: backward and forward. As the compatibility name implies, you must consider changes in one direction. In backward compatibility, you updated the consumers first, as records could arrive in either the new or old format. In forward compatibility, you first updated the producers to ensure the records from that time are in the new format. The last compatibility strategy to explore is the full compatibility mode.

3.4.3   *Full compatibility*

You can add or remove fields in full compatibility mode, but there is one catch: any changes you make must be to optional fields only. An optional field is one where you provide a default value in the schema definition should the original deserialized record not provide that specific field.

NOTE   Avro and JSON Schema support explicitly providing default values. With Protocol Buffers, version 3 (the version used in the book), every field automatically has a default based on its type. For example, number types are 0, strings are ", collections are empty, etc.

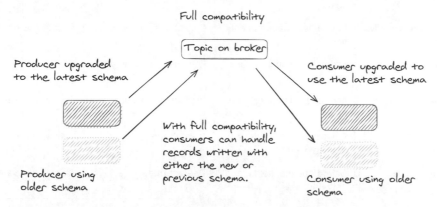

Figure 3.11   Full compatibility allows producers to send with the previous or new schema, and consumers can handle the records with either the new schema or the previous one.

Since the fields in the updated schema are optional, these changes are compatible with existing producer and consumer clients. As such, the upgrade order, in this case, is up to you. Consumers will continue to work with records produced with the new or old schema.

3.4.4   *No compatibility*

Specifying a compatibility of NONE instructs Schema Registry to do just that: no compatibility checks. Not using compatibility checks means someone can add new fields, remove existing fields, or change the field type. Any changes are accepted.

Not providing any compatibility checks provides a great deal of freedom. But the tradeoff is you're vulnerable to breaking changes that might go undetected until the worst possible time: in production.

It could be that every time you update a schema, you upgrade all producers and consumers simultaneously. Another possibility is to create a new topic for clients to use. Applications can use the new topic without concerns containing records from the older, incompatible schema.

Now you've learned how to migrate a schema to use a new version with changes within the different schema compatibility modes. For review, table 3.2 is a quick summary table of the various compatibility types.

But there's more you can do with schemas. Much like working with objects, you can share common code to reduce duplication and make maintenance more manageable. You can do the same with schema references.

Table 3.2 Schema compatibility mode summary

Mode	Changes allowed	Client update order	Retro guaranteed compatibility
Backward	Delete fields, add optional fields	Consumers, producers	Prior version
Backward transitive	Delete fields, add optional fields	Consumers, producers	All previous versions
Forward	Add fields, delete optional fields	Producers, consumers	Prior version
Forward transitive	Add fields, delete optional fields	Producers, consumers	All previous versions
Full	Delete optional fields, add optional fields	Doesn't matter	Prior version
Full transitive	Delete optional fields, add optional fields	Doesn't matter	All previous versions

3.5 Schema references

A schema reference is just what it sounds like, referring to another schema inside the current schema. Reuse is a core principle in software engineering, as the ability to use something you've already built solves two issues. First, you could save time by not rewriting some existing code. Second, when you need to update the original work (which always happens), all the downstream components using the original get automatically updated.

When would you want to use a schema reference? Suppose you have an application providing information on commercial businesses and universities. To model the business, you have a `Company` schema; for the universities, you have a `College` schema. A company has executives, and the college has professors. You want to represent both with a nested `Person` domain object. The schemas would look something like the following listing.

Listing 3.28 College schema

```
"namespace": "bbejeck.chapter_3.avro",
  "type": "record",
  "name": "CollegeAvro",
  "fields": [
    {"name": "name", "type": "string"},
    {"name": "professors", "type":              <-- Array of professors
    {"type": "array", "items": {
      "namespace": "bbejeck.chapter_3.avro",
      "name":"PersonAvro",                      <-- The item type in the array is a Person object.
      "fields": [
        {"name": "name", "type":"string"},
        {"name": "address", "type": "string"},
        {"name": "age", "type": "int"}
```

```
        ]
      }},
        "default": []
      }
   ]
}
```

So you can see here you have a nested record type in your college schema, which is not uncommon. Now let's look at the company schema.

Listing 3.29   Company schema

```
{
   "namespace": "bbejeck.chapter_3.avro",
   "type": "record",
   "name": "CompanyAvro",
   "fields": [
      {"name": "name", "type": "string"},
      {"name": "executives", "type":                   ◀── Array of
      {"type": "array", "items": {                          executives
         "type":"record",
         "namespace": "bbejeck.chapter_3.avro",
         "name":"PersonAvro",                         ◀── Item type is a
         "fields": [                                       PersonAvro
            {"name": "name", "type":"string"},
            {"name": "address", "type": "string"},
            {"name": "age", "type": "int"}
         ]
      }},
        "default": []
      }
   ]
}
```

Again, you have a nested record for the type contained in the schema array. It's natural to model the executive or professor type as a person, as it allows you to encapsulate all the details into an object. But as you can see here, there's duplication in your schemas. If you need to change the person schema, you need to update every file containing the nested person definition. Additionally, as you add more definitions, the size and complexity of the schemas can get unwieldy quickly due to all the nesting of types.

It would be better to put a reference to the type when defining the array. So let's do that next. We'll put the nested `PersonAvro` record in its schema file, `person.avsc`.

I won't show the file here, as nothing changes. I'm putting the definition you see here in a separate file. Now let's take a look at how you'd update the `college.avsc` and `company.avsc` schema files.

Listing 3.30   Updated College schema

```
{
   "namespace": "bbejeck.chapter_3.avro",
   "type": "record",
```

```
    "name": "CollegeAvro",
    "fields": [
      {"name": "name", "type": "string"},
      {"name": "professors", "type":
      {"type": "array", "items": "bbejeck.chapter_3.avro.PersonAvro"},
        "default": []
      }
    ]
}
```

**This is the new part; it's a
reference to the person object**

NOTE   The referring schema you provide must be the same type when using
schema references. For example, you can't give a reference to an Avro
schema or JSON Schema inside the Protocol Buffers schema; the reference
must be another Protocol Buffers schema.

Here, you've cleaned things up by referencing the object created by the `person.avsc`
schema. Now let's look at the updated company schema.

Listing 3.31   Updated Company schema

```
{
    "namespace": "bbejeck.chapter_3.avro",
    "type": "record",
    "name": "CompanyAvro",
    "fields": [
      {"name": "name", "type": "string"},
      {"name": "executives", "type":
        {
          "type": "array", "items": "bbejeck.chapter_3.avro.PersonAvro"},
          "default": []
        }
    ]
}
```

**This is the new part;
it's a reference to the
person object.**

Now both schemas refer to the same object created by the person schema file. For
completeness let's take a look at how you implement a schema reference in both
JSON Schema and Protocol Buffers. First we'll look at the JSON Schema version.

Listing 3.32   Company schema reference in JSON Schema

```
{
    "$schema": "http://json-schema.org/draft-07/schema#",
    "title": "Exchange",
    "description": "A JSON schema of a Company using Person refs",
    "javaType": "bbejeck.chapter_3.json.CompanyJson",
    "type": "object",
    "properties": {
      "name": {
        "type": "string"
      },
      "executives": {
        "type": "array",
```

```
    "items": {
      "$ref": "person.json"          ◄──   The reference to
    }                                       the Person object
  }                                         schema
  }
}
```

The concept with references in JSON Schema is the same, but you provide an explicit `$ref` element pointing to the referenced schema file. It's assumed that the referenced file is located in the same directory as the referring schema.

Now let's take a look at the equivalent reference with Protocol Buffers.

Listing 3.33   Company Schema reference in Protocol Buffers

```
syntax = "proto3";

package bbejeck.chapter_3.proto;          Import statement
                                          for the referenced
import "person.proto";          ◄──       schema

option java_outer_classname = "CompanyProto";
option java_multiple_files = true;

message Company {
  string name = 1;                        Refers to the
  repeated Person executives = 2;   ◄──   Person proto
}
```

With Protocol Buffers, you have a minor extra step of providing an import referring to the proto file containing the referenced object.

But now the question is how the (de)serializers will know how to serialize and deserialize the object into the correct format. You've removed the definition from inside the file, so you need to get a reference to the schema as well. Fortunately, Schema Registry allows schema references.

You need to register a schema for the person object first. When you register the schema for the college and company schemas, you reference the already registered person schema.

Using the Gradle schema-registry plugin makes this a simple task. The following listing shows how you would configure it for using schema references.

Listing 3.34   Gradle plugin reference configuration

```
register {                                  Registers the person schema

    subject('person','src/main/avro/person.avsc', 'AVRO')   ◄──
    subject('college-value','src/main/avro/college.avsc', 'AVRO')
        .addReference("bbejeck.chapter_3.avro.PersonAvro", "person", 1)   ◄──
    subject('company-value','src/main/avro/company.avsc', 'AVRO')

                                          Registers the college schema and adds
                                               a reference to the person schema
```

```
        .addReference("bbejeck.chapter_3.avro.PersonAvro", "person", 1)    ◁─┐
    }
```

**Registers the company schema and
adds a reference to the person schema**

So you first registered the `person.avsc` file, but in this case, the subject is simply `person` because, in this case, it's not associated directly with any one topic. Then, you registered the college and company schemas using the `<topic name> - value` pattern, as the college and company schemas are tied to topics with the same names and use the default subject name strategy (`TopicNameStrategy`). The `addReference` method takes three parameters:

- A name for the reference. Since you're using Avro, it's the fully qualified name of the schema. For Protobuf, it's the name of the proto file, and for JSON Schema, it's the URL in the schema.
- The subject name for the registered schema.
- The version number for the reference.

Now, with the references in place, you register the schemas, and your producer and consumer client can serialize and deserialize the objects with the references correctly.

There are examples in the source code for running a producer and consumer with the schema references in action. Since you've already run the `./gradlew streams :registerSchemasTask` for the main module, you've already set up your references. To see using schema references in action, you can run the code in the following listing.

Listing 3.35   Tasks for schema references in action

```
./gradlew streams:runCompanyProducer
./gradlew streams:runCompanyConsumer

./gradlew streams:runCollegeProducer
./gradlew streams:runCollegeConsumer
```

3.6   *Schema references and multiple events per topic*

We've covered the different subject strategies `RecordNameStrategy` and `TopicRecord-NameStrategy` and how they allow the production of records of various types for a topic. But with the `RecordNameStrategy`, any topic you have must use the same schema version for the given type. If you want to change or evolve the schema, all topics must use the new schema. Using the `TopicRecordNameStrategy` allows for multiple events in a topic, and it scopes the schema to a single topic, allowing you to evolve the schema independently of other topics.

But you can't control the number of different types produced to the topic with both approaches. If someone wants to produce a record of another kind that is not desired, you don't have any way to enforce this policy.

However, by using schema references, there is a way to produce multiple event types to a topic and restrict the types of records produced to the topic. Using `TopicName-Strategy` in conjunction with schema references allows all records in the topic to be constrained by a single subject. In other words, schema references will enable you to have multiple types, but only those the schema refers to. This is best understood by walking through an example scenario.

Imagine you are an online retailer and have developed a system for precisely tracking packages you ship to customers. You have a fleet of trucks and planes that take packages anywhere in the country. Each time a package is handled along its route, it's scanned into your system, generating one of three possible events represented by these domain objects: `PlaneEvent`, `TruckEvent`, or `DeliveryEvent`.

These are distinct events, but they are closely related. Also, since the order of these events is essential, you want them produced on the same topic so you have all related events together and in the proper sequence of their occurrence. I'll cover more about how combining related events in a single topic helps with sequencing in chapter 4 when we cover clients. Now, assuming you've already created schemas for `PlaneEvent`, `TruckEvent`, and `DeliveryEvent`, you could create a schema like in the following listing to contain the different event types.

Listing 3.36 Avro schema `all_events.avsc` with multiple events

```
[
    "bbejeck.chapter_3.avro.TruckEvent",      ←—  An Avro union type for
    "bbejeck.chapter_3.avro.PlaneEvent",           the different events
    "bbejeck.chapter_3.avro.DeliveryEvent"
]
```

The `all_events.avsc` schema file is an Avro union, an array of the possible event types. You use a union when a field or, in this case, a schema could be of more than one type.

Since you're defining all the expected types in a single schema, your topic can now contain multiple types, but it's limited to only those listed in the schema. When using schema references in this format with Avro, it's critical to always set `auto.register.schemas=false` and `use.latest.version=true` in your Kafka producer configuration. Here's why you need to use these configurations with the given settings.

When the Avro serializer goes to serialize the object, it won't find the schema since it's in the union schema. As a result, it will register the schema of the individual object, overwriting the union schema. So, setting the auto registration of schemas to `false` avoids overwriting the schema problem. In addition, by specifying `use.latest.version=true`, the serializer will retrieve the latest version of the schema (the union schema) and use that for serialization. Otherwise, it would look for the event type in the subject name, and since it won't find it, a failure will result.

> **TIP** When using the `oneOf` field with references in Protocol Buffers, the referenced schemas are automatically registered recursively, so you can use the `auto.register.schemas` configuration set to `true`. You can also do the same with JSON Schema `oneOf` fields.

Let's now take a look at how you'd register the schema with references.

Listing 3.37   Registering the `all_events` schema with references

Registers all_events schema   **Registers the individual schemas**
                                                   referenced in the all_events.avsc file

```
subject('truck_event',
        'src/main/avro/truck_event.avsc', 'AVRO')
subject('plane_event','src/main/avro/plane_event.avsc', 'AVRO')
subject('delivery_event','src/main/avro/delivery_event.avsc', 'AVRO')

subject('inventory-events-value',
        'src/main/avro/all_events.avsc','AVRO')
  .addReference("bbejeck.chapter_3.avro.TruckEvent",
                        "truck_event", 1)
  .addReference("bbejeck.chapter_3.avro.PlaneEvent", "plane_event", 1)
  .addReference("bbejeck.chapter_3.avro.DeliveryEvent", "delivery_event", 1)
```

Adds the references of the individual schemas

As you saw in section 3.5 with Avro, you need to register the individual schemas before the schema with the references. After that, you can register the main schema with the references.

When working with Protobuf, there isn't a `union` type but a `oneOf`, which is essentially the same thing. However, with Protobuf, you can't have a `oneOf` at the top level; it must exist in a Protobuf message. For your Protobuf example, imagine you want to track customer interactions, logins, searches, and purchases as separate events. But since they are closely related, sequencing is essential, so you want them in the same topic. The following listing shows the Protobuf file containing the references.

Listing 3.38   Protobuf file with references

```
syntax = "proto3";

package bbejeck.chapter_3.proto;

import "purchase_event.proto";
import "login_event.proto";
import "search_event.proto";

option java_multiple_files = true;
option java_outer_classname = "EventsProto";

message Events {

  oneof type {
    PurchaseEvent purchase_event = 1;
    LoginEvent login_event = 2;
    SearchEvent search_event = 3;
  }
  string key = 4;
}
```

Imports the individual Protobuf messages

The oneOf field, which could be one of the three types listed

You've seen a Protobuf schema earlier in the chapter, so I won't review all the parts here. Still, critical to this example is the `oneOf` field `type`, which could be `Purchase-Event`, `LoginEvent`, or `SearchEvent`. When you register a Protobuf schema, it has enough information present to register all of the referenced schemas recursively, so it's safe to set the `auto.register` configuration to `true`. You can structure your Avro references similarly.

Listing 3.39 Avro schema with references using an outer class

```
{
  "type": "record",
  "namespace": "bbejeck.chapter_3.avro",       Outer class
  "name": "TransportationEvent",        ◁——┘   name

  "fields" : [                              Field named
    {"name": "event", "type"[       ◁——     "event"
      "bbejeck.chapter_3.avro.TruckEvent",    ◁——┐  Avro union for
      "bbejeck.chapter_3.avro.PlaneEvent",         the field type
      "bbejeck.chapter_3.avro.DeliveryEvent"
    ]}
  ]
}
```

So, the main difference between this Avro schema and the previous Avro schema with references is that this one has an outer class, and the references are now a field in the class. Also, when you provide an outer class with Avro references like you have done here, you can now set the `auto.register` configuration to `true`, although you still need to register the schemas for the referenced objects ahead of time as Avro, unlike Protobuf, does not have enough information to register the referenced objects recursively.

There are some additional considerations regarding using multiple types with producers and consumers. I'm referring to the generics you use on the Java clients and how you can determine the appropriate action on an object depending on its concrete class name. These topics are better suited for when we cover clients, so we'll cover those subjects in the next chapter.

Now, you've learned about the different schema compatibility strategies, how to work with schemas, and how to use references. In all the examples you've run, you've used the built-in serializers and deserializers provided by Schema Registry. The following section will cover the configuration for producer and consumer (de)serializers. But we'll only cover the configurations related to the (de)serializers, not general producer and consumer configuration, which we'll discuss in the next chapter.

3.7 *Schema Registry (de)serializers*

At the beginning of the chapter, I noted that when producing records in Kafka, you need to serialize the records for transport over the network and storage in Kafka. Conversely, when consuming records, you deserialize them so you can work with objects.

You must configure the producer and consumer with the classes required for the serialization and deserialization process. Schema Registry provides a serializer, deserializer, and a Serde (used in Kafka Streams) for all three supported types (Avro, Protobuf, JSON).

Providing the serialization tools is a strong argument for using Schema Registry, which I addressed earlier in the chapter. Freeing developers from writing serialization code speeds up development and increases standardization across an organization. Also, using a standard set of serialization tools reduces errors and the chance that one team implements a custom serialization framework.

> **NOTE** What's a Serde? A Serde class contains a serializer and deserializer for a given type. You will use Serdes when working with Kafka Streams because you cannot access the embedded producer and consumer. Hence, providing a class containing the correct serializer and deserializer makes sense. In chapter 6, you'll see Serdes in action when we start working with Kafka Streams.

In the following sections, I will discuss the configuration for using Schema Registry–aware serializers and deserializers. One important thing to remember is you don't configure the serializers directly. You set the serializer configuration when configuring the `KafkaProducer` or `KafkaConsumer`. If the following sections aren't entirely clear, that's OK because we'll cover clients (producers and consumers) in the next chapter.

3.7.1 *Avroserializers and deserializers*

`KafkaAvroSerializer` and `KafkaAvroDeserializer` classes can be used to serialize and deserialize Avro records. When configuring a consumer, you'll need to include an additional property, `KafkaAvroDeserializerConfig.SPECIFIC_AVRO_READER_CONFIG=true`, indicating that you want the deserializer to create a `SpecificRecord` instance. Otherwise, the deserializer returns a `GenericRecord`.

Listing 3.40 shows how to add these properties to the producer and consumer. Note the following example only shows the configurations required for the serialization. I've left out the other configurations for clarity. We'll cover the configuration of producers and consumers in chapter 4.

Listing 3.40 Required configuration for Avro

```
// producer properties
producerProps.put(ProducerConfig.KEY_SERIALIZER_CLASS_CONFIG,          The serializer
    StringSerializer.class);                                           for the key
producerProps.put(ProducerConfig.VALUE_SERIALIZER_CLASS_CONFIG,        The serializer
    KafkaAvroSerializer.class);                                        for the value
producerProps.put(AbstractKafkaSchemaSerDeConfig.SCHEMA_REGISTRY_URL_CONFIG,
    "http://localhost:8081");                                          Sets the URL for
                                                                       the serializer
//consumer properties are set separately on the consumer
props.put(ConsumerConfig.KEY_DESERIALIZER_CLASS_CONFIG,
    StringDeserializer.class);                                         The deserializer
props.put(ConsumerConfig.VALUE_DESERIALIZER_CLASS_CONFIG,              for the key
    KafkaAvroDeserializer.class);
props.put(KafkaAvroDeserializerConfig.SPECIFIC_AVRO_READER_CONFIG,
    true);
```

The deserializer for the value

Indicates to construct a specific record instance

```
props.put(AbstractKafkaSchemaSerDeConfig.SCHEMA_REGISTRY_URL_CONFIG,
    "http://localhost:8081");
```
⟵ **Sets the URL for the deserializer**

Next, let's take a look at the configuration for working with Protobuf records.

3.7.2   Protobuf

To work with Protobuf records, you can use `KafkaProtobufSerializer` and `Kafka-ProtobufDeserializer` classes. When using Protobuf with a schema registry, it's probably a good idea to specify both `java_outer_classname` and `java_multiple_files` to `true` in the Protobuf schema. If you end up using the `RecordNameStrategy` with Protobuf, you must use these properties so the deserializer can determine the type when creating an instance from the serialized bytes.

Earlier in the chapter, we discussed that when using Schema Registry–aware serializers, those serializers will attempt to register a new schema. If your Protobuf schema references other schemas via imports, the referenced schemas are also registered. Only Protobuf provides this capability; Avro and JSON do not register referenced schemas automatically. Again, if you don't want autoregistration of schemas, deactivate it with the following configuration: `auto.shema.registration = false`.

Let's look at a similar example of providing the relevant Schema Registry configurations for working with Protobuf records.

Listing 3.41   Required configuration for Protobuf

The key deserializer

```
// producer properties
producerProps.put(ProducerConfig.KEY_SERIALIZER_CLASS_CONFIG,
    StringSerializer.class);
producerProps.put(ProducerConfig.VALUE_SERIALIZER_CLASS_CONFIG,
    KafkaProtobufSerializer.class);
producerProps.put(AbstractKafkaSchemaSerDeConfig.SCHEMA_REGISTRY_URL_CONFIG,
    "http://localhost:8081");

// consumer properties again set separately on the consumer
props.put(ConsumerConfig.KEY_DESERIALIZER_CLASS_CONFIG,
    StringDeserializer.class);
props.put(ConsumerConfig.VALUE_DESERIALIZER_CLASS_CONFIG,
    KafkaProtobufDeserializer.class);
props.put(KafkaProtobufDeserializerConfig.SPECIFIC_PROTOBUF_VALUE_TYPE,
    AvengerSimpleProtos.AvengerSimple.class);
props.put(AbstractKafkaSchemaSerDeConfig.SCHEMA_REGISTRY_URL_CONFIG,
    "http://localhost:8081");
```

The key serializer

The Protobuf value serializer

Provides the URL for Schema Registry for the consumer

The Protobuf value deserializer

The location of Schema Registry for the producer

The specific class the deserializer should instantiate

As with the Avro deserializer, you must instruct it to create a specific instance. But in this case, you configure the actual class name instead of setting a Boolean flag indicating you want a particular class. If you leave out the specific value type configuration, the deserializer returns a type of `DynamicRecord`. We cover `DynamicRecord` in section C.2.5.

The `bbejeck.chapter_3.ProtobufProduceConsumeExample` class in the book source code demonstrates producing and consuming a Protobuf record. Now, we'll move on to the final example of the configuration of Schema Registry's supported types, JSON schemas.

3.7.3   *JSON Schema*

Schema Registry provides the `KafkaJsonSchemaSerializer` and `KafkaJsonSchema-Deserializer` for working with JSON schema objects. The configuration should feel familiar to both Avro and the Protobuf configurations.

> **NOTE**   Schema Registry also provides `KafkaJsonSerializer` and `KafkaJson-Deserializer` classes. While the names are very similar, these (de)serializers are for working with Java objects for conversion to and from JSON without a JSON Schema. While the names are close, ensure you use the serializer and deserializer with `Schema` in the name. We'll talk about the generic JSON serializers in the next section.

Listing 3.42   Required configuration for JSON Schema

Provides the URL for Schema Registry for the producer

```
// producer configuration
producerProps.put(AbstractKafkaSchemaSerDeConfig.SCHEMA_REGISTRY_URL_CONFIG,
    "http://localhost:8081");
producerProps.put(ProducerConfig.KEY_SERIALIZER_CLASS_CONFIG,
    StringSerializer.class);                              ← The key serializer
producerProps.put(ProducerConfig.VALUE_SERIALIZER_CLASS_CONFIG,
    KafkaJsonSchemaSerializer.class);                     ← The JSON Schema value serializer

// consumer configuration
props.put(AbstractKafkaSchemaSerDeConfig.SCHEMA_REGISTRY_URL_CONFIG,
    "http://localhost:8081");                             ← Provides the URL for Schema Registry for the producer
props.put(ConsumerConfig.KEY_DESERIALIZER_CLASS_CONFIG,
    StringDeserializer.class);
props.put(ConsumerConfig.VALUE_DESERIALIZER_CLASS_CONFIG,
    KafkaJsonSchemaDeserializer.class);                   ← Specifies the JSON Schema value deserializer
props.put(KafkaJsonDeserializerConfig.JSON_VALUE_TYPE,
    SimpleAvengerJson.class);       ← Configures the specific classes this deserializer will create
```

The key deserializer

Here, you can see a similarity with the Protobuf configuration: you need to specify the class the deserializer should construct from the serialized form in the last line of this example. If you leave out the value type, the deserializer returns a `Map`, the generic form of a JSON Schema deserialization. The same applies to keys. If your key is a JSON schema object, you'll need to supply a `KafkaJsonDeserializerConfig.JSON_KEY_TYPE` configuration for the deserializer to create the exact class.

There is a simple producer and consumer example for working with JSON schema objects in the `bbejeck.chapter_3.JsonSchemaProduceConsumeExample` in the source code for the book. As with the other basic producer and consumer examples, there are sections demonstrating how to work with the specific and generic return types. There's a discussion of the structure of the JSON schema generic type in section C.3.5.

Now, we've covered the different serializers and deserializers for each type of serialization supported by Schema Registry. Although using Schema Registry is recommended, it's not required. The following section outlines how to serialize and deserialize your Java objects without Schema Registry.

3.8 Serialization without Schema Registry

At the beginning of this chapter, I stated that your event objects, specifically their schema representations, are a contract between the producers and consumers of the Kafka event streaming platform. Schema Registry provides a central repository for those schemas, enforcing these schema contracts across your organization. Additionally, Schema Registry provides serializers and deserializers, which offer a convenient way of working with data without writing custom serialization code.

Does this mean using Schema Registry is required? No, not at all. Sometimes, you may not have access to Schema Registry or don't want to use it. Writing custom serializers and deserializers is easy. Remember, producers and consumers are decoupled from the (de)serializer implementation; you only provide the class name as a configuration setting. However, it's good to remember that when you use Schema Registry, you can use the same schemas across Kafka Streams, Connect, and ksqlDB.

So, to create your serializer and deserializer, you create classes that implement the `org.apache.kafka.common.serialization.Serializer` and `org.apache.kafka.common.serialization.Deserializer` interfaces. With the `Serializer` interface, there is only one method you must implement: `serialize`. For the `Deserializer`, it's the `deserialize` method. Both interfaces have additional default methods (`configure`, `close`) you can override if necessary. The following listing shows a section of a custom serializer using the `jackson-databindobjectMapper` (see details are omitted for clarity).

Listing 3.43 Serialize method of a custom serializer

```
@Override
public byte[] serialize(String topic, T data) {
    if (data == null) {
        return null;
    }
    try {
        return objectMapper.writeValueAsBytes(data);
    } catch (JsonProcessingException e) {
        throw new SerializationException(e);
    }
}
```

Converts the given object to a byte array

Here, you call `objectMapper.writeValueAsBytes()`, which returns a serialized representation of the passed-in object. Now let's look at an example of the deserializing counterpart (some details are omitted for clarity).

Listing 3.44 Deserialize method of a custom deserializer

```
@Override
public T deserialize(String topic, byte[] data) {
    try {
        return objectMapper.readValue(data, objectClass);   ⟵
    } catch (IOException e) {
        throw new SerializationException(e);
    }
}
```

Converts the bytes back to an object specified by the objectClass parameter

The `bbejeck.serializers` package contains the serializers and deserializers shown here and additional ones for Protobuf. You can use these serializers/deserializers in any examples in this book but remember that they don't use Schema Registry. Or they can serve as examples of how to implement your own (de)serializers.

In this chapter, we've covered how event objects, specifically their schemas, represent contracts between producers and consumers. We discussed how Schema Registry stores these schemas and enforces an implied contract across the Kafka platform. Finally, we covered the supported serialization formats of Avro, Protobuf, and JSON. In the next chapter, you'll move up further in the event streaming platform to learn about Kafka clients, the `KafkaProducer` and `KafkaConsumer`. If you think of Kafka as your central nervous system for data, the clients are its sensory inputs and outputs.

Summary

- Schemas represent a contract between producers and consumers. Even if you don't use explicit schemas, you have an implied one with your domain objects, so developing a way to enforce this contract between producers and consumers is critical.
- Schema Registry stores all your schemas, enforcing data governance and providing versioning and three schema compatibility strategies: backward, forward, and full. The compatibility strategies offer assurance that the new schema will work with its immediate predecessor but not necessarily older ones. You must use backward transitive, forward transitive, and full transitive for full compatibility across all versions. Schema Registry provides a convenient REST API for uploading, viewing, and testing schema compatibility.
- Schema Registry supports three serialization formats: Avro, Protocol Buffers, and JSON Schema. It also provides integrated serializers and deserializers you can plug into your `KafkaProducer` and `KafkaConsumer` instances for seamless support for all three supported types. The provided (de)serializers cache sche-

mas locally and only fetch them from Schema Registry when they can't locate a schema in the cache.

- Using code generation with tools like Avro and Protobuf or open source plugins supporting JSON Schema helps speed development and eliminate human error. Plugins that integrate with Gradle and Maven also allow testing and the uploading of schemas in the development build cycle.

Kafka clients 4

> **This chapter covers**
> - Producing records with `KafkaProducer`
> - Understanding message delivery semantics
> - Consuming records with `KafkaConsumer`
> - Learning about Kafka's exactly-once streaming
> - Using the `Admin` API for programmatic topic management
> - Handling multiple event types in a single topic

This chapter is where the "rubber hits the road." We take what you've learned over the previous two chapters and apply it here to start building event streaming applications. We'll begin by working individually with the producer and consumer clients to understand how each works.

4.1 Introducing Kafka clients

In their simplest form, clients operate like this: producers send records (in a produce request) to a broker, the broker stores them in a topic, consumers send a fetch request, and the broker retrieves records from the topic to fulfill that request (figure 4.1). When we talk about the Kafka event streaming platform, it's common to mention producers and consumers. After all, it's a safe assumption that you produce data for someone else to consume. But it's essential to understand that the producers and consumers are unaware of each other; there's no synchronization between these two clients.

Figure 4.1   **Producers send batches of records to Kafka in a produce request.**

`KafkaProducer` has one task: sending records to the broker. The records themselves contain all the information the broker needs to store them.

   `KafkaConsumer`, on the other hand, only reads or consumes records from a topic (figure 4.2). Also, as we mentioned in chapter 1, covering the Kafka broker, the broker handles the storage of the records. Consuming records does not affect how long the broker retains them.

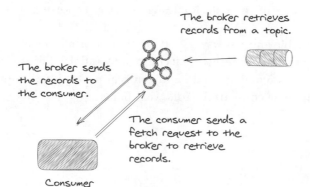

Figure 4.2   **Consumers send fetch requests to consume records from a topic, and the broker retrieves those records to fulfill the request.**

In this chapter, you'll take a `KafkaProducer`, dive into the essential configurations, and walk through examples of producing records to the Kafka broker. Learning how `KafkaProducer` works is important because that's the crucial starting point for building event streaming applications: getting the records into Kafka.

   Next, you'll learn how to use the `KafkaConsumer`. Again, we'll cover the vital configuration settings, and from working with some examples, you'll see how an event streaming application works by continually consuming records from the Kafka broker. You've started your event streaming journey by getting your data into Kafka, but when you start consuming the data, you begin building valuable applications.

Then, we'll go into working with the `Admin` interface. As the name implies, it's a client that allows you to perform administrative functions programmatically.

From there, you'll get into more advanced subject matter, such as the idempotent producer configuration, which guarantees per-partition, exactly-once message delivery, and the Kafka transnational API for exactly-once delivery across multiple partitions.

When you complete this chapter, you'll know how to build event streaming applications using the `KafkaProducer` and `KafkaConsumer` clients. Additionally, you'll understand how they work to recognize when you have a good use case for including them in your application. You should also come away with a good sense of configuring the clients to ensure your applications are robust and can handle situations when things don't go as expected.

So, with this overview in mind, we will embark on a guided tour of how the clients do their jobs. First, we'll discuss the producer, and then we'll cover the consumer. Along the way, we'll take some time going into deeper details, and then we'll return and continue with the tour.

4.2   *Producing records with the KafkaProducer*

You've seen the `KafkaProducer` in chapter 3 when we covered Schema Registry, but I didn't go into how the producer works. Let's do that now.

Say you work on a medium-sized wholesale company's data ingest team. You get transaction data delivered via a point-of-sale service, and several different departments within the company want access to the data for things such as reporting, inventory control, detecting trends, etc.

Your task is to build a reliable and fast way of making that information available to anyone within the company that wants access. The company Vandelay Industries uses Kafka to handle all of its event streaming needs, and you realize this is your opportunity to get involved. The sales data contains the following fields:

- Product name
- Per-unit price
- Quantity of the order
- The timestamp of the order
- Customer name

At this point in your data pipeline, you don't need to do anything with the sales data other than send it into a Kafka topic, which makes it available for anyone in the company to consume (figure 4.3).

To ensure everyone is on the same page with the data structure, you've modeled the records with a schema and published it to Schema Registry. All that's left for you to do is write the `KafkaProducer` code to take the sales records and send them to Kafka.

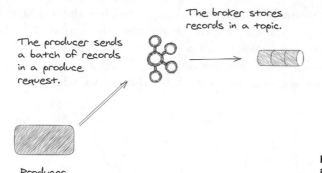

The broker stores records in a topic.

The producer sends a batch of records in a produce request.

Producer

Figure 4.3 Sending the data into a Kafka topic

The following listing shows what your code looks like. Some details are left out for clarity. The source code can be found at bbejeck.chapter_4.sales.SalesProducerClient.

Listing 4.1 KafkaProducer

Creates the KafkaProducer instance using a try-with-resources statement so the producer closes automatically when the code exits

```
try (
Producer<String, ProductTransaction> producer = new KafkaProducer<>(
  producerConfigs)) {
    while(keepProducing) {
     Collection<ProductTransaction> purchases =
       salesDataSource.fetch();
     purchases.forEach(purchase -> {
        ProducerRecord<String, ProductTransaction> producerRecord =
          new ProducerRecord<>(topicName, purchase.getCustomerName(),
            purchase);
        producer.send(producerRecord,
          (RecordMetadata metadata, Exception exception) -> {
            if (exception != null) {
                LOG.error("Error producing records ", exception);
            } else {
              LOG.info("Produced record at offset
                     {} with timestamp {}",
                     metadata.offset(), metadata.timestamp());
            }
        });
     });
   });
```

The data source providing the sales transaction records

Creates the ProducerRecord from the incoming data

In the success case, logs the offset and timestamp of the record stored in the topic

Logs if an exception occurred with the produce request

Sends the record to the Kafka broker and provides a lambda for the Callback instance

Notice KafkaProducer takes a Map of configuration items (In section 4.2.1 we'll discuss some of the more critical KafkaProducer configurations). We use a data generator to simulate the delivery of sales records. You take the list of ProductTransaction objects and use the Java stream API to map each object into a ProducerRecord object.

For each ProducerRecord created, you pass it as a parameter to the KafkaProducer .send() method. However, the producer does not immediately send the record to the broker. Instead, it attempts to batch up records. By using batches, the producer makes fewer requests, which helps with the performance of both the broker and the producer client. The `KafkaProducer.send()` call is asynchronous to allow for the continual addition of records to a batch. The producer has a separate thread (the I/O thread) that sends records when the batch is full or when it decides to transmit the batch (figure 4.4).

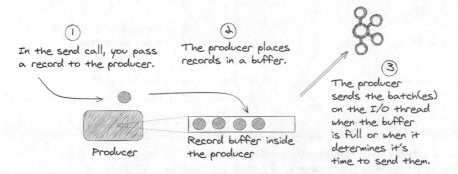

Figure 4.4   The producer batches records and sends them to the broker when the buffer is full or it's time to send them.

There are two signatures for the `send` method. The version you use in the code here accepts a `ProducerRecord` and `Callback` objects as parameters. However, since the `Callback` interface only contains one method, also known as a functional interface, we can use a lambda expression instead of a concrete implementation. The producer I/O thread executes the `Callback` when the broker acknowledges the record as persisted.

The `Callback.onCompletion` method, again represented here as a lambda, accepts two parameters, `RecordMetadata` and `Exception`. The `RecordMetadata` object contains metadata of the record the broker has acknowledged. It's worth noting that if you set `acks=0`, the `RecordMetadata.offset` field will have a value of -1. The offset is -1 because the producer doesn't wait for acknowledgment from the broker, so it can't report the offset assigned to the record. The exception parameter is non-null if an error occurred.

Since the producer I/O thread executes the callback, it's best to refrain from doing any heavy processing, as that would hold up the sending of records. The other overloaded `KafkaProducer.send()` method only accepts a `ProducerRecord` parameter and returns a `Future<RecordMetadata>`. The `Future.get()` method blocks until the broker acknowledges the record (request completion). Note that executing the `get` method throws an exception if an error occurs during the send.

Generally speaking, it's better to use the `send` method with the `Callback` parameter as it's a bit cleaner to have the I/O thread handle the results of the send asynchronously versus having to keep track of every `Future` resulting from the send calls.

At this point, we've covered the fundamental behavior of a `KafkaProducer`, but before we move on to consuming records, we should take a moment to discuss other essential subjects involving the producer: configurations, delivery semantics, partition assignment, and timestamps.

4.2.1   Producer configurations

The following are some important producer configurations:

- `bootstrap.servers`—One or more host:port configurations specifying a broker for the producer to connect to. Here, we have a single value because this code runs against a single broker in development. In a production setting, you could list every broker in your cluster in a comma-separated list.
- `key.serializer`—The serializer for converting the key into bytes. In this example, the key is a `String`, so we can use the `StringSerializer` class. The `org.apache.kafka.common.serialization` package contains serializers for `String`, `Integer`, `Double`, etc. You could also use Avro, Protobuf, or JSON Schema for the key and use the appropriate serializer.
- `value.serializer`—The serializer for the value. Here, we're using an object generated from an Avro schema. Since we're using Schema Registry, we'll use the `KafkaAvroSerializer` we saw in chapter 3. But the value could also be a string, integer, etc., and you would use one of the serializers from the `org.apache.kafka.common.serialization` package.
- `acks`—The number of acknowledgments required to consider the produce request successful. The valid values are `0`, `1`, and `all`. The `acks` configuration is one of the most important to understand as it directly affects data durability. Let's go through the different settings here.
  - Zero (`acks=0`)—Using a value of `0` means the producer will not wait for any acknowledgment from the broker about persisting the records. The producer considers the send successfully immediately after transmitting it to the broker. You could consider using `acks=0` as "fire and forget." Using this setting has the highest throughput but the lowest data durability guarantee.
  - One (`acks=1`)—A setting of `1` means the producer waits for notification from the lead broker for the topic partition that it successfully persisted the record to its log. But the producer doesn't wait for acknowledgment from the leader that any of the followers persisted in the record. While you have a little more assurance on the record's durability in this case, it will be lost if the lead broker fails before the followers replicate the record.
  - All (`acks=all`)—This setting gives the highest guarantee of data durability. In this case, the producer waits for acknowledgment from the lead broker that it successfully persisted the record to its log, and the following in-sync brokers could also persist the record. This setting has the lowest throughput but the highest durability guarantees. When using the `acks=all` setting, it's

best to set the `min.insync.replicas` configuration for your topics to a value higher than the default of 1. For example, with a replication factor of 3, setting `min.insyc.replicas=2` means the producer will raise an exception if there are not enough replicas available for persisting a record. We'll go into more detail on this scenario later in this chapter.

- `delivery.timeout.ms`—This is an upper bound on the time you want to wait for a response after calling `KafkaProducer.send()`. Since Kafka is a distributed system, failures in delivering records to the broker will occur. But in many cases, these errors are temporary and hence retryable. For example, the producer may encounter trouble connecting due to a network partition. But network partitions can be a temporary issue, so the producer will retry sending the batch, and in many cases, the resending of records succeeds. But after a certain point, you'll want the producer to stop trying and throw an error, as prolonged connectivity problems mean a problem that needs attention. Note that if the producer encounters a fatal error, the producer will throw an exception before this timeout expires.

- `retries`—When the producer encounters a nonfatal error, it will retry sending the record batch. The producer will continue to retry until the `delivery.timeout.ms` timeout expires. The default value for `retries` is `INTEGER.MAX_VALUE`. Generally, you should leave the retries configuration at the default value. If you want to limit the number of retries a producer makes, you should reduce the value of the `delivery.timeout.ms` configuration. With errors and retries, records may arrive out of order on a partition. Consider the producer sends a batch of records, but an error forces a retry. But in the meantime, the producer sent a second batch that encountered no errors. The first batch succeeds in the subsequent retry, but now it's appended to the topic after the second batch. To avoid this issue, you can set the configuration `max.in.flight.requests.per.connection=1`. Another approach to avoid out-of-order batches is to use the `idempotent producer`, which we'll discuss in section 4.4.1.

Now that you have learned about the concept of retries and record acknowledgments, let's look at message delivery semantics.

4.2.2 *Kafka delivery semantics*

Kafka provides three different delivery semantic types: at least once, at most once, and exactly once. Let's discuss each of them here:

- *At least once*— With at least once, records are never lost but may be delivered more than once. From the producer's perspective, this can happen when a producer sends a batch of records to the broker. The broker appends the records to the topic-partition, but the producer does not receive the acknowledgment in time. In this case, the producer resends the batch of records. From the consumer's point of view, you have processed incoming records, but an error occurs before the consumer can commit. Your application reprocesses data from the last committed offset, including records already processed, so there are duplicates. Kafka provides at-least-once delivery by default.

- *At most once*—Records are successfully delivered but may be lost with an error. From the producer's standpoint, enabling `acks=0` would be an example of at-most-once semantics. Since the producer does not wait for acknowledgment as soon as it sends the records, it has no notion if the broker received or appended them to the topic. From the consumer perspective, it commits the offsets before processing any records, so in the event of an error, it will not start reprocessing missed records as the consumer has already committed the offsets. To achieve at most once, producers set `acks=0`, and consumers commit offsets before processing.

- *Exactly once*—With exactly-once semantics, records are neither delivered more than once nor lost. Kafka uses transactions to achieve exactly-once semantics. If Kafka aborts a transaction, the consumer effectively ignores the aborted data, its internal position continues to advance, and the stored offsets aren't visible to any consumer configured with `read_committed`.

Both of these concepts are critical elements of Kafka's design. Partitions determine the level of parallelism and allow Kafka to distribute the load of a topic's records to multiple brokers in a cluster. The broker uses timestamps to determine which log segments it will delete. In Kafka Streams, timestamps drive records' progress through a topology (we'll return to timestamps in chapter 9).

4.2.3 Partition assignment

When it comes to assigning a partition to a record, there are four possibilities:

1 You can provide a valid partition number.
2 If you don't give the partition number, but there is a key, the producer sets the partition number by taking the hash of the key modulo the number of partitions.
3 Without providing a partition number or key, `KafkaProducer` sets the partition by alternating the partition numbers for the topic.

The approach to assigning partitions without keys has changed somewhat over time. Before Kafka 2.4, the default partitioner assigned partitions on a round-robin basis. That meant the producer assigned a partition to a record; it would increment the partition number for the next record. This round-robin approach sends multiple, smaller batches to the broker. This approach causes more load on the broker due to more requests. Figure 4.5 will help clarify what is going on.

But now, when you don't provide a key or partition for the record, the partitioner assigns the partition per batch. When the producer flushes its buffer and sends records to the broker, the batch is for a single partition, resulting in a single request. Let's take a look at figure 4.6 to visualize how this works.

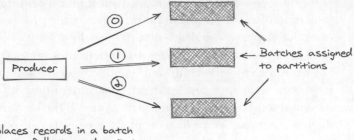

The producer places records in a batch
for partitions in the following order: 0, 1,
and 2. Then the order starts
over again at 0.

Figure 4.5   Round-robin partition assignment

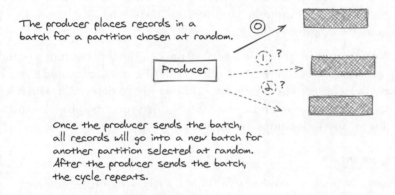

The producer places records in a
batch for a partition chosen at random.

Once the producer sends the batch,
all records will go into a new batch for
another partition selected at random.
After the producer sends the batch,
the cycle repeats.

Figure 4.6   Sticky partition assignment

After sending the batch, the partitioner randomly selects a partition and assigns it to the next batch. Over time, there should still be an even distribution of records across all partitions, but it's done one batch at a time.

Sometimes, the provided partitioners may not suit your requirements, and you'll need finer-grained control over partition assignment. For those cases, you can write a custom partitioner.

4.2.4   *Writing a custom partitioner*

Let's revisit the producer application from section 4.1. The key is the customer's name, but some orders don't follow the typical process and end up with a customer name of CUSTOM. You'd prefer to restrict those orders to a single partition 0 and have all other orders on partition 1 or higher. So, in this case, you'll need to write a custom partitioner that can look at the key and return the appropriate partition number.

The following listing shows that a custom partitioner does just that. The `Custom-OrderPartitioner`, from src/main/java/bbejeck/chapter_4/sales/CustomOrder-Partitioner.java, examines the key to determine which partition to use (some details are omitted for clarity).

Listing 4.2 `CustomOrderPartitioner` custom partitioner

```
public class CustomOrderPartitioner implements Partitioner {

@Override
public int partition(String topic,
                     Object key,
                     byte[] keyBytes,
                     Object value,
                     byte[] valueBytes,
                     Cluster cluster) {                    Retrieves the number
                                                           of partitions for
    Objects.requireNonNull(key, "Key can't be null");      the topic
    int numPartitions = cluster.partitionCountForTopic(topic);   ◁
    String strKey = (String) key;
    int partition;
                                          If the name of the
    if (strKey.equals("CUSTOM")) {        customer is CUSTOM,
        partition = 0;            ◁       returns 0
    } else {
        byte[] bytes = strKey.getBytes(StandardCharsets.UTF_8);
        partition = Utils.toPositive(Utils.murmur2(bytes)) %
                                    (numPartitions - 1) + 1;    ◁
    }
    return partition;                     Determines the partition to use in
  }                                        the noncustom order case
}
```

To create your partitioner, you implement the `Partitioner` interface, which has three methods: `partition`, `configure`, and `close`. I'm only showing the `partition` method here as the other two don't do anything in this implementation. The logic is straightforward: if the customer name equates to "CUSTOM", return zero for the partition. Otherwise, you determine the partition as usual, with a slight twist. First, we subtract one from the number of candidate partitions since the 0 partition is reserved. Then, we shift the partition number by 1, which ensures we always return 1 or greater for the non-custom order case.

NOTE This example does not represent a typical use case and is presented only to demonstrate how you can provide a custom partitioner. In most cases, it's best to go with one of the provided ones.

You've just seen how to construct a custom partitioner. Next, we'll wire it up with our producer.

4.2.5   *Specifying a custom partitioner*

Now that you've written a custom partitioner, let's configure the producer to use it instead of the default partitioner. You specify a different partitioner when configuring the Kafka producer:

```
producerConfigs.put(ProducerConfig.PARTITIONER_CLASS_CONFIG,
 CustomOrderPartitioner.class);
```

The bbejeck.chapter_4.sales.SalesProducerClient uses the `CustomOrderPartitioner`, but you can comment-out the line if you don't want to use it. You should note that since the partitioner config is a producer setting, you must configure each one you want to use the custom partitioner.

4.2.6   *Timestamps*

The `ProducerRecord` object contains a timestamp field of type `Long`. If you don't provide a timestamp, the `KafkaProducer` adds one to the record, which is the current time of the system the producer is running on. Timestamps are an essential concept in Kafka. The broker uses them to determine when to delete records by taking the oldest timestamp in a segment and comparing it to the current time. The broker removes the segment if the difference exceeds the configured retention time. Kafka Streams and ksqlDB also rely heavily on timestamps, but I'll defer those discussions until we get to their respective chapters.

Kafka may use two possible timestamps depending on the configuration of the topic. In Kafka, topics have a configuration, `message.timestamp.type`, which can either be `CreateTime` or `LogAppendTime`. A configuration of `CreatTime` means the broker stores the record with the timestamp provided by the producer. Suppose you configure your topic with `LogAppendTime`. In that case, the broker overwrites the timestamp in the record with its current wall-clock (i.e., system) time when the broker appends the record in the topic. In practice, the difference between these timestamps should be small. Another consideration is that you can embed the event's timestamp in the record value payload when creating it.

This wraps up our discussion on the producer-related issues. Next, we'll move on to the mirror image of producing records to Kafka, consuming records.

4.3   *Consuming records with the KafkaConsumer*

So you're back on the job at Vandelay Industries, and you now have a new task. Your producer application is up and running, happily pushing sales records into a topic. But now you're asked to develop a `KafkaConsumer` application to serve as a model for consuming records from a Kafka topic.

The `KafkaConsumer` sends a fetch request to the broker to retrieve records from its subscribed topics (figure 4.7). The consumer makes a `poll` call to get the records. But each time the consumer polls, it doesn't necessarily result in the broker fetching records. Instead, it could be retrieving records cached by a previous call.

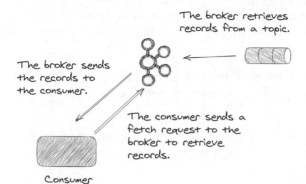

The broker retrieves records from a topic.

The broker sends the records to the consumer.

The consumer sends a fetch request to the broker to retrieve records.

Consumer

Figure 4.7 Consumers send fetch requests to consume records from a topic, and the broker retrieves those records to fulfill the request.

NOTE Producer and consumer clients are available in other programming languages, but in this book, we'll focus on the Java clients available in the Apache Kafka distribution. To see a list of clients available in other languages check out take a look at this resource: http://mng.bz/Y7qK.

Let's get started by looking at the code for creating a `KafkaConsumer` instance. Some details are left out for clarity.

Listing 4.3 `KafkaConsumer` code found in bbejeck.chapter_4.sales.SalesConsumerClient

```
try (
  final Consumer<String, ProductTransaction> consumer = new KafkaConsumer<>(
    consumerConfigs)) {                                    ⟵ Creates the new
    consumer.subscribe(topicNames);                           consumer instance
    while (keepConsuming) {
      ConsumerRecords<String, ProductTransaction> consumerRecords =
        consumer.poll(Duration.ofSeconds(5));
      consumerRecords.forEach(record -> {                  ⟵
        ProductTransaction pt = record.value();
        LOG.info("Sale for {} with product {} for a total sale of {}",
          record.key(),
          pt.getProductName(),                              Does some processing
          pt.getQuantity() * pt.getPrice());                with each of the
      });                                                   returned records
    }
  }
```

Subscribes to topic(s)

Polls for records

In this code example, you're creating a `KafkaConsumer`, again using the try-with-resources statement. After subscribing to a topic or topics, you begin processing records returned by the `KafkaConsumer.poll` method. When the `poll` call returns records, you start processing them. In this example case, we're simply logging out the details of the sales transactions.

TIP Whenever using a `KafkaProducer` or `KafkaConsumer`, you must close them to ensure you clean up all the threads and socket connections. The try-with-resources (http://mng.bz/GZxR) in Java ensures that resources created in the `try` portion are closed at the end of the statement. Using the try-with-resources statement is a good practice as it's easy to overlook adding a `close` call on either a producer or a consumer.

You'll notice that just like with the producer, you create a `Map` of configurations and pass them as a parameter to the constructor. The following are some of the more prominent ones:

- `bootstrap.servers`—One or more host:port configurations specifying a broker for the consumer to connect to. We have a single value, but this could be a comma-separated list.
- `max.poll.interval.ms`—The maximum time a consumer can take between calls to `KafkaConsumer.poll()`; otherwise, the consumer is considered nonactive and triggers a rebalance. We'll talk more about the consumer group coordinator and relabances in this section.
- `group.id`—An arbitrary string value used to associate individual consumers as part of the same consumer group. Kafka uses the concept of a consumer group to map multiple consumers logically as one consumer.
- `enable.auto.commit`—A Boolean flag that sets whether the consumer will automatically commit offsets. If you put this to `false`, your application code must manually commit the offsets of records you considered successfully processed.
- `auto.commit.interval.ms`—The time interval for automatically committing offsets.
- `auto.offset.reset`—When a consumer starts, it will resume consuming from the last committed offset. If offsets aren't available, this configuration specifies where to begin consuming records, either the earliest or latest available offset. The latest means the offset of the next record that arrives after the consumer starts.
- `key.deserializer.class`—The class name of the deserializer the consumer uses to convert record key bytes into the expected object type for the key.
- `value.deserializer.class`—The class name of the deserializer the consumer uses to convert record value bytes into the expected object type for the value. Here, we're using the provided `KafkaAvroDeserializer` for the value, which requires the `schema.registry.url` configuration we have in our configuration.

The code we use in our first consumer application is simple, but that's not the main point. Your business logic (i.e., how you handle processing) will be different on a case-by-case basis.

It's more important to grasp how the `KafkaConsumer` works and the implications of the different configurations. With this understanding, you'll be better able to know how to write the code to perform the desired operations on the consumed records. So, as we did in the producer example, we will detour from our narrative and go a little deeper into the implications of these different consumer configurations.

4.3.1   *The poll interval*

Let's first discuss the role of `max.poll.interval.ms`. It will be helpful to look at figure 4.8 to see the poll interval configuration in action to get a full understanding,

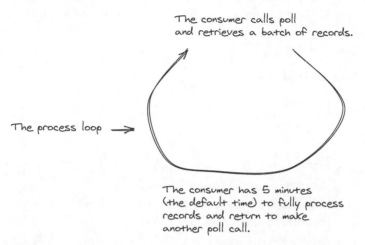

The consumer calls poll
and retrieves a batch of records.

The process loop →

The consumer has 5 minutes
(the default time) to fully process
records and return to make
another poll call.

Figure 4.8   The `max.poll.interval.ms` **configuration specifies how long a consumer may take between calls to** `KafkaConsumer.poll()` **before the consumer is considered inactive and removed from the consumer group.**

In the illustration here, the consumer processing loop starts with a call to `KafkaConsumer.poll(Duration.ofSeconds(5))`; the time passed to the `poll(Duration)` call is the maximum time the consumer waits for new records, in this case, 5 seconds. When the `poll(Duration)` call returns, if any records are present, the `for` loop over the `ConsumerRecords` executes your code over each one. Had no records been returned, the outer `while` loop goes back to the top for another `poll(Duration)` call.

Going through this illustration, iterating over all the records and execution for each record must be completed before the `max.poll.interval.ms` time elapses. By default, this value is 5 minutes, so if your processing takes longer, that individual consumer is considered dead, and a rebalance ensues. I know I've mentioned a few new terms in group coordinator and rebalancing; we'll cover them in the next section when we look at the `group.id` configuration.

If your processing takes longer than the `max.poll.interval.ms`, you have a couple of options. The first approach would be to validate what you're doing when processing the records and look for ways to speed up the processing. If you find no changes to make to your code, the next step could be to reduce the maximum number of records the consumer retrieves from a `poll` call. You can do this by setting the `max.poll.records` configuration to a setting less than the default `500`. I don't have any recommendations; you'll have to experiment to determine a good number.

4.3.2   *The group id configuration*

The `group.id` configuration takes us into a deeper conversation about consumer groups in Kafka. Kafka consumers use a `group.id` configuration, which Kafka uses to map all consumers with the same `group.id` into the same consumer group. A consumer group is a way to treat all group members as one consumer logically. Figure 4.9 illustrates how group membership works.

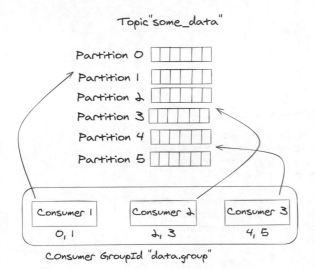

Figure 4.9   **Consumer groups allow the assignment of topic partitions across multiple consumers.**

So, going off figure 4.9, there is one topic with six partitions. There are three consumers in the group, so each consumer has an assignment of two partitions. Kafka guarantees that only a single consumer maintains an assignment for a given topic partition. To have more than one consumer assigned to a single topic partition would lead to undefined behavior.

Life with distributed systems means that failures aren't to be avoided but embraced with sound practices to deal with them as they occur. So what happens with our scenario here if one of the consumers in the group fails, whether from an exception or missing a required timeout, as we discussed with the `max.poll.interval.ms` timeout? The answer is the Kafka rebalance protocol, depicted in figure 4.10.

What we see in figure 4.10 is that consumer 2 fails and can no longer function. So rebalancing takes the topic partitions owned by consumer 2 and reassigns one topic partition to each of the other active consumers in the group. If consumer 2 becomes active again (or another consumer joins the group), another rebalance occurs and reassigns topic partitions from the active members. Each group member will be responsible for two topic-partitions again.

NOTE   The number of partitions limits the number of active consumers you can have. From our example here, you can start up to six consumers in the

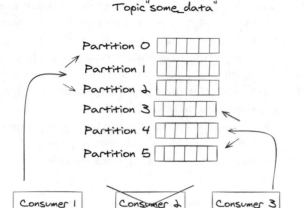

Topic "some_data"

Partition 0
Partition 1
Partition 2
Partition 3
Partition 4
Partition 5

Consumer 1
0, 1, 2

Consumer 2

Consumer 3
3, 4, 5

Consumer 2 fails and drops out of the group. Its
partitions are reassigned to consumer 1 and consumer 3.

Figure 4.10   The Kafka rebalance protocol reassigns topic partitions from failed consumers to currently active ones.

group, but any more beyond six will be idle. Also, note that different groups don't affect each other; each is treated independently.

So far, I've discussed how not making a `poll()` call within the specified timeout will cause a consumer to drop out of the group, triggering a rebalance and assigning its topic partition assignments to other consumers. But if you recall, the default setting for `max.poll.interval.ms` is 5 minutes. Does this mean it takes up to 5 minutes for the potentially dead consumer to get removed from the group and its topic partitions reassigned? The answer is no. Let's look at the poll interval illustration again, but we'll update it to reflect session timeouts (figure 4.11).

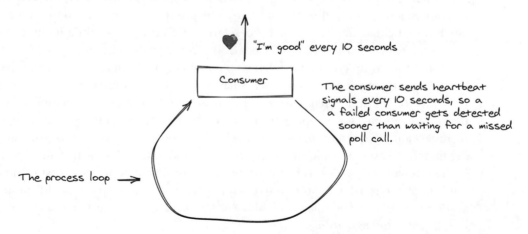

"I'm good" every 10 seconds

Consumer

The consumer sends heartbeat signals every 10 seconds, so a a failed consumer gets detected sooner than waiting for a missed poll call.

The process loop →

Figure 4.11   In addition to needing to call `poll` within the timeout, a consumer must send a heartbeat every 10 seconds.

There is another configuration timeout, `session.timeout.ms`, with a default value of 45 seconds. Each `KafkaConsumer` runs a separate thread for sending heartbeats indicating it's still alive. Should a consumer fail to send a heartbeat within 45 seconds, it's marked as dead and removed from the group, triggering a rebalance. This two-level approach for confirming consumer liveliness is essential to ensure all consumers are functioning. It also allows the reassignment of their topic partitions to other group members to ensure continued processing should one of them fail.

Let's discuss the new terms *group coordinator, rebalancing,* and *group leader* I just spoke about to give you a clear picture of how group membership works. Let's start with a visual representation of how these parts are tied together (figure 4.12).

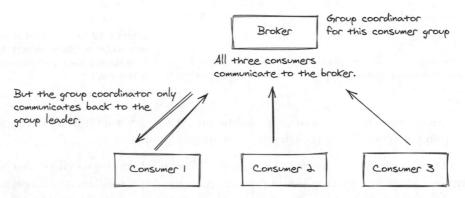

Figure 4.12   Group coordinator is a broker assigned to track a subset of consumer groups, and the group leader is a consumer communicating with the group coordinator.

The group coordinator is a broker handling membership for a subset of all available consumer groups. No single broker will act as the group coordinator; the responsibility is spread among the different brokers. The group coordinator monitors the membership of a consumer group via requests to join a group or when a member fails to communicate (either a poll or heartbeat) within the given timeouts.

When the group coordinator detects a membership change, it triggers a rebalance for the existing members. A rebalance is having all group members rejoin so that group resources (topic partitions) can be evenly distributed to the other members. When a new member joins, some topic partitions are removed from some or all existing group members and assigned to the new member. The opposite process occurs when an existing member leaves: its topic partitions are reassigned to the other active members.

The rebalance process is pretty straightforward, but it comes at a cost of time lost waiting for the rebalance process to complete, known as a "stop-the-world" or an *eager rebalance*. But with the release of Kafka 2.4, you can use a new rebalance protocol called *cooperative rebalancing*.

Let's look at both protocols, beginning with eager rebalancing.

EAGER REBALANCING

When the group coordinator detects a change in membership, it triggers a rebalance. This is true of both rebalance protocols we're going to discuss.

Once the rebalancing process starts, each group member gives up ownership of all its assigned topic partitions. Then, they send a `JoinGroup` request to the controller. Part of the request includes the topic partitions that the consumer is interested in, the ones they just relinquished control of. As a consequence of the consumers giving up their topic partitions, processing stops (figure 4.13).

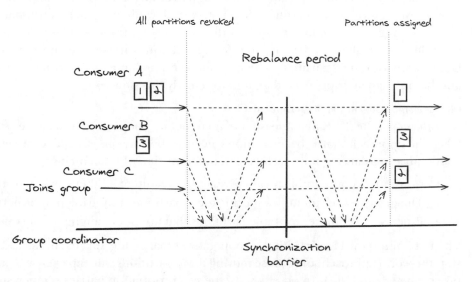

Figure 4.13 Rebalancing with the eager or "stop-the-world" approach. Processing occurs on all partition stops until reassigned, but most of the partitions end up with the original consumer.

The controller collects all of the topic partition information from the group and sends out the `JoinGroup` response, but the group leader receives all of the included topic partition information.

> **NOTE**   Remember from our discussion of the broker in chapter 2: all actions are a request/response process.

The group leader takes this information and creates topic partition assignments for all members of the group. Then, the group leader sends assignment information to the coordinator in a `SyncGroup` request. Note that the other members of the group also send `SyncGroup` requests but don't include any assignment information. After the group controller receives the assignment information from the leader, all members of the group get their new assignment via the `SyncGroup` response.

Now, with their topic partition assignments, all group members begin processing again. Again, note that no processing occurred when group members sent the `Join-Group` request until the `SyncGroup` response arrived with their assignments. This gap in processing is known as a synchronization barrier and is required as it's vital to ensure that each topic partition only has one consumer owner. If a topic partition had multiple owners, undefined behavior would result.

> **NOTE** During this entire process, consumer clients don't communicate with each other. All the consumer group members communicate only with the group coordinator. Additionally, only one group member, the leader, sets the topic partition assignments and sends them to the coordinator.

While the eager rebalance protocol redistributes resources and ensures only one consumer owns a given topic-partition, it comes with downtime as each consumer is idle from the initial `JoinGroup` request and the `SyncGroup` response. This cost might be negligible for smaller applications, but for applications with several consumers and a large number of topic partitions, the cost of downtime increases. Fortunately, there's another rebalancing approach that aims to remedy this situation.

INCREMENTAL COOPERATIVE REBALANCING

Introduced in the 2.4 Kafka release, the incremental cooperative rebalance protocol takes the approach that rebalances don't need to be so expensive. The incremental cooperative rebalancing approach takes a different view of rebalancing:

1 Consumers don't automatically give up ownership of all their topic partitions.
2 The group leader identifies specific topic partitions requiring new ownership.
3 Processing continues for topic partitions that are not changing ownership.

The last point is the big win with the cooperative rebalancing approach. Instead of the stop-the-world approach, only those moving topic partitions will experience a pause in processing (figure 4.14). In other words, the synchronization barrier is much smaller.

I'm skipping over some details, so let's walk through the incremental cooperative rebalancing protocol process. Like before, when the group controller detects a change in group membership, it triggers a rebalance. Each group member encodes their current topic-partition subscriptions in a `JoinGroup` request, but each member retains ownership for now.

The group coordinator assembles all the subscription information. In the `Join-Group` response, the group leader looks at the assignments and determines which topic-partitions need to migrate to new ownership. The leader removes any topic partitions requiring new ownership from the assignments and sends the updated subscriptions to the coordinator via a `SyncGroup` request. Again, each group member sends a `SyncGroup` request, but only the leader's request contains the subscription information.

> **NOTE** All group members receive a `JoinGroup` response, but only the response to the group leader contains the assignment information. Likewise, each group member issues a `SyncGroup` request, but only the leader encodes a new assignment. In the `SyncGroup` response, all members receive their possible updated assignment.

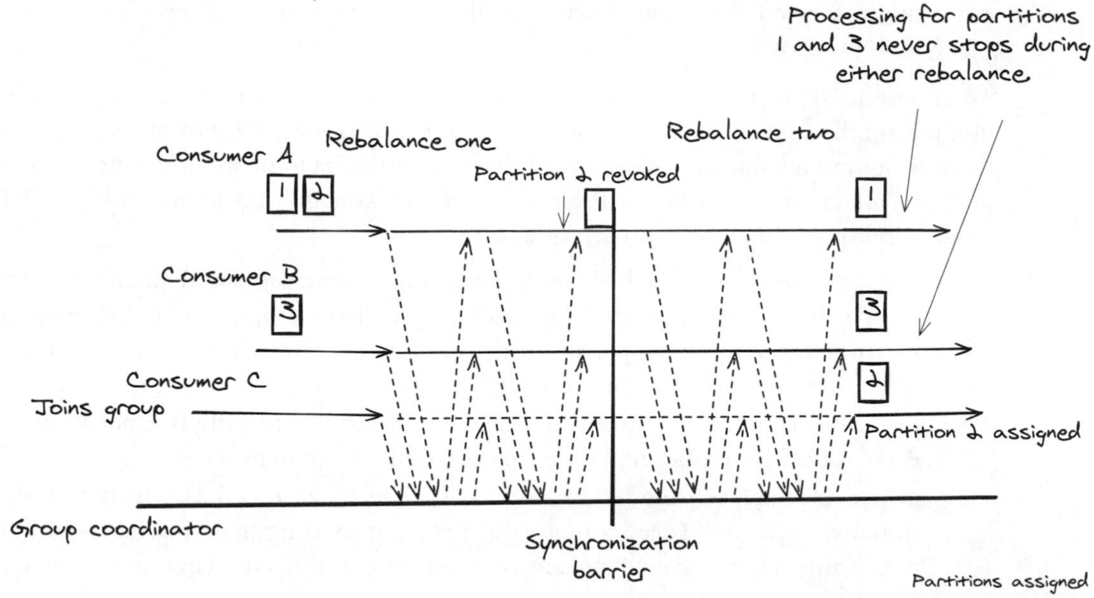

Figure 4.14 Rebalancing with cooperative approach processing continues and only stops for partitions marked for reassignment.

The group members take the SyncGroup response and potentially calculate a new assignment, either revoking topic partitions not included or adding ones in the new assignment but not the previous one. Topic partitions included in both the old and new assignments require no action.

Members then trigger a second rebalance, but only topic-partitions changing ownership are included. This second rebalance acts as the synchronization barrier as in the eager approach, but since it only comprises topic partitions receiving new owners, it is much smaller. Additionally, nonmoving topic partitions continue to process records!

After discussing the different rebalance approaches, we should cover some broader information about the available partition assignment strategies and how you apply them.

4.3.3 Applying partition assignment strategies

We've already discussed that a broker serves as a group coordinator for some subset of consumer groups. Since two different consumer groups could have differing ideas of distributing resources (topic partitions), the responsibility for which approach to use is entirely on the client side.

> **NOTE** For Kafka Connect and Kafka Streams, which are abstractions built on top of Kafka producers and consumers, use cooperative rebalance protocols, and I recommend staying with the default settings. This discussion about

partitioners is to inform you of what's available for applications directly using a `KafkaConsumer`.

To choose the partition strategy you want the `KafkaConsumer` instances in a group to use, you set the `partition.assignment.strategy` by providing a list of supported partition assignment strategies. All of the available petitioners implement the `Consumer-PartitionAssignor` interface. Here's a list of the available assignors with a brief description of the functionality each one provides:

- `RangeAssignor`—The default setting. `RangeAssignor` uses an algorithm to sort the partitions in numerical order and assigns them to consumers by dividing the number of available partitions by the number of consumers. This strategy gives partitions to consumers in lexicographical order.
- `RoundRobinAssignor`—Takes all available partitions and assigns a partition to each available member of the group in a round-robin manner.
- `StickyAssignor`—Attempts to assign partitions as balanced as possible. Additionally, `StickyAssignor` attempts to preserve existing assignments as much as possible during a rebalance. `StickyAssignor` follows the eager rebalancing protocol.
- `CooperativeStickyAssignor`—Follows the same assignment algorithm as `StickyAssignor`. The difference is that `CooperativeStickyAssignor` uses the cooperative rebalance protocol.

While it's difficult to provide concrete advice as each use case requires careful analysis of its unique needs, in general, for newer applications, one should favor using the `CooperativeStickyAssignor` for the reasons outlined in the section on incremental cooperative rebalancing.

> **TIP**   If you are upgrading from a version of Kafka 2.3 or earlier, you need to follow a specific upgrade path found in the 2.4 upgrade documentation (https://kafka.apache.org/documentation/#upgrade_240_notable) to upgrade to the cooperative rebalance protocol safely.

We've concluded our coverage of consumer groups and how the rebalance protocol works. Next, we'll cover a different configuration, static membership, where there's no initial rebalance when a consumer leaves the group.

4.3.4   *Static membership*

In the previous section, you learned that when a consumer instance shuts down, it sends a leave group request to the group controller. Or if it's considered unresponsive by the controller, it gets removed from the consumer group. Either way, the result is the same: the controller triggers a rebalance to reassign resources (topic partitions) to the remaining group members.

While this protocol is what you want to keep your applications robust, there are some situations where you'd prefer slightly different behavior. For example, let's say you have several consumer applications deployed. You might do a rolling upgrade or restart whenever you need to update the applications (figure 4.15).

In a rolling upgrade, each application is shut down upgraded and then restarted.

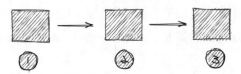

Each shutdown sends a "leave group" request; then, restarting issues a "join group" request. So each restart results in two rebalances for all instances in the group.

Figure 4.15 Rolling upgrades trigger multiple relabances.

You'll stop instance 1, upgrade and restart it, and then move on to instance number 2 and continue until you've updated every application. By doing a rolling upgrade, you don't lose nearly as much processing time if you shut down every application simultaneously. But this rolling upgrade triggers two rebalances for every instance: one when the application shuts down and another when it starts back up. Or consider a cloud environment where an application node can drop off at any moment only to have it back up and running once its failure is detected.

Even with the improvements brought by cooperative rebalancing, it would be advantageous not to have a rebalance triggered automatically for these transient actions. The concept of static membership was introduced in the 2.3 version of Apache Kafka. We'll use the illustration in figure 4.16 to discuss how static membership works.

At a high level, with static membership, you set a unique ID in the consumer configuration, `group.instance.id`. The consumer provides this ID to the controller when it joins a group, and the controller stores this unique group ID. When a consumer leaves the group, it does not send a leave group request. When it rejoins, it presents this unique membership ID to the controller. The controller looks it up and returns the original assignment to this consumer without rebalancing! The tradeoff for using static membership is that you'll need to increase the `session.timeout.ms` configuration to a value higher than the default of 10 seconds, as once a session timeout occurs, the controller kicks the consumer out of the group and triggers a rebalance.

Your value should be long enough to account for transient unavailability and not trigger a rebalance but not so long that an actual failure gets handled correctly with a rebalance. So, if you can sustain 10 minutes of partial unavailability, set the session timeout to 8 minutes. While static membership can be a good option for those running KafkaConsumer applications in a cloud environment, it's essential to consider the performance implications before opting to use it. Note that you must have Kafka brokers and clients on version 2.3.0 or higher to take advantage of static membership.

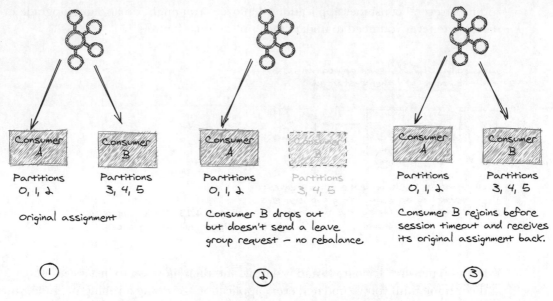

Figure 4.16   **Static members don't issue leave group requests when dropping out of a group, and a static ID allows the controller to remember them.**

Next, we'll cover a crucial subject committing the offsets of messages when using `KafkaConsumer`.

4.3.5   *Committing offsets*

In chapter 2, we discussed how the broker assigns a number to incoming records, called an offset. The broker increments the offset by one for each incoming record. Offsets are vital because they identify a record's logical position in a topic. `KafkaConsumer` uses offsets to know where it last consumed a record. For example, if a consumer retrieves a batch of records with offsets from 10 to 20, the starting offset of the next batch of records the consumer wants to read starts at offset 21.

To ensure the consumer continues to progress across restarts or failures, it needs to periodically commit the offset of the last record it has successfully processed. Kafka consumers provide a mechanism for automatically committing offsets. You enable automatic offset commits by setting the `enable.auto.commit` configuration to `true`. This configuration is turned on by default, but I've listed it here to discuss how automatic commits work. Also, we'll want to discuss the concept of a consumer's position versus its latest committed offset. A related configuration, `auto.commit.interval.ms`, specifies how much time needs to elapse before the consumer should commit offsets. It is based on the system time of the consumer.

But first, let's show how automatic commits work (figure 4.17).

Following figure 4.17, the consumer retrieves a batch of records from the `poll (Duration)` call. Next, the code takes the `ConsumerRecords`, iterates over them, and

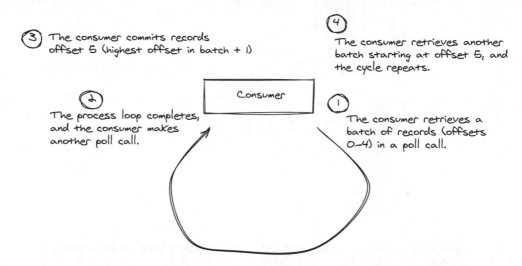

③ The consumer commits records offset 5 (highest offset in batch + 1)

④ The consumer retrieves another batch starting at offset 5, and the cycle repeats.

Consumer

② The process loop completes, and the consumer makes another poll call.

① The consumer retrieves a batch of records (offsets 0–4) in a poll call.

Figure 4.17   With automatic commits enabled when returning to the top of the `poll` loop, the consumer commits the highest offset +1 of the previous batch if the auto-commit interval has passed.

does some processing. After that, the code returns to the top of the `poll` loop and attempts to retrieve more records. But before retrieving records, if the consumer has auto-commit enabled and the amount of time elapsed since the last auto-commit check is greater than the `auto.commit.interval.ms` interval, the consumer commits the offsets of the records from the previous batch. By committing the offsets, we are marking these records as consumed, and under normal conditions, the consumer won't process these records again. I'll describe what I mean by this statement later.

What does it mean to commit offsets? Kafka maintains an internal topic named `__consumer_offsets`, that stores the committed offsets for consumers. When we say a consumer commits, it's not storing the offsets for each record; it's the highest offset per partition plus one.

For example, in figure 4.17, let's say the records returned in the batch contained offsets 0–4. So when the consumer commits, it will be offset 5 (figure 4.18).

So, the committed position is the offset successfully processed (plus one), indicating the starting record for the next batch it will retrieve. In figure 4.18, it's 5. Should the consumer in this example fail or you restart the application, the consumer would consume records starting at offset 5 again since it couldn't commit before the failure or restart.

Consuming from the last committed offset means you are guaranteed not to miss processing a record due to errors or application restarts. But it also means that you may process a record more than once (figure 4.19).

Suppose you processed some of the records with offsets larger than the latest one committed, but your consumer failed to commit for whatever reason. In that case, when you resume processing, you start with records from the committed offset, so

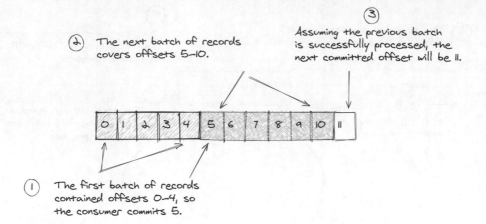

Figure 4.18   A consumer's committed position is the largest offset it has consumed so far, plus one.

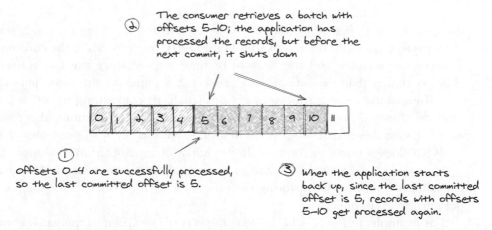

Figure 4.19   Restarting a consumer after processing without a commit means reprocessing some records.

you'll reprocess some of the records. This potential for reprocessing is known as "at least once." We covered at-least-once delivery in the delivery semantics in section 4.2.2.

To avoid reprocessing records, you could manually commit offsets immediately after retrieving a batch of records, giving you at-most-once delivery. But you risk losing some records if your consumer encounters an error after committing and before it can process the records.

COMMITTING CONSIDERATIONS

When enabling auto-commit with a Kafka consumer, you must ensure you've fully processed all the retrieved records before the code returns to the top of the `poll` loop. In practice, this should present no issue, assuming you are working with your records synchronously, meaning your code waits for the completion of processing of each record. However, suppose you were to hand off records to another thread for asynchronous processing or set the records aside for later processing. In that case, you also run the risk of not processing all consumed records before you commit (figure 4.20). Let me explain how this could happen.

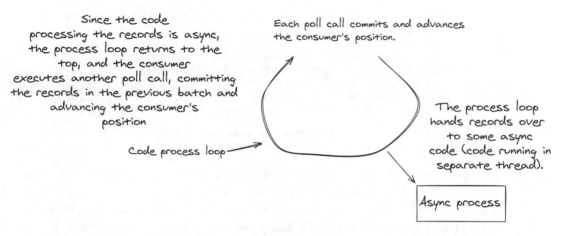

Figure 4.20 **Asynchronous processing with auto-committing can lead to potentially lost records.**

When you hand off the records to an asynchronous process, the code in your `poll` loop won't wait for the successful processing of each record. When your application calls the `poll()` method again, it commits the current position—that is, the highest offset plus one for each topic-partition consumed in the previous batch. But your async process may not complete working with all the records up to the highest offset at the time of the commit. Suppose your consumer application experienced a failure or a shutdown for any reason when it resumes processing. In that case, it will start from the last committed offset, which skips over the unprocessed records in the previous run of your application.

To avoid prematurely/maturely committing records before you consider them fully processed, you'll want to turn off auto-commits by setting `enable.auto.commit` to `false`. But why would you need to use asynchronous processing requiring manual committing? When you consume records, you do some processing that takes a long time (up to 1 second) to process each record. Your topic has a high traffic volume, so you want to stay caught up. You decide that as soon as you consume a batch of records, you'll hand them off to an async process so the consumer can immediately return to the `poll` call to retrieve the next batch.

Using an approach like this is called *pipelining*. But you must ensure you only commit the offsets for successfully processed records. This means turning off auto-committing and creating a way to only commit records your application considers fully processed. The following example code shows one example approach you could take. Note that I'm only showing the key details here, and you should consult the source code to see the entire example (bbejeck.chapter_4.pipelining.PipliningConsumerClient).

Listing 4.4   Consumer code

After you've retrieved a batch of records, you hand
off the batch of records to the async processor.

```
ConsumerRecords<String, ProductTransaction> consumerRecords = consumer.poll(
  Duration.ofSeconds(5));
if (!consumerRecords.isEmpty()) {
    recordProcessor.processRecords(consumerRecords);
    Map<TopicPartition, OffsetAndMetadata> offsetsAndMetadata =
      recordProcessor.getOffsets();                              ← Checks for offsets
    if (offsetsAndMetadata != null) {                              of completed
       consumer.commitSync(offsetsAndMetadata);   ←               records
    }
}
```

If the Map is not empty, you commit the
offsets of the records processed so far.

The key point with this consumer code is that the `RecordProcessor.processRe-`
`cords()` call returns immediately, so the next call to `RecordProcessor.getOffsets()`
returns offsets from a previous batch of records that are fully processed. What I want to emphasize here is how the code hands over new records for processing and then collects the offsets of records already fully processed for committing. Let's take a look at the processor code to see this is done. For the full code, see bbejeck.chapter_4.piplining.ConcurrentRecordProcessor.

Listing 4.5   Asynchronous processor code

Does the actual work on the consumed records

Creates the Map for collecting
the offset for committing

Iterates over the TopicPartition objects

```
Map<TopicPartition, OffsetAndMetadata> offsets = new HashMap<>();
  consumerRecords.partitions().forEach(topicPartition -> {            ←
      List<ConsumerRecord<String,ProductTransaction>> topicPartitionRecords =
    consumerRecords.records(topicPartition);
      topicPartitionRecords.forEach(this::doProcessRecord);
      long lastOffset = topicPartitionRecords.get(
  topicPartitionRecords.size() - 1).offset();
      offsets.put(topicPartition, new OffsetAndMetadata(lastOffset + 1));  ←
  });
  ....
  offsetQueue.offer(offsets);        ←       Puts the entire
                                             Map of offsets in
Gets the last offset for all records of a given TopicPartition     a queue
Gets records by TopicPartition for processing
```

Stores the offset to
commit for the
TopicPartition

The takeaway with the code here is that by iterating over records using `TopicPartition`, it's easy to create the map entry for the offsets to commit. Once you've iterated over all the records in the list, you only need to get the last offset. You, the observant reader, might ask yourself, "Why does the code add 1 to the last offset?" When committing offsets, it's always the offset of the *next* record you'll retrieve. For example, if the last offset is 5, you want to commit 6. Since you've already consumed offsets 0–5, you only want to consume records from offset 6 forward.

Then, you use the `TopicPartition` from the top of the loop as the key and the `OffsetAndMetadata` object as the value. When the consumer retrieves the offsets from the queue, it's safe to commit those offsets as the application has fully processed the records. The main point of this example is that you can ensure you only commit records you consider complete if you need to asynchronously process records outside of the `Consumer.poll` loop. It's important to note that this approach only uses a *single thread* and consumer for the record processing, which means the code still processes the records in order, so committing the offsets as they are handed back is safe.

> **NOTE** For a fuller example of threading and `KafkaConsumer`, you should consult "Introducing the Confluent Parallel Consumer," by Anthony Stubbs (http://mng.bz/z8xX) and https://github.com/confluentinc/parallel- consumer.

WHEN OFFSETS AREN'T AVAILABLE

I mentioned earlier that Kafka stores offsets in an internal topic named `__consumer_offsets`. But what happens when a consumer can't find its offsets? Take the case of starting a new consumer against an existing topic. The new `group.id` will not have any commits associated with it. So, the question becomes where to start consuming if offsets aren't found for a given consumer. The `KafkaConsumer` provides a configuration, `auto.offset.reset` that allows you to specify a relative position to start consuming when no offsets are available.

There are three settings:

1 `earliest`—Resets the offset to the earliest one
2 `latest`—Resets the offset to the latest one
3 `none`—Throws an exception to the consumer

With a setting of `earliest` the implications are that you'll start processing from the head of the topic, meaning you'll see all the records currently available. Using a setting of `latest` means you'll only start receiving records that arrive at the topic once your consumer is online, skipping all the previous records currently in the topic. The setting of `none` means that an exception gets thrown to the consumer, and depending on whether you are using any `try`/`catch` blocks, your consumer may shut down.

The choice of which setting to use depends entirely on your use case. Once a consumer starts, you only care about reading the latest data, or it may be too costly to process all records.

Whew! That was quite a detour, but well worth the effort to learn some of the critical aspects of working with the `KafkaConsumer`. So far we've covered how to build streaming applications using a `KafkaProducer` and `KafkaConsumer`. We've discussed situations where you're using at-least-once processing. But there are situations where you need to guarantee that you process records exactly once. For this, use the exactly-once semantics offered by Kafka.

4.4   *Exactly-once delivery in Kafka*

The 0.11 release of Apache Kafka saw `KafkaProducer` introduce exactly once message delivery. There are two modes for the `KafkaProducer` exactly-once message semantics: the idempotent producer and the transactional producer.

> **NOTE**   Idempotence means you can perform an operation multiple times, and the result will stay the same as it was after the first application of the operation.

The idempotent producer guarantees that the producer will deliver messages in order and only once to a topic partition. The transactional producer allows you to produce messages to multiple topics atomically, meaning all messages across all topics succeed together or none at all. In the following sections, we'll discuss the idempotent and the transactional producer.

4.4.1   *The idempotent producer*

You only need to set the configuration `enable.idempotence=true` to use the idempotent producer, and it's now the default value. Some other configuration factors come into play:

1 `max.in.flight.requests.per.connection` must not exceed a value of 5 (the default value is 5).
2 `retries` must be greater than 0 (the default value is Integer.MAX_VALUE).
3 `acks` must be set to `all`.

Consider the following listing. Some details are omitted for clarity.

Listing 4.6   KafkaProducer configured for idempotence

```
Map<String, Object> producerProps = new HashMap<>();
//Standard configs
producerProps.put(ProducerConfig.BOOTSTRAP_SERVERS_CONFIG, "somehost:9092");
producerProps.put(ProducerConfig.KEY_SERIALIZER_CLASS_CONFIG, ...);
producerProps.put(ProducerConfig.VALUE_SERIALIZER_CLASS_CONFIG, ...);

//Configs related to idempotence                              Sets acks to "all"
producerProps.put(ProducerConfig.ACKS_CONFIG, "all");     ◁──┘
producerProps.put(
  ProducerConfig.ENABLE_IDEMPOTENCE_CONFIG, true);        ◁──┘ Enables idempotence
producerProps.put(
```

```
ProducerConfig.RETRIES_CONFIG,
        ➡ Integer.MAX_VALUE);
producerProps.put(
ProducerConfig.MAX_IN_FLIGHT_REQUESTS_PER_CONNECTION,
        ➡ 5);
```

Sets retries to Integer.MAX_VALUE. This is the default value, shown here for completeness.

Setting max in-flight requests per connection to 5. This is the default value, shown here for completeness

In our earlier discussion about KafkaProducer, we outlined a situation where record batches within a partition can end up out of order due to errors and retries, and setting the max.inflight.requests.per.connection configuration to 1 allows you to avoid this situation. Using the idempotent producer removes the need to adjust that configuration. In section 4.2.2 on message delivery semantics, we also discussed that you would need to set retries to 0 to prevent possible record duplication, risking possible data loss.

Using the idempotent producer avoids both the records out of order and possible record duplication with retries. If you require strict ordering within a partition and no duplicated delivery of records, then using the idempotent producer is necessary.

NOTE As of the 3.0 release of Apache Kafka, the idempotent producer settings are the default, so you'll get the benefits of using it out of the box with no additional configuration needed.

The idempotent producer uses two concepts to achieve its in-order and only-once semantics: unique producer IDs (PIDs) and message sequence numbers. The idempotent producer gets initiated with a PID. Since each creation of an idempotent producer results in a new PID, idempotence for a producer is only guaranteed during a single producer session. For a given PID, a monotonically sequence ID (starting at 0) gets assigned to each batch of messages. There is a sequence number for each partition the producer sends records to (figure 4.21).

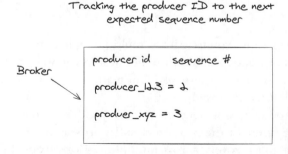

Figure 4.21 The broker keeps track of sequence numbers for each PID and topic partition it receives.

The broker maintains a listing (in-memory) of sequence numbers per topic partition per PID. If the broker receives a sequence number not *exactly one greater* than the sequence number of the last committed record for the given PID and topic partition, it will reject the produce request (figure 4.22).

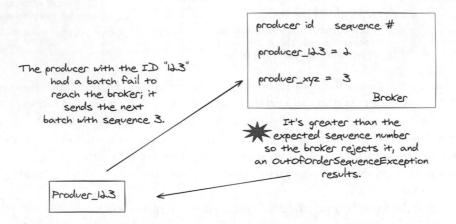

The producer with the ID "123" had a batch fail to reach the broker; it sends the next batch with sequence 3.

producer id   sequence #

producer_123 = 2

producer_xyz = 3

Broker

It's greater than the expected sequence number so the broker rejects it, and an OutOfOrderSequenceException results.

Producer_123

Figure 4.22   The broker rejects produce requests when the message sequence number doesn't match the expected one.

If the number is less than the expected sequence number, it's a duplication error that the producer ignores. If the number exceeds the expected one, the produce request results in an OutOfOrderSequenceException. For the idempotent producer, the OutOfOrderSequenceException is not a fatal error, and retries will continue. When there is a retriable error, if there is more than one in-flight request, the broker will reject the subsequent requests, and the producer will put them back to resend them to the broker.

So, if you require strict ordering of records within a partition, using the idempotent producer is necessary. But what do you do if you need to write to multiple topic-partitions atomically? In that case, you would use the transactional producer, which we'll cover next.

4.4.2   *Transactional producer*

Using the transactional producer allows you to write to multiple topic partitions atomically; all of the writes succeed, or none of them do. When would you want to use the transactional producer? In any scenario where you can't afford to have duplicate records, like in the financial industry, for example.

To use the transaction producer, you need to set the producer configuration transactional.id to a unique value for the producer. Kafka brokers use the transactional.id to enable transaction recovery across multiple sessions from the same producer instance. Since the ID needs to be unique for each producer and applications can have numerous producers, it's a good idea to come up with a strategy where the ID for the producers represents the segment of the application it's working on.

NOTE   Kafka transactions are a deep subject and could take up an entire chapter. Therefore, I won't go into detail about the design of transactions. For readers interested in more information, see the original KIP (KIP stands for Kafka Improvement Process): http://mng.bz/9Qqq.

Enabling a producer to use transactions automatically upgrades it to an idempotent producer. You can use the idempotent producer without transactions, but you can't do the opposite, using transactions without the idempotent producer. Let's dive into an example. We'll take our previous code and make it transactional.

Listing 4.7   KafkaProducer basics for transactions

Sets a unique ID for the producer. Note that it's up to the user to provide this unique ID.

```
HashMap<String, Object> producerProps = new HashMap<>();

producerProps.put("transactional.id", "set-a-unique-transactional-id");

Producer<String, String> producer = new KafkaProducer<>(producerProps);
producer.initTransactions();                        ← Calls initTransactions

try {
    producer.beginTransaction();
    producer.send(topic, "key", "value");
    producer.commitTransaction();
} catch (ProducerFencedException | OutOfOrderSequenceException
  | AuthorizationException e) {
    producer.close();
} catch (KafkaException e) {
    producer.abortTransaction();
    // safe to retry at this point
}
```

The beginning of the transaction, but does not start the clock for transaction timeouts

Commits the transaction after sending all the records

Handles fatal exceptions; your only choice at this point is to close the producer and re-instantiate the producer instance.

Sends record(s): in practice, you'd send more than one, but it's shortened here for clarity.

Handling a non-fatal exception, you can begin a new transaction with the same producer and try again.

After creating a transactional producer instance, you must execute the `init-Transactions()` method. The `initTransaction` sends a message to the transaction coordinator (the transaction coordinator is a broker handling transactions for producers) so it can register the `transactional.id` for the producer to manage its transactions. The transaction coordinator is a broker managing transactions for producers.

If the previous transaction has started, this method blocks until completion. Internally, it also retrieves some metadata, including an `epoch`, which this producer uses in future transactional operations.

Before sending records, you call `beginTransaction()`, which starts the transaction for the producer. Once the transaction begins, the transaction coordinator will only wait for a period defined by the `transaction.timeout.ms` (1 minute by default), and without an update (a commit or abort), it will proactively abort the transaction. However, the transaction coordinator does not start the transaction timeout clock until the broker sends records. Then, after the code completes processing and producing the records, you commit the transaction.

You should notice a subtle difference in error handling between the transactional example and the previous nontransactional one. With the transactional produce, you don't have to check for an error either with a `Callback` or checking the returned `Future`. Instead, the transactional producer throws them directly for your code to handle.

It's important to note that any exceptions in the first `catch` block are fatal, and you must close the producer. To continue working, you'll have to create a new instance. But any other exception is retryable, and you'll need to abort the current transaction and start over.

Of the fatal exceptions, we've already discussed the `OutOfOrderSequenceException` in section 4.4.1, and the `AuthorizationException` is self-explanatory. We should quickly discuss the `ProducerFencedException`. Kafka has a strict requirement that there is only one producer instance with a given `transactional.id`. When a new transactional producer starts, it "fences" off any previous producer with the same ID and must close. However, there is another scenario where you can get a `ProducerFencedException` without starting a new producer with the same ID (figure 4.23).

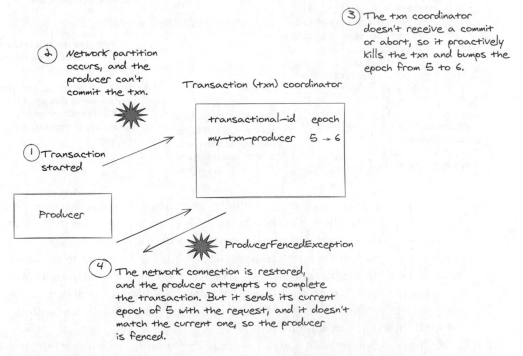

Figure 4.23   Transactions proactively aborted by the Transaction Coordinator cause an increase in the epoch associated with the transaction ID.

When you execute the `producer.initTransactions()` method, the transaction coordinator increments the producer epoch. The producer epoch is a number the transaction coordinator associates with the transactional ID. When the producer makes any transactional request, it provides the epoch and its transaction ID. If the epoch in the request doesn't match the current epoch, the transaction coordinator rejects the request, and the producer is fenced.

The timeout expires if the current producer can't communicate with the transaction coordinator. In that case, as we discussed before, the coordinator proactively aborts the transaction and increments the epoch for that ID. When the producer attempts to work again after the break in communication, it finds itself fenced, and you must close the producer and restart at that point.

> **NOTE**   There is an example for transactional producers in the form of a test located at src/test/java/bbejeck/chapter_4/TransactionalProducerConsumerTest.java in the source code.

So far, I've only covered how to produce transactional records, so let's move on to consuming them.

4.4.3   Consumers in transactions

Kafka consumers can subscribe to multiple topics simultaneously, with some containing transactional records and others not. But for transactional records, you'll only want to consume ones from successful transactions. Fortunately, it's only a matter of a simple configuration. To configure your consumers for transactional records, you set the `isolation.level` configuration to `read_committed`. See the following code listing (some details are omitted for clarity).

Listing 4.8   KafkaConsumer configuration for transactions

```
HashMap<String, Object> consumerProps = new HashMap<>();
consumerProps.put(ConsumerConfig.BOOTSTRAP_SERVERS_CONFIG, "localhost:9092");
consumerProps.put(ConsumerConfig.GROUP_ID_CONFIG, "the-group-id");

consumerProps.put(ConsumerConfig.ISOLATION_LEVEL_CONFIG,
    "read_committed");                                          ⟵─┐ Sets the isolation
                                                                  │ configuration for
                                                                  │ the consumer
consumerProps.put(ConsumerConfig.KEY_DESERIALIZER_CLASS_CONFIG,
    StringDeserializer.class);
consumerProps.put(ConsumerConfig.VALUE_DESERIALIZER_CLASS_CONFIG,
    IntegerDeserializer.class);
```

Setting this configuration guarantees your consumer only retrieves successfully committed transactional records. If you use the `read_uncommitted` setting, the consumer retrieves successful and aborted transactional records. The consumer is guaranteed to retrieve nontransactional records with either configuration set. There is a difference in the highest offset a consumer can retrieve in the `read_committed` mode. Let's follow along with figure 4.24 to demonstrate this concept.

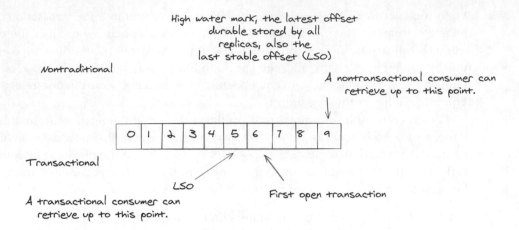

Figure 4.24   High water mark vs. last stable offset in a transactional environment

In Kafka, there is a concept of the last stable offset (LSO), an offset where all offsets below it have been "decided." There's another concept known as the high water mark. The high water mark is the most significant offset successfully written to all replicas. In a nontransactional environment, the LSO is the same as the high water mark; its records are considered decided or durable and written immediately. But with transactions, an offset can't be considered decided until the transaction is either committed or aborted, so this means the LSO is the offset of the first open transaction minus 1.

In a nontransactional environment, the consumer can retrieve up to the high water mark in a `poll()` call. But with transactions, it will only retrieve up to the LSO.

NOTE   The test, located at src/test/java/bbejeck/chapter_4/Transactional-ProducerConsumerTest.java, also contains a couple of tests demonstrating consumer behavior with both `read_committed` and `read_uncommitted` configurations.

So far, we've covered how to use a producer and a consumer separately. But there's one more case to consider: using a consumer and producer together within a transaction.

4.4.4   *Producers and consumers within a transaction*

When building applications to work with Kafka, it's a common practice to consume from a topic, perform some transformations on the records, and then produce the results back to Kafka in a different topic. Records are considered consumed when the consumer commits the offsets. If you recall, committing offsets is simply writing to a topic (__consumer_offsets).

If you are doing a consume–transform–produce cycle, you'd want to make sure that committing offsets is part of the transaction as well. Otherwise, you could end up in a situation where you've committed offsets, but the transaction fails. Then, restarting the application skips the recently processed records as the consumer committed the offsets.

Imagine you have a stock reporting application and need to provide broker compliance reporting. You must send the compliance reports only once, so the best approach is to consume the stock transactions and build the compliance reports within a transaction. Following this approach guarantees you send the reports only once. Consider the following listing, found in src/test/java/chapter_4/Transactional-ConsumeTransformProduceTest.java (some details are left out for clarity).

Listing 4.9 Example of the consume–transform–produce with transactions

Creates the HashMap to hold the offsets to commit

```
Map<TopicPartition, OffsetAndMetadata> offsets = new HashMap<>();
producer.beginTransaction();                                        Starts the
consumerRecords.partitions().forEach(topicPartition -> {            transaction
    consumerRecords.records(topicPartition).forEach(record -> {
        lastOffset.set(record.offset());
        StockTransaction stockTransaction = record.value();
        BrokerSummary brokerSummary = BrokerSummary.newBuilder()

        producer.send(new ProducerRecord<>(outputTopic, brokerSummary));
    });
    offsets.put(topicPartition,                                     Transforms the
      new OffsetAndMetadata(lastOffset.get() + 1L));                StockTransaction
});                                                                 object into a
try {                                                               BrokerSummary
    producer.sendOffsetsToTransaction(offsets,
      consumer.groupMetadata());                                    Stores the
    producer.commitTransaction();                                   TopicPartition and
}                                                                   OffsetAndMetadata
                                          Commits the               in the map
                        Commits the       offsets for the
                        transaction       consumed
                                          records in the
                                          transaction
```

The most significant difference between this code and a nontransactional consume–transform–produce application is that we keep track of the TopicPartition objects and the offset of the records. We do this because we need to provide the offsets of the records we just processed to the KafkaProducer.sendOffsetsToTransaction method. In consume–transform–produce applications with transactions, the producer sends offsets to the consumer group coordinator, ensuring that the offsets are part of the transaction. The offsets are not committed if the transaction fails or gets aborted. By having the producer commit the offsets, you don't need any coordination between the producer and consumer in the cases of rolled-back transactions.

So far, we've covered using producer and consumer clients for sending and receiving records to and from a Kafka topic. Another type of client uses the Admin API, which allows you to perform programmatically topic and consumer group-related administrative functions.

4.5   *Using the Admin API for programmatic topic management*

Kafka provides an administrative client for inspecting topics, brokers, access control lists (ACLs), and configuration. While there are several functions you can use for the admin client, I'm going to focus on the administrative tasks for working with topics and records. You'll usually have an operations team responsible for managing your Kafka brokers in production. What I'm presenting here are things you can do to facilitate testing or prototyping an application using Kafka.

To create topics with the admin client is simply a matter of creating the admin client instance and then executing the command to create the topic(s).

Listing 4.10   Creating a topic

```
Map<String, Object> adminProps = new HashMap<>();
adminProps.put("bootstrap.servers", "localhost:9092");

try (Admin adminClient = Admin.create(adminProps)) {          ◁── Creates the Admin instance. Note the use of a try with resources block.

    final List<NewTopic> topics = new ArrayList<>)();          ◁── The list to hold the NewTopic objects

    topics.add(new NewTopic("topic-one", 1, 1));               ◁── Creates the NewTopic objects and adds them to the list
    topics.add(new NewTopic("topic-two", 1, 1));

    adminClient.createTopics(topics);                          ◁── Executes the command to create the topics
}
```

NOTE I'm referring to an admin client, but the type is the interface `Admin`. There is an abstract class, `AdminClient`, but its use is discouraged over using the `Admin` interface instead. An upcoming release may remove the `AdminClient` class.

This code can be handy when prototyping and building new applications by ensuring the topics exist before running the code. Let's expand this example and show how to list topics and, optionally, delete one.

Listing 4.11   More topic operations

In this example, you list all the non-internal topics in the cluster. Note that if you wanted to include the internal topics, you would provide a ListTopicOptions object, which you would call the ListTopicOptions.listInternal(true) method.

```
Map<String, Object> adminProps = new HashMap<>();
adminProps.put("bootstrap.servers", "localhost:9092");

try (Admin adminClient = Admin.create(adminProps)) {          ◁── Prints the current topics found

    Set<String> topicNames = adminClient.listTopics().names.get();
    System.out.println(topicNames);
```

```
adminClient.deleteTopics(Collections.singletonList("topic-two"));
}
```

You delete a topic and list all of the topics again, but you should not see the recently deleted topic in the list.

An additional note is that the `Admin.listTopics()` returns a `ListTopicResult` object. To get the topic names, you use `ListTopicResult.names()`, which returns a `Kafka-Future<Set<String>>`, so you use the `get()` method, which blocks until the admin client request completes. This command completes immediately since we're using a broker container running on your local machine.

You can execute several other methods with the admin client, such as deleting records and describing topics. But the way you execute them is very similar, so I won't list them here, but look at the source code (src/test/java/bbejeck/chapter_4/Admin-ClientTest.java) to see more examples of using the admin client.

> **TIP** Since we're working on a Kafka broker running in a Docker container on your local machine, we can execute all the admin client topics and record operations risk free. However, you should exercise caution when working in a shared environment to ensure you don't create problems for other developers. Additionally, remember you might not have the opportunity to use the admin client commands in your work environment. You should never attempt to modify topics on the fly in production environments.

That wraps up our coverage of using the `Admin` API. In our next and final section, we'll discuss considerations when producing multiple event types for a topic.

4.6 *Handling multiple event types in a single topic*

Let's say you're building an application to track activity on a commerce website. You need to track clickstream events such as logins, searches, and purchases. Conventional wisdom says that the different events (logins, searches) and purchases could go into separate topics as they are independent events. But you can learn by examining how these related events occurred in sequence.

But you'll need to consume the records from the different topics and then try to stitch the records together in the proper order. Remember, Kafka guarantees record order within a partition of a topic but not across partitions of the same topic, not to mention other topics.

Is there another approach you can take? The answer is yes; you can produce those different event types to the same topic. Assuming you provide a consistent key across the event types, you will receive the various events in order on the same topic partition.

At the end of chapter 3, I covered how to use multiple event types in a topic, but I deferred to show an example with producers and consumers. Now, we'll explain how you can produce and consume multiple event types safely with Schema Registry.

In chapter 3, specifically section 3.6 on schema references and multiple events per topic, I discussed using Schema Registry to support multiple event types in a single topic. I didn't go through an example using a producer or consumer at that point, as it fits better in this chapter. So that's what we're going to cover now.

> **NOTE** Since chapter 3 covered Schema Registry, I will not do any review in this section. I may mention some terms introduced in that chapter, so you should refer back to refresh your memory if needed.

Let's start with the producer side.

4.6.1   *Producing multiple event types*

We'll use this Protobuf schema in the following listing:

```
syntax = "proto3";

package bbejeck.chapter_4.proto;

import "purchase_event.proto";
import "login_event.proto";
import "search_event.proto";

option java_multiple_files = true;
option java_outer_classname = "EventsProto";

message Events {
    oneof type {
      PurchaseEvent purchase_event = 1;
      LogInEvent login_event = 2;
      SearchEvent search_event = 3;
    }
    string key = 4;
  }
```

What happens when you generate the code from the Protobuf definition? You get an `Events` object that contains a single field `type` that accepts one of the three possible event objects (a Protobuf `oneof` field). Some details are omitted for clarity.

Listing 4.12   Example of creating KafkaProducer using Protobuf with a `oneof` field

```
...
producerConfigs.put(ProducerConfig.KEY_SERIALIZER_CLASS_CONFIG,
  StringSerializer.class);
producerConfigs.put(ProducerConfig.VALUE_SERIALIZER_CLASS_CONFIG,
  KafkaProtobufSerializer.class);                           ◁──┐ Configures the
...                                                             producer to use the
                                                                Protobuf serializer
Producer<String, Events> producer = new KafkaProducer<>(
    producerConfigs));                  ◁──┐ Creates the KafkaProducer instance
```

Since Protobuf doesn't allow the `oneof` field as a top-level element, the events you produce always have an outer message container. As a result, your producer code looks the same for the case when you're sending a single event type. So, the generic type for `KafkaProducer` and `ProducerRecord` is the class of the Protobuf outer message (in this case, `Events`). In contrast, if you were to use an Avro union for the schema like the example in the following code listing, it can be a top level element on its own.

Listing 4.13 Avro schema of a union type

```
[
  "bbejeck.chapter_3.avro.TruckEvent",
  "bbejeck.chapter_3.avro.PlaneEvent",
  "bbejeck.chapter_3.avro.DeliveryEvent"
]
```

Your producer code will change to use a common interface type of all generated Avro classes, as in the following example (some details omitted for clarity).

Listing 4.14 `KafkaProducer` instantiation with Avro union type schema

Configures the producer not to autoregister schemas   **Specifies to use the Kafka Avro serializer**

```
producerConfigs.put(ProducerConfig.KEY_SERIALIZER_CLASS_CONFIG,
  StringSerializer.class);
producerConfigs.put(ProducerConfig.VALUE_SERIALIZER_CLASS_CONFIG,
  KafkaAvroSerializer.class);
producerConfigs.put(AbstractKafkaSchemaSerDeConfig.AUTO_REGISTER_SCHEMAS,
  false);
producerConfigs.put(AbstractKafkaSchemaSerDeConfig.USE_LATEST_VERSION,
  true);
```

Sets the use latest schema version to true

```
Producer<String, SpecificRecord> producer = new KafkaProducer<>(
  producerConfigs())
```

Instantiates the producer

Because you don't have an outer class, in this case, each event in the schema is a concrete class of either `TruckEvent`, `PlaneEvent`, or a `DeliveryEvent`. To satisfy the `Kafka-Producer` generics, you need to use the `SpecificRecord` interface as every Avro-generated class implements it. As we covered in chapter 3, it's crucial when using Avro schema references with a union as the top-level entry is to turn off autoregistration of schemas and to enable using the latest schema version.

Let's move to the other side of the equation, consuming multiple event types.

4.6.2 *Consuming multiple event types*

Depending on your approach, you may need to instantiate `KafkaConsumer` with a generic type of a common base class or interface that all of the records implement when consuming from a topic with multiple event types.

Let's consider using Protobuf first. Since you will always have an outer wrapper class, that's the class you'll use in the generic type parameter, the value parameter in this example (some configuration details are omitted for clarity).

Listing 4.15   Configuring the consumer for working with multiple event types in Protobuf

```
consumerProps.put(ConsumerConfig.VALUE_DESERIALIZER_CLASS_CONFIG,
   KafkaProtobufDeserializer.class);                                    ⊲── Uses Protobuf
consumerProps.put(                                                          deserializer
   KafkaProtobufDeserializerConfig.SPECIFIC_PROTOBUF_VALUE_TYPE,
      Events.class);                                     ⊲── Sets the Protobuf
                                                             deserializer to return
Consumer<Events> consumer = new KafkaConsumer<>(             a specific type
   consumerProps);           ⊲──┐
                                  Creates KafkaConsumer
```

As you've seen before, you are setting up your consumer; you're configuring the deserializer to return a specific type (in this case, the `Events` class). With Protobuf, when you have a `oneof` field, the generated Java code includes methods to help you determine the field type with `hasXXX` methods. In our case, the `Events` object contains the following three methods:

```
hasSearchEvent()
hasPurchaseEvent()
hasLoginEvent()
```

The Protobuf-generated Java code also contains an enum named `<oneof field name>Case`. In this example, we've named the `oneof` field `type` so it's named `TypeCase`, and you access it by calling `Events.getTypeCase()`. You can use the enum to determine the underlying object succinctly (some details omitted for clarity):

```
switch (event.getTypeCase()) {                      Individual case statement
    case LOGIN_EVENT -> {                           based on the enum
        logins.add(event.getLoginEvent());     ⊲──
    }                                                  Retrieves the event
    case SEARCH_EVENT -> {                             object using getXXX
        searches.add(event.getSearchEvent());          methods for each
    }                                                  potential type in the
    case PURCHASE_EVENT ->  {                          oneof field
        purchases.add(event.getPurchaseEvent());
    }
}
```

Which approach you use to determine the type is personal choice.

Next, let's see how you would set up your consumer for multiple types with the Avro union schema (some configuration details are omitted for clarity).

Listing 4.16   Configuring the consumer for working with union schema with Avro

```
consumerProps.put(ConsumerConfig.VALUE_DESERIALIZER_CLASS_CONFIG,
   KafkaAvroDeserializer.class);
consumerProps.put(KafkaAvroDeserializerConfig.SPECIFIC_AVRO_READER_CONFIG,
```

Usies the Avro deserializer

```
true);

Consumer<SpecificRecord> consumer = new KafkaConsumer<>(consumerProps);
```

Specifies the deserializer to return a specific Avro type   **Creates KafkaConsumer**

You specify the `KafkaAvroDeserializer` for the deserializer configuration. We also covered before how Avro is slightly different from Protobuf and JSON Schema in that you tell it to return the specific class type, but you don't provide the class name. So when you have multiple event types in a topic and are using Avro, the consumer needs to use the `SpecificRecord` interface again in the generics shown in listing 4.16.

So, by using the `SpecificRecord` interface when you start retrieving records from the `Consumer.poll` call, you'll need to determine the concrete type to do any work with it. See the following listing (some details omitted for clarity).

Listing 4.17 Determining the concrete type of a record using Avro union schemas

```
SpecificRecord avroRecord = record.value();
if (avroRecord instanceof PlaneEvent) {
    PlaneEvent planeEvent = (PlaneEvent) avroRecord;
    ....
} else if (avroRecord instanceof TruckEvent) {
    TruckEvent truckEvent = (TruckEvent) avroRecord;
    ....
} else if (avroRecord instanceof DeliveryEvent) {
    DeliveryEvent deliveryEvent = (DeliveryEvent) avroRecord;
    ....
}
```

The approach here is similar to what you did with Protobuf, but this is at the class level instead of the field level. You could also model your Avro approach to something similar to Protobuf and define a record that contains a field representing the union. See the following example.

Listing 4.18 Avro with embedding the union field in a record

```
{
  "type": "record",
  "namespace": "bbejeck.chapter_4.avro",     |  Outer class
  "name": "TransportationEvent",             ◄─┘ definition

  "fields" : [
    {"name": "txn_type", "type": [           ◄──  Avro union type
      "bbejeck.chapter_4.avro.TruckEvent",         at the field level
      "bbejeck.chapter_4.avro.PlaneEvent",
      "bbejeck.chapter_4.avro.DeliveryEvent"
    ]}
  ]
}
```

In this case, the generated Java code provides a single method, `getTxnType()`, but it has a return type of `Object`. As a result, you'll need to use the same approach of checking for the instance type as you previously did when using a union schema, essentially pushing the task of determining the record type from the class level to the field level.

> **NOTE** Java 16 introduces pattern matching with the `instanceof` keyword that removes the need for casting the object after the `instanceof` check.

Summary

- Kafka producers send records in batches to topics on the Kafka broker and will continue to retry sending failed batches until the `delivery.timeout.ms` configuration expires. You can configure a Kafka producer to be an idempotent producer, meaning it guarantees to send records only once and in order for a given partition. Kafka producers also have a transactional mode that guarantees exactly-once delivery of records across multiple topics. You enable the Kafka transactional API in producers by using the configuration `transactional.id`, which must be a unique ID for each producer. When using consumers in the transactional API, you want to set the `isolation.level` to `read_committed` so you only consume committed records from transactional topics.
- Kafka consumers read records from topics. Multiple consumers with the same group ID get topic partition assignments and work together as one logical consumer. Should one group member fail, its topic partition assignment(s) are redistributed to other group members via a process known as rebalancing. Consumers periodically commit the offsets of consumed records, so when restarted after a shutdown, they pick up processing where they left off.
- Kafka producers and consumers offer three different delivery guarantees at least once, at most once, and exactly once. At-least-once delivery means no records are lost, but you may receive duplicates due to retries. At-most-once delivery means that you won't receive duplicate records, but records could be lost due to errors. Exactly-once delivery means you won't receive duplicates and won't lose any records due to errors.
- Static membership provides stability in environments where consumers frequently drop off only to return online within a reasonable time.
- `CooperativeStickyAssignor` provides the much improved rebalance behavior. The cooperative rebalance protocol is the best choice in most cases as it significantly reduces the amount of downtime during a rebalance.
- The `Admin` API provides a way to create and manage topics, partitions, and records programmatically.
- When you have different event types, but the events are related and processing them in order is essential, placing the multiple event types in a single topic is worth considering until the `delivery.timeout.ms` configuration expires. You can configure a Kafka producer to be an idempotent producer, meaning it guarantees it will send records only once and in order for a given partition.

Kafka producers also have a transactional mode that guarantees exactly-once delivery of records across multiple topics. You enable the Kafka transactional API in producers by using the configuration `transactional.id`, which must be a unique ID for each producer. When using consumers in the transactional API, you want to set the `isolation.level` to `read_committed`, so you only consume committed records from transactional topics.

Kafka Connect

5

This chapter covers
- Getting started with Kafka Connect
- Applying Single Message Transforms
- Building and deploying your own connector
- Making your connector dynamic with a monitoring thread
- Creating a custom transformation

This chapter will teach you how to move events into and out of Apache Kafka quickly. While Kafka can function as a central nervous system for data, it primarily provides a decoupled and centralized approach to data access. Still, other essential services, like full-text searching, report generation, and data analysis, can only be serviced by applications specific to those purposes. No single technology or application can satisfy all the needs of a business or organization.

We've established in earlier chapters that Kafka simplifies the architecture for a technology company by ingesting events once, and any group within the organization can consume events independently. In the previously mentioned cases where the consumer is another application, you'll need to write a consumer specifically for that application. You'll repeat a lot of code if you have more than one. A better approach would be to have an established framework that you can deploy that handles either getting data from an application into Kafka or getting data out of Kafka into an external application. That framework exists and is a critical component of Apache Kafka: Kafka Connect.

5.1 An introduction to Kafka Connect

Kafka Connect is part of the Apache Kafka project. It provides the availability to integrate Kafka with other systems like relational databases, search engines, NoSQL stores, Cloud object stores, and data warehouse systems. Connect lets you quickly stream large amounts of data in and out of Kafka. The ability to perform this streaming integration is critical to legacy systems in today's event streaming applications.

Using Connect, you can achieve a two-way flow between existing architectures and new applications. For example, you can migrate incoming event data from Kafka to a typical Model–View–Controller (MVC) application by using a connector to write results to a relational database. So you could consider Kafka Connect as a sort of "glue" that enables you to integrate different applications with new sources of event data seamlessly.

One concrete example of Kafka Connect is capturing changes in a database table as they occur, called *change data capture* (CDC). CDC exports changes to a database table (INSERT, UPDATE, and DELETE) to other applications. Using CDC, you store the database changes in a Kafka topic, making them available to downstream consumers with low latency. One of the best parts of this integration is that the old application doesn't need to change; it's business as usual.

Now, you could implement this type of work yourself using producer and consumer clients. But that statement oversimplifies all the work you would be required to put into getting that application production-ready. Not to mention that each system you consume from or produce to would require different handling.

Consuming changes from a relational database differs from consuming from a NoSQL store, and producing records to ElasticSearch differs from producing records to Amazon S3 storage. It makes sense to go with a proven solution with several off-the-shelf components ready to use. There are hundreds of connectors available from Confluent (https://www.confluent.io/hub/); while some are commercial and require payment, over 100 connectors are freely available for you to use.

Connect runs in a separate process from a Kafka cluster, either as a single application or several instances as a distributed application. Connect exposes the connectors as plugins that you configure on the classpath. No coding is involved in running a connector; you supply a JSON configuration file to the connect server. Having said that, if no connector is available to cover your use case, you can also implement your own connector. We'll cover creating a custom connector in the last section of this chapter.

Regarding Connect, it's important to understand there are two kinds of connectors: sources and sinks. A source connector will consume data from an external source such as a Postgres or MySql database, MongoDB, or an S3 bucket into Kafka topics. A sink connector does the opposite, producing event data from a Kafka topic into an external application like ElasticSearch or Google BigQuery. Additionally, due to Kafka's design, you can simultaneously have more than one sink connector exporting data from a Kafka topic. This pattern of multiple sink connectors means that you can take an existing application and, with a combination of source and sink connectors, share its data with other systems without making any changes to the original application.

Finally, Connect provides a way to modify data coming into or out of Kafka. Single Message Transforms (SMTs) allow you to alter the format of records from the source system when reading them into a Kafka. Or, if you need to change the format of a record to match the target system, you can also use an SMT there. For example, when importing customer data from a database table, you may want to mask sensitive data fields with the `MaskField` SMT. There's also the ability to transform the format of the data.

Let's say you're using the Protobuf format in your Kafka cluster. You have a database table that feeds a topic via a source connector, and for the target topic, you also have a sink connector writing records out to a Redis key-value store. Neither the database nor Redis works with the Protobuf format in either case. But by using a value converter, you can seamlessly transform the incoming records into a Protobuf format from the source connector. In contrast, the sink connector will use another value converter to change the outgoing records back into the plain text from Protobuf.

To wrap up our introduction, in this chapter, you'll learn how to deploy and use Kafka Connect to integrate external systems with Kafka, enabling you to build event-streaming data pipelines. You'll learn how to apply SMTs to make changes to incoming or outgoing records and how to develop custom transformations. Finally, we'll cover how to build your own connector if no existing one meets your needs.

5.2   *Integrating external applications into Kafka*

So far, we've learned what Connect is and that it acts as a "glue" that can bind different systems together, but let's see this in action with an example.

Imagine you are responsible for coordinating new student signups for college orientation. The students indicate which orientation session they will attend through a web form the college has used for some time. It used to be that when the student arrived for orientation, other departments, housing, food service, and guidance would need to meet with the students. To use the information the students applied on the form, a staff member would print the information and hand out paper copies to each department's staff member.

However, this process was error-prone and could be time-consuming as other staff members saw the student information for the first time, and figuring out the logistics for each student on the spot takes time. It would be better if a process were in place to share the information as soon as students signed up. You've learned that the university has recently embarked on a technological modernization approach and has adopted Kafka at the center of this effort.

The Office of Student Affairs, the department responsible for getting orientation information, is set on its current application structure, a basic web application feeding a PostgreSQL database. The reluctance of the department to change at first presents an issue, but then you realize there's a way to integrate their data into the new event streaming architecture.

You can send orientation registrant information directly into a Kafka topic using a JDBC source connector. Then, the other departments can set up consumer applications to receive the data immediately upon registration.

NOTE The integration point is a Kafka topic when you use Kafka Connect to bring in data from other sources. This means *any* application using `Kafka-Consumer` (including Kafka Streams) can use the imported data.

Figure 5.1 shows how this integration between the database and Kafka works. In this case, you'll use Kafka Connect to monitor a database table and stream updates into a Kafka topic.

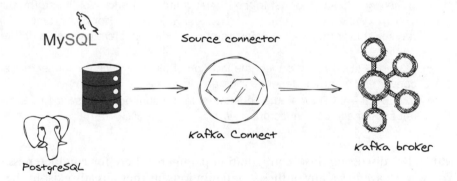

Figure 5.1 Kafka Connect integrating a database table and a Kafka topic

Now that you've decided to use Kafka Connect to help you integrate the orientation student signups with the new event streaming platform rolled out at the university, let's dive in with a simple working example in the next section. After that, we'll go into more detail on how Connect works and the main concepts.

5.3 Getting started with Kafka Connect

Kafka Connect runs in two flavors: distributed and standalone mode. Running in distributed mode makes sense for most production environments because you can use the parallelism and fault tolerance available when running multiple Connect instances. Standalone mode is suitable for development on a local machine or your laptop.

Setting up Kafka Connect

We'll use a connect Docker image instead of downloading Kafka Connect and running it locally on your laptop. Remember, Kafka Connect runs separately from a Kafka broker, so setting up a new `docker-compose` file and adding a connect service simplifies development and allows you to focus on learning.

When setting up Kafka Connect, you'll provide two levels of configuration. One set of configurations is for the connect server (or worker), and the other is for the individual connector. I've just mentioned a new term, connect worker, that we still need to discuss, but we'll cover what a worker is along with other core connect concepts in the next section.

Let's look at some of the key configuration parameters you'll work with for Kafka Connect:

- `key.converter`—Class of the converter that controls serialization of the key from Connect format to the format written to Kafka. In this case, you'll use the built-in `org.apache.kafka.connect.storage.StringConverter`. The value specified here sets a default converter for connectors created on this worker, but individual connectors can override this value.
- `value.converter`—Class of the converter controlling serialization of the value from Connect format to the format written to Kafka. You'll use the built-in `org.apache.kafka.connect.json.JsonConverter` for this example. Like the key converter, this is a default setting, and connectors can use a different setting.
- `plugin.path`—Tells Connect the location of plugins, such as connectors and converters, and their dependencies on the worker.
- `group.id`—An ID for all the consumers in the connected cluster; this setting is used for source connectors only.

NOTE I'm discussing these configuration parameters here for completeness. You won't have to set any of these configurations, as they already exist in the docker-compose file under the connect service. The connect parameters in the file will take a form similar to CONNECT_PLUGIN_PATH, which will set the `plugin.path` configuration. The docker-compose file will also start a Postgres DB instance and automatically populate a table with values needed to run the example.

I want to explain briefly what a converter is. Connect uses a converter to switch the form of the data captured and produced by a source connector into Kafka format or convert from Kafka format to an expected format of an external system, as in the case of a sink connector (figure 5.2).

Since you can change this setting for individual connectors, any connector can work in any serialization format. For example, one connector could use Protobuf while another could use JSON.

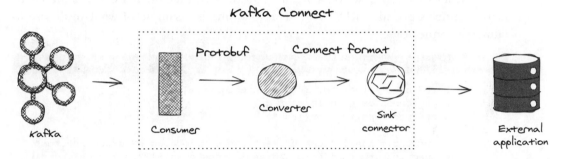

Figure 5.2   Connect converter changes the data format before the data gets into Kafka or after it leaves Kafka.

Next, look at some configurations you'll provide for a connector. In this case, we're using a JDBC connector for our first example, so you'll need to provide the information needed for the connector to connect to the database, like username and password, and you'll also need to specify how the connector will determine which rows to import into Kafka. Let's take a look at a few of the more important configs for the JDBC connector:

- `connector.class`—Class of the connector.
- `connection.url`—URL used to connect to the database.
- `mode`—Method the JDBC source connector uses to detect changes.
- `timestamp.column.name`—Name of the column tracked for detecting changes.
- `topic.prefix`—Connect writes each table to a topic named *topic.prefix+ Table name.*

Most of these configurations are straightforward, but we need to discuss two of them, `mode` and `timestamp.column.name`, in more detail because they have an active role in how the connector runs. The JDBC source connector uses `mode` to detect which rows it needs to load.

For this example, you'll use the `timestamp` setting for the `mode`, which relies on a column containing a timestamp, named in the `timestamp.column.name` config. While it should be evident that INSERT sets the timestamp, we've also added a trigger to the database Docker image that refreshes the timestamp with UPDATE statements. By using a timestamp, the connector will pull any values from the database table whose timestamp is greater than the last one from the previous import.

Another value for the mode could be `incrementing`, which relies on a column with an auto-incrementing number that focuses only on new inserts. The JDBC connector determines which rows to import using several configuration items. Under the covers, the connector will query the database, producing the results into Kafka. In this chapter, I won't go into more detail on the JDBC connector. It's not that we've covered everything there is to know. The information on the JDBC connector could fill an entire chapter. The more important point is that you need to provide configurations for the individual connector, and most offer a rich set to control the connector's behavior.

Now, let's look at how you start an individual connector. It's easily achieved by using connects provided REST API. The following is a sample of what you'll use to launch a connector.

Listing 5.1   REST API call to start a connector

```
curl -i -X PUT http://localhost:8083/connectors/
  jdbc_source_connector_example/config \
  -H "Content-Type: application/json" \
  -d '{
        "connector.class": "io.confluent.connect.jdbc.JdbcSourceConnector",
        "connection.url": "jdbc:postgresql://postgres:5432/postgres",
        "connection.user": "postgres",
        "connection.password": "postgres",
        "mode":"timestamp",
        "timestamp.column.name":"ts",
        "topic.prefix":"postgres_",
        "value.converter":"org.apache.kafka.connect.json.JsonConverter",
        "value.converter.schemas.enable": "false",
        "tasks.max":"1"
     }'
```

Specifies to use a JsonConverter, which converts incoming records to JSON

Disables the schema for the records

Sets the maximum number of tasks for the connector

So, this REST call will start the JDBC connector running. There are a few configurations I'd like to call your attention to here. You set `value.converter`, which will convert the incoming records from the database into JSON format. But you'll see a `value.converter.schemas.enable` configuration set to `false`, meaning the converter will not preserve the schema from the connector in the message's contents.

Remember, when using a JSON converter, the schema is attached to each incoming record, which can increase its size significantly. We can turn off the inferred schemas since we are producing records *into* Kafka. But when consuming from a Kafka topic to write to an external system, depending on the connector, you must enable the schema inferring so Connect can understand the byte arrays stored in Kafka. A better approach would be to use Schema Registry, and then you could use Avro, Protobuf, or JSONSchema for the value converter. We covered schemas and Schema Registry in chapter 3, so I won't review those details again here.

You also see a `tasks.max` setting. To fully explain this configuration, let's provide some additional context. So far, we've learned that you use Kafka Connect to pull data from an external system into Kafka or to push data from Kafka out to another application. You just reviewed the JSON required to start a connector, and it does not do the pulling or pushing of data. Instead, the connector instance is responsible for starting a number of tasks whose job it is to move the data.

Earlier in the chapter, we mentioned two types of connectors: `SourceConnector` and `SinkConnector`. They use two corresponding kinds of tasks: `SourceTask` and `SinkTask`. It's the connector's primary job to generate task configurations. When running in distributed mode, the Connect framework code will distribute and start them across the different workers in the Connect cluster. Each task instance will run in its own thread. Note that setting the `tasks.max` doesn't guarantee the total number of tasks that will run; the connector will determine the correct number it needs *up to* the maximum number. At this point, it would be helpful to look at an illustration of the relationship between workers' connectors and tasks in figure 5.3.

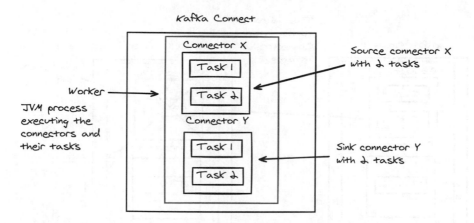

Figure 5.3   Connect in standalone mode; all tasks reside in one worker.

Here, we're looking at Kafka Connect in standalone mode; there's a single worker, a JVM process responsible for running the connector(s) and their task(s). Now let's take a look at distributed mode in figure 5.4.

As you can see here, in distributed mode, the tasks are spread out to other workers in the Connect cluster. Not only does distributed mode allow for higher throughput due to spreading out the load, but it also provides the ability to continue processing in the face of a connector failure. Let's look at one more diagram to illustrate what this means in figure 5.5.

From looking at this illustration, if a Kafka Connect worker stops running, the connector task instances on that worker will get assigned to other workers in the cluster. So, while standalone mode is excellent for prototyping and getting up and running

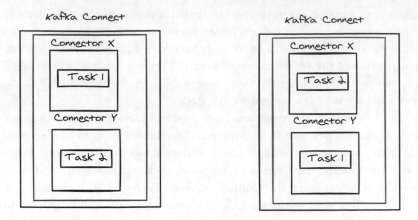

Figure 5.4   Connect in distributed mode; tasks get spread around to other connector instances.

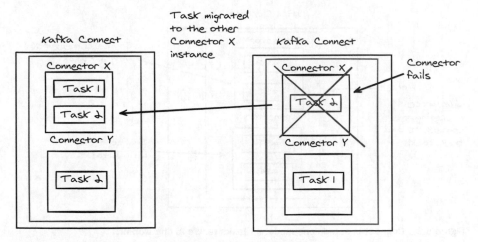

Figure 5.5   Connect in distributed mode provides fault tolerance.

quickly with Kafka Connect, running in distributed mode is recommended for production systems as it provides fault tolerance; tasks from failed workers get assigned to the remaining workers in the cluster. Note that in the distributed mode, you'll need to issue a REST API call to start a particular connector on each machine in the Connect cluster.

Now, let's get back to our example. You have your connector running, but there are a few things you'd like to do differently. First, there are no keys for the incoming records. No keys are a problem because records for the same student may end up on different partitions. Since Kafka only guarantees ordering within a partition, multiple updates made by a student could get processed out of order. Second, the students input their social security numbers. It's essential to limit the exposure of that data, so it would be great to alter or mask it before it gets into Kafka. Fortunately, Connect offers a simple but powerful solution: transforms.

5.4 Applying Single Message Transforms

Simply put, SMTs can perform lightweight changes to records before they get into Kafka or on their way out to external systems, depending on whether you're using a source or sink connector. The main point about transforms is that the work done should be simple, meaning they work on only one record at a time (no joins or aggregations), and the use case should be applicable solely for connectors and not custom producer or consumer applications. It's better to use Kafka Streams or ksqlDB for anything more complex, as they are purpose-built for complex operations. Figure 5.6 depicts the role of the transformation.

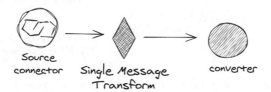

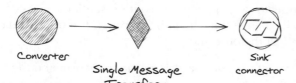

Figure 5.6   A connect converter changes the data format before the data gets into Kafka or after it leaves Kafka.

As you can see here, the role of an SMT is to sit between the connector and the converter. For a source connector, it will apply its operation the record *before* it gets to the converter, and in the case of a sink connector, it will perform the transformation *after* the converter. In both cases, SMTs operate on data in the same format, so most SMTs will work equally well on a sink or a source connector.

Connect provides several SMTs out of the box that tackle a wide range of use cases. For example, there are SMTs for filtering records, flattening nested structures, or removing a field. I won't list them all here, but for the complete list of SMTs, consult the Kafka Connect documentation (http://mng.bz/84VP).

We'll use three transforms: `ValueToKey`, `ExtractField`, and `MaskField`. Working with the provided SMTs requires adding some JSON configuration; no code is required. You can write your own transform if you have a use case where the provided transformations won't offer what you need. We'll cover creating a custom SMT a little later in the chapter. The following listing provides the complete JSON you'll use to add the necessary transforms for your connector.

Listing 5.2   The JSON transform configuration

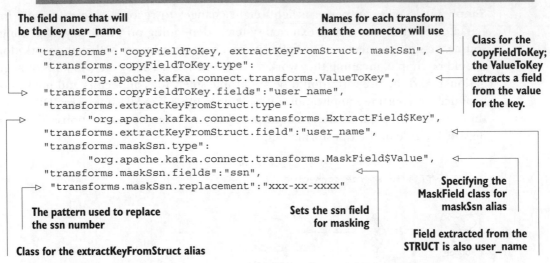

The field name that will be the key user_name

Names for each transform that the connector will use

Class for the copyFieldToKey; the ValueToKey extracts a field from the value for the key.

```
"transforms":"copyFieldToKey, extractKeyFromStruct, maskSsn",
"transforms.copyFieldToKey.type":
     "org.apache.kafka.connect.transforms.ValueToKey",
"transforms.copyFieldToKey.fields":"user_name",
"transforms.extractKeyFromStruct.type":
     "org.apache.kafka.connect.transforms.ExtractField$Key",
"transforms.extractKeyFromStruct.field":"user_name",
"transforms.maskSsn.type":
     "org.apache.kafka.connect.transforms.MaskField$Value",
"transforms.maskSsn.fields":"ssn",
"transforms.maskSsn.replacement":"xxx-xx-xxxx"
```

The pattern used to replace the ssn number

Sets the ssn field for masking

Specifying the MaskField class for maskSsn alias

Field extracted from the STRUCT is also user_name

Class for the extractKeyFromStruct alias

Most of the JSON for the transform configuration should be straightforward. There's a comma-separated list of names, each representing a single transform.

TIP   The order of names in the `transforms` entry is not arbitrary. The names are in order of how the connector will apply them, so it's essential to consider how each transformation will change the data going through the transformation chain and how it will affect the outcome.

Then, for each name in the list, you provide the class of the transform and the field to which it will apply. But there is one thing I'd like to point out that might not be readily apparent. With the `copyFieldToKey` transform, you've indicated that you want to use the user_name column for the key of each resulting Kafka record. But the result produces a single field in STRUCT that looks like the following listing.

Listing 5.3   Struct

```
Struct {"user_name" : "artv"}
```

But you want the value of the struct field user_name for the key, so you also apply the `ExtractField` transform. In the configs, you need to specify the transform that will extract the key like this: `ExtractField$Key`. Connect applies the second transform, and the key ends up with the raw single value applied to the key for the incoming record.

I want to point out something about the transformations that could go unnoticed here. You can chain multiple transforms together to operate on the same field; this is demonstrated in our example by first copying a field to the key and then extracting a value from the intermediate result of the field copy operation. But there is a balance you'll need to strike here; if you find yourself building a transform chain that starts to go beyond 2, consider using Kafka Streams to perform the transformation work, as it will be more efficient.

The final transform you utilize is `MaskField`, which you'll apply to the field containing the student's Social Security number. Again, you'll see in the configuration how you specified you want to apply the masking of the value with the `Mask-Field$Value` setting. In this case, you specify the replacement for the Social Security number as a string of x characters for each number, resulting in a value of xxx-xx-xxxx. With the `MaskField` transform, you also have the option of not specifying a specific replacement, and it will use an empty value based on the type of the field it's replacing—an empty string for string field or 0 for numerical ones.

Now, you've completed a fully configured connector that will poll the database for changes or updates and import them into Kafka, let's make your relational database part of your event streaming platform!

> **NOTE**   We've talked about the JDBC connector in this section. There are a few corner cases where the JDBC connector won't get the latest changes into Kafka. I won't go into those here, but I recommend looking at the Debezium connector for integrating relational databases with Apache Kafka (http://mng.bz/E9JJ). Instead of using an incrementing value or timestamp field, Debezium uses the database changelog to capture changes that need to go into Kafka.

To run the example we just detailed in this section, consult the README file in the chapter 5 directory of the source code for this book. There's also a sink connector as part of the example, but we won't cover it here as you deploy it similarly, provide some JSON configurations, and issue a REST call to the connect worker.

So far, you've set up a source connector, and you'll also want to send events from Kafka into an external system. To accomplish that, you'll use a sink connector, specifically an elastic search sink connector.

5.5   *Adding a sink connector*

One additional feature you suggested adding was the ability for incoming students to search for a potential roommate based on the input data gathered for the orientation process. When students come for orientation, part of the process can be to enter a few keywords and get hits from other students who have the same preferences.

When you pitched the idea, everyone was enthusiastic, but questions on how to get the data seemed too complicated. The team wanted to avoid setting up a new pipeline for gathering information, so they put the idea on hold. But now, with the adoption of Kafka and Connect for importing the incoming student information, exposing a search application with Elastic Search (https://www.elastic.co/) is as simple as providing a sink connector.

Fortunately, there's already an elastic search sink connector, so all you need to do is install the jar files to your running Connect cluster, and you're all set. So, getting started with the elastic sink connector is simply a matter of making a REST API call to get the connector up and running.

Listing 5.4   REST call to install an elastic connector

```
$ curl -i -X PUT localhost:8083/connectors/
    student-info-elasticsearch-connector/config \
    -H "Content-Type: application/json" \
    -d '{
          "connector.class":
        "io.confluent.connect.elasticsearch.ElasticsearchSinkConnector",
          "connection.url": "http://elasticsearch:9200",
          "tasks.max": "1",
          "topics": "postgres_orientation_students",      ← Topic to import records from
          "type.name": "_doc",
          "value.converter": "org.apache.kafka.connect.json.JsonConverter",
          "value.converter.schemas.enable": "false",
          "schema.ignore": "true",
          "key.ignore": "false",
          "errors.tolerance":"all",                        ← Error tolerance setting / Sets the replication factor to 1 for development setting
          "errors.deadletterqueue.topic.name":"orientation_student_dlq",  ← The name of the dead letter queue
          "errors.deadletterqueue.context.headers.enable": "true",  ← Enables headers so error description is provided
          "errors.deadletterqueue.topic.replication.factor": "1"
    }'
```

For the most part, the configurations here are similar to what you saw for the JDBC connector, except for the configurations specific to the connector. There is the topic name the sink connector will use to read records and write them to elastic search.

But there are three configurations, starting with `errors`, that we haven't seen before, and we should discuss them now. Since a sink connector is attempting to write event records to an external system, there's a good possibility for errors. After all, working with distributed applications means embracing failure and providing the mechanisms for how you want your application to respond to them.

> **NOTE**   Listing 5.4 sets the replication factor for the dead letter queue (DLQ) topic. In a single-node cluster, such as our Docker development environment, you'll want to put this to 1. Otherwise, the connector won't start since there aren't enough brokers for the default replication factor of 3.

The `errors.tolerance` configuration in listing 5.4 specifies how the connector will respond when it gets an error. Here, you've used a setting of `all`, meaning that the connector will keep running regardless of errors during conversion and transformation while writing records to the sink. Then, you'd need to review the logs for that specific connector to determine what went wrong and how to proceed. While the `all` setting allows the connector to continue operating, you still want to know when a record fails. The configuration creates a dead letter queue (DLQ), where Connect can set aside the records it can't deliver.

However, when enabling a DLQ topic, Connect only stores the failed record. To get the reason why the failure occurred, you'll need to allow storage of some additional information in the record header, which you've done. We previously covered record headers, so you can refer to chapter 4 to review information on reading with record headers.

As with anything in life, there are tradeoffs to consider when using a DLQ. While a setting of `none` for `errors.tolerance` and shutting down on error sounds harsh, if this were a production system, you'd probably find out quickly that something was amiss and needed to be fixed. Contrast this with the setting of `all`, which will continue regardless of any errors the connector may encounter; it's critical to track any errors because running indefinitely with an error condition could arguably be worse than shutting down. In other words, "if a tree falls in a forest and no one is there to hear it, does it make a sound?" If errors arrive in your DLQ, but no one is looking at it, it's the same as no errors occurring.

So, with enabling a DLQ, you'll want to set up some monitoring (i.e., `Kafka-Consumer`) that can alert on any errors and possibly take action, like shutting down the problematic connector in the face of continuous problems. The source code for the book will have a basic implementation demonstrating this type of functionality.

So now you've learned about using a source connector to bring event data into Kafka and a sink connector to export from Kafka to an external system. While this is a simple workflow, this example is not representative of the full usefulness of Kafka Connect. First, there could be several sink connectors writing the imported records out to external systems, not just one, and the use of Kafka to create an event streaming platform could be increased by effectively bringing all external systems into one central data flow.

In practice, you can have several different source connectors importing data, and there would likely be more processing of those incoming records and not a simple data pipe demonstrated in this chapter. For example, consider that any number of client applications could rely on the topic with data a source connector provides. Each application could produce unique results for another topic. Then, any number of sink connectors could produce the updated records back out of external systems in your architecture.

Connect plays an integral part in gluing together external data applications into the Kafka event streaming platform. Using Connect, any application with a connector can plug into it and use Kafka as the central hub for all incoming data.

But what if you have a source or a sink candidate that doesn't have an existing connector? While there are hundreds of existing connectors, the one you need might not exist. But there's good news: a Connect API makes it possible for you to implement your own connector, and that's what you'll do in the next section.

5.6   *Building and deploying your own connector*

In this section, we will walk through the steps you'll take to implement your connector. Let's say you work for a fintech firm, and along with providing institutional-grade financial data to investors and investment firms, your company is branching out and wants to provide analysis based on a subscription model.

To offer the analysis data, the company has set up a few departments to ingest real-time stock ticker data. Getting an entire feed of live data is costly, and the cost for each department to have its feed would be prohibitively expensive. But you realize that if you created a source connector, you could consume the feed once, and every department could set up its client application to consume the feed from your Kafka cluster in near real time. So, with your idea firmly in mind, let's move on to implementing your connector.

5.6.1   *Implementing a connector*

To develop your connector, you will work the following interfaces: `Connector` and `Task`. Specifically, you will extend the abstract class `SourceConnector`, which extends the `Connector` class. You'll need to implement several abstract methods, but we're not going to go over each one, just the most significant ones. You can consult the source code from the book to look at the entire implementation.

Let's start with the configuration. The `Connector` class only does a little when it comes to moving data. Its primary responsibility is properly configuring each `Task` instance as it's the `Task` that gets data into or out of Kafka. So you'll create a `Stock-TickerSourceConnectorConfig` class that you'll instantiate directly inside your connector. The configuration class contains a `ConfigDef` instance to specify the expected configurations.

> **Listing 5.5   Setting up a `ConfigDef` instance**

```
public class StockTickerSourceConnector extends SourceConnector {
private static final ConfigDef CONFIG_DEF = new ConfigDef()      ⟵── Creates the
          .define(API_URL_CONFIG,                                      ConfigDef
                ConfigDef.Type.STRING,                                 instance
                ConfigDef.Importance.HIGH,
                "URL for the desired API call")
          .define(TOPIC_CONFIG,                              ⟵── Adds the topic name
                ConfigDef.Type.STRING,                           to the configuration
                ConfigDef.Importance.HIGH,
                "The topic to publish data to")

... followed by the rest of the configurations
}
```

Defines the API URL config — (annotation pointing to `.define(API_URL_CONFIG,`)

Here, you can see we're adding the configurations using a fluent style where we chain the method calls `define` one after another. Note that there are more configurations we're adding, but it's not necessary to show all of them here; you get the point of how they all get added. By building the `ConfigDef`, the connector will "know" what configurations to expect. The `ConfigDef.define` method allows you to provide a default value. So, if you don't set a configuration, the default value gets populated in its place. But if you don't give a default value, the connector starts it will throw a `ConfigException` and shuts down. The next part of the connector we'll want to look at is how it determines the configuration for each `Task` instance.

Listing 5.6 Configuring the tasks

```
@Override
public List<Map<String, String>> taskConfigs(int maxTasks) {
 List<Map<String, String>> taskConfigs = new ArrayList<>();
 List<String> symbols = monitorThread.symbols();
 int numTasks = Math.min(symbols.size(), maxTasks);
 List<List<String>> groupedSymbols =
       ConnectorUtils.groupPartitions(symbols, numTasks);
 for (List<String> symbolGroup : groupedSymbols) {
    Map<String, String> taskConfig = new HashMap<>();
    taskConfig.put(TOPIC_CONFIG, topic);
    taskConfig.put(API_URL_CONFIG, apiUrl);
    taskConfig.put(TOKEN_CONFIG, token);
    taskConfig.put(TASK_BATCH_SIZE_CONFIG, Integer.toString(batchSize));
    taskConfig.put(TICKER_SYMBOL_CONFIG, String.join(",", symbolGroup));
    taskConfigs.add(taskConfig);
 }
return taskConfigs;
}
```

- **Gets the list of symbols from the monitoring thread** → `List<String> symbols = monitorThread.symbols();`
- **Partitions the stock symbols by the number of tasks** → `ConnectorUtils.groupPartitions(symbols, numTasks);`
- **Creates the Map for configurations for each task** → `Map<String, String> taskConfig = new HashMap<>();`
- **Stores the symbols as a comma-separated string** → `taskConfig.put(TICKER_SYMBOL_CONFIG, String.join(",", symbolGroup));`

You'll notice you're getting the list of symbols from an instance variable named `monitor-Thread`. I want to defer talking about this field until we get to monitoring later in this chapter (section 5.6.2). For now, it's enough to know this method returns the list of symbols needed to run the ticker.

If you recall from earlier in the chapter, one of the configurations you set for the connector is `max.tasks`. This configuration determines the maximum number of tasks the connector may spin up to move data. Our stock API service can retrieve information about those companies using a comma-separated list of up to 100 ticker symbols. You'll want to partition the ticker symbols into separate lists to ensure Connect evenly distributes the work across all tasks.

For example, if you have specified 2 as the maximum number of tasks and there are 10 ticker symbols, the connector will partition them into two lists of 5 ticker symbols each. This grouping of ticker symbols uses the `ConnectionUtils.groupPartitions` utility method. That's about as much as we want to cover for the connector implementation, so we'll move on to implementing the task.

Since you're creating a `SourceConnector`, you'll extend the abstract class `Source-Task` to build your task implementation, `StockTickerSourceTask`. There's a handful of methods on the class you'll need to override, but we'll focus on the `poll()` method since that is at the heart of what a `SourceTask` does: it gets data from an external source and loads it into Kafka.

Let's review the behavior of a `SourceTask` at a high level. Once you start a connector, it will configure the correct number of tasks and start them. Once the thread executes, the `SourceTask.poll()` is called periodically to retrieve records from the configured external application. There's no overall configuration on how often to execute the `poll` method. While you don't have explicit control over how often to call your task's `poll` method, you can add some throttling inside your task implementation to wait a desired amount of time before executing your `SourceTask` logic.

There's a basic implementation of throttling in our `SourceTask` example. Why would you want to add throttling to the source task execution? In our case, we will issue a request to an HTTP endpoint for a stock service API. Many available APIs limit how often you can hit the service or give a total number of requests per day. So, by adding throttling, you can ensure the API calls stay within the plan limits.

Also, you'll want to handle the case when there are no records to return. It would be a good idea to wait a small amount of time, 1 to 2 seconds, for example, to allow the source to have something new to return. It's essential, however, to return control to the calling thread by returning `null` so the worker can respond to requests to pause or shutdown.

Now, let's look at the action of the `StockTickerSourceTask.poll` method. We will view the code in sections to ensure we fully understand what is going on with the polling logic.

Listing 5.7   Start of the `poll` method for the source task

Gets the amount of time to wait for the next polling   **Calculates the next time a poll should occur**

```
public List<SourceRecord> poll() throws InterruptedException {
    final long nextUpdate = lastUpdate.get() + timeBetweenPoll;
    final long now = sourceTime.milliseconds();
    final long sleepMs = nextUpdate - now;

    if (sleepMs > 0) {
        LOG.trace("Waiting {} ms to poll API feed next", sleepMs);
        sourceTime.sleep(sleepMs);
    }
```

Gets the current time

At the very top of the method, we calculate the time our next API call should occur; remember, we want to stay within the limit of calls we can make to the service. If the current time is less than our last call, plus the interval we want between calls, we'll have the polling thread sleep.

Now, once the wait time passes, the next part of the method that executes is the core logic of retrieving the stock results (some details are omitted for clarity).

Listing 5.8   Core polling logic

```
HttpRequest request = HttpRequest.newBuilder()
                .uri(uri)
```

Builds the HttpRequest object

```
                                        .GET()
Converts the returned                   .headers("Content-Type", "text/plain;charset=UTF-8")
JSON into a JsonNode                    .build();
        HttpResponse<String> response;
                                                          Sends the request for
        try {                                             stock ticker results
            response = httpClient.send(request,
                        HttpResponse.BodyHandlers.ofString());          ◁
            AtomicLong counter = new AtomicLong(0);
            JsonNode apiResult = objectMapper.readTree(response.body());
            ArrayNode tickerResults = (ArrayNode) apiResult.get(resultNode);
            LOG.debug("Retrieved {} records", tickerResults.size());
            Stream<JsonNode> stockRecordStream =
                        StreamSupport.stream(tickerResults.spliterator(), false);

            List<SourceRecord> sourceRecords = stockRecordStream.map(entry -> {
                Map<String, String> sourcePartition =
                    Collections.singletonMap("API", apiUrl);
                Map<String, Long> sourceOffset =
                    Collections.singletonMap("index", counter.getAndIncrement());
                Schema schema = getValueSchema(entry);          ◁         Creates the schema
                Map<String, Object> resultsMap = toMap(entry);  ◁         of the ticker record
                return new SourceRecord(sourcePartition,
                                        sourceOffset,
   Maps each of the returned              topic,                         Converts the JSON
   JSON ticker results into a             null,                          record to a Map
   SourceRecord                           schema,
                                        toStruct(schema, resultsMap));  ◁

            }).toList();                                                 Builds a Struct for
                                                                         the SourceRecord
            lastUpdate.set(sourceTime.milliseconds());          ◁        value

        return sourceRecords;          ◁
        }                                      Sets the last
                                            update timestamp
                        Returns the
                        SourceRecord toList
```

The logic for the polling is clear. You create an `HttpRequest` object and then submit it to the ticker API endpoint; then, you read the string response body that the API endpoint returns. The results are JSON, with the ticker symbol information in a nested array. You extract the array and flatten the results, mapping each to a `SourceRecord` so that each entry will become a record sent to Kafka.

There's one part of this conversion that we need to cover here. The `SourceRecord` constructor accepts two parameters, a source partition and a source offset, both represented by a `Map` instance. The notion of a source partition and offset might seem a bit out of context, considering the source for a `SourceConnector` will not be Kafka. However, the concept of a source partition indicates a general location description of where the connector sources the data—a database table name, file name, or, in our case, the API URL. We build the schema for the record returned from the ticker API call. Then, you convert the JSON record into a `Map`. We need to do these two steps to create a `Struct` for the value of the `SourceRecord`. Additionally, we pass in the generated schema to the `SourceRecord` instance, which may end up in the Kafka topic if you've configured the connector to include the schema.

With that generalized definition of a source partition, the source offset is the position of the individual record in the retrieved result. If you remember, in Kafka, a consumer commits the offset of the last fully consumed record, so if it shuts down for any reason, it can resume operations from that committed offset.

Earlier in the section, your custom connector instance used a `monitorThread` variable to get the list of ticker symbols to track. In the next section, I'll explain this variable and why the connector is using it.

5.6.2   *Making your connector dynamic with a monitoring thread*

In this connector example, assuming the list of ticker symbols is static would be reasonable. But what if you wanted or needed to change them? Of course, you can use the Connect REST API to update the configurations, but that means you must keep track of any changes and manually update the connector. But you can provide a monitoring thread with your custom connector to keep track of any changes and update the connector tasks automatically.

To be clear, a monitoring thread is not unique to the connector; you'll implement a class that extends a regular `java.lang.Thread` class. Conceptually, you'll start the thread when your connector starts, and it will contain the logic needed to check for any changes and reconfigure the connector tasks.

Imagine you have a separate microservice that handles which information to include in the source connector configuration. The microservice returns a comma-separated list of the ticker symbols. So you'll need your monitoring thread to periodically issue an `HttpRequest` to the service and compare the response against the current list of symbols. If there's a change, it will trigger a reconfiguration of the connector tasks. A better example of why you'd want to use a monitoring thread is the JDBC connector. You can use the JDBC connector to import an entire relational database, possibly consisting of many tables. In any organization, a relational database is not a static resource; it will change. So you'll want to automatically pick up those changes to ensure the connector is importing the latest data into Kafka.

Let's start the monitoring thread analysis by looking at the class declaration and constructor.

Listing 5.9   Class declaration and constructor of monitoring thread

```
public StockTickerSourceConnectorMonitorThread(
    final ConnectorContext connectorContext,          ◁──  ConnectorContext reference passed
    final int monitorThreadCheckInterval,   ◁───┐           in from the SourceConnector
┌─▷ final HttpClient httpClient,                 │
│
│   The HttpClient for making requests                An integer representing
│   to get updated ticker symbols                     how often to check (in
                                                      milliseconds) for changes
```

```
        final String serviceUrl) {
            this.connectorContext = connectorContext;
            this.monitorThreadCheckInterval = monitorThreadCheckInterval;
            this.httpClient = httpClient;
            this.serviceUrl = serviceUrl;
    }
```
The URL for the microservice

The constructor parameters are self-explanatory, but `ConnectorContext` deserves a quick discussion. Any class extending the abstract `Connector` class will have access to the `ConnectorContext`, so you don't need to be concerned about where it comes from. You'll use it to interact with the connect runtime, requesting task reconfiguration when the source changes.

Now, for the behavior of the monitoring thread, you'll override the `Thread.run()` method and provide the logic you want to execute should there be any relevant changes, as in the following listing (some details are omitted for clarity).

Listing 5.10 Overridden run method of the monitoring thread

Guard condition on while loop using a java.util.concurrent.CountDownLatch object

```
public void run() {
  while (shutDownLatch.getCount() > 0) {
    try {
        if (updatedSymbols()) {

            connectorContext.requestTaskReconfiguration();
        }
    boolean isShutdown = shutDownLatch.await(monitorThreadCheckInterval,
                            TimeUnit.MILLISECONDS);

    if (isShutdown) {
        return;
    }
    ....
```

Checks for any changes to symbols

The symbols have changed, so requests a reconfiguration

Waits the specified interval

If the CountDownLatch counted down, returns and breaks the loop

All the `StockTickerSourceConnectorMonitorThread` does is check whether the stock symbols have changed and, if they have, request a configuration change for the connector task(s). After checking for changes, the `shutDownLatch` waits for the designated time specified by the `monitorThreadCheckInterval` instance variable set when the connector created the thread. If the `shutDownLatch` counts down while waiting, the `isShutdown` variable returns `true`, and the monitoring thread stops running. Otherwise, it continues to monitor for changes.

> **NOTE** The `CountDownLatch` is a class from the `java.util.concurrent` package and is a synchronization tool that allows you to have one or more threads wait until a particular condition. I won't go into any more details here, but consult the Javadoc for a full explanation (http://mng.bz/j1N8).

To conclude the discussion on the monitoring thread, let's quickly review how it determines whether there are any changes (some details are omitted for clarity).

Listing 5.11 Logic used to detect changes to symbols

```
List<String> maybeNewSymbols = symbols();              ⟵───┤ Retrieves the
 boolean foundNewSymbols = false;                             symbols
 if (!Objects.equals(maybeNewSymbols, this.tickerSymbols)) {  ⟵───┐ Compares
     tickerSymbols = new ArrayList<>(maybeNewSymbols);    ⟵───┐    retrieved
     foundNewSymbols = true;                                  │    symbols to
}                                          Updates the instance│    the current
return foundNewSymbols;                    list of symbols          set
```

To determine whether there are changes, the `symbols()` method issues an `HttpRequest` to the microservice and returns a `List<String>` contained in the response. If the retrieved list's contents differ from the current one, we update the instance list and set the `foundNewSymbols` Boolean to `true`, which triggers a reconfiguration of the task when the method returns.

So this wraps up our coverage of the custom `SourceConnector`; the code presented here isn't production-ready, but it gives you a good understanding of how you would implement your connector. There are instructions in the book's source code that describe how to run this custom connector locally with a Docker image.

The API returns a rich set of fields for the stock ticker symbol, but you may only interested in storing a subset of those fields in Kafka. Of course, you could extract the fields of interest right at the source when the connector task retrieves the API results, but if you switched API services with a different JSON constructor, it would mean you'd need to change the logic of your connector. A better approach would be to use a transform to extract the fields you need before the records make it to Kafka. You'll need a custom transformation to take arbitrary fields from a JSON object and return one with just those fields, and that's what we will cover in the next section: creating a custom transformation.

5.6.3 *Creating a custom transformation*

Although Kafka Connect provides several transformations out of the box, those transformations may not handle every case. While the need for a custom connector is less likely due to the number of connectors available for popular external systems, it's more likely that you'll need to write a custom transformation.

The stock API feed we used for our custom connector produces a result containing 70+ fields of different metrics for each stock symbol. While each is useful, we only want to keep a fraction—5 or 6 at most. So, you will create a transformation that keeps only the fields you specify via a configured comma-separated list of field names.

To complete the transformation, you will implement the `Transformation` interface. There are a few methods on the interface you'll need to implement, but we'll focus on one in particular, the `apply` method, as that's where all the action of dropping the fields you're not interested in happens.

But implementing a `Transformation` object has a bit of a twist. If you remember earlier in the chapter, when you launch a connector, one of the configurations you set specifies whether you want to include the schema for each record. So, you'll need to consider that when implementing your custom SMT. We'll see how this works in the upcoming section. Additionally, you'll need to allow a user to apply this transformation to the key or the value, and you'll see how we handle that.

Let's get into the implementation of the `Transformation` now (some details are omitted for clarity).

Listing 5.12   Implementation of the custom transformation

The apply method where the transformation takes place

Class declaration

```
public abstract class MultiFieldExtract<R extends ConnectRecord<R>>
                                  implements Transformation<R> {

    @Override
    public R apply(R connectRecord) {
      if (operatingValue(connectRecord) == null) {
        return connectRecord;
      } else if (operatingSchema(connectRecord) == null) {
        return applySchemaless(connectRecord);
      } else {
        return applyWithSchema(connectRecord);
      }
    }
}
```

When the underlying value is null, there's nothing to do but return the original record.

If the schema is null, we apply the transformation to just the value.

Otherwise, you'll need to adjust the schema to match the new value structure.

You'll probably notice that our `Transformation` implementation is an `abstract` class. If you go back to section 5.3, you'll remember when you configured SMTs, you needed to specify that the transform was either for the key or the value with a configuration like `MaskField$Value`(in Java, the `$` indicates an inner class). So, you declare the transformation class as abstract since it will have, by convention, three abstract methods: `operatingSchema`, `operatingValue`, and `newRecord`. You'll implement these methods with two inner classes, `Key` and `Value`, representing the transformation for the respective part of the Connect record. We won't go into any more details here, so we can continue moving forward and discuss the action in the `apply` method.

You'll see the simplest case in listing 5.12: the Connect record's underlying value is `null`. Remember, in Kafka, it's acceptable to have a `null` key or value. Keys are optional, and with compacted topics, a `null` value represents a tombstone and indicates to the log cleaner to delete the record from the topic. For the rest of this section, I will assume that we're only working with values.

Next, we check whether the record has an embedded schema. Again, in section 5.2, we discussed the `value.converter.schemas.enable` configuration, which, if enabled, embeds the schema for each record coming through the connector. If this branch evaluates to `true`, we'll use the `applySchemaless` method to complete the transformation.

Listing 5.13   Transforming without a schema

```
private R applySchemaless(R connectRecord) {
    final Map<String, Object> originalRecord =
                requireMap(operatingValue(connectRecord), PURPOSE);
    final Map<String, Object> newRecord = new LinkedHashMap<>();
    List<Map.Entry<String,Object>> filteredEntryList =
      originalRecord.entrySet().stream()
      .filter(entry -> fieldNamesToExtract.contains(entry.getKey()))
      .toList();

    filteredEntryList.forEach(entry -> newRecord.put(entry.getKey(),
                                          entry.getValue()));
    return newRecord(connectRecord, null, newRecord);
    }
```

Since there's no schema, we can create a `Map` with `String` keys (for the field name) and `Object` for the values for the current record. Then, you create an empty map for the new record and filter the existing records by checking whether the list of configured field names contains it. If it is in the list, the key and value are placed in the map, representing the new record. If the record does have a schema, we follow a very similar process, but we first have to adjust the schema only to contain the fields we're interested in keeping.

Listing 5.14   Transforming with a schema

```
private R applyWithSchema(R connectRecord) {
final Struct value =
        requireStruct(operatingValue(connectRecord), PURPOSE);

Schema updatedSchema = schemaUpdateCache.get(value.schema());
if(updatedSchema == null) {
    updatedSchema = makeUpdatedSchema(value.schema());
    schemaUpdateCache.put(value.schema(), updatedSchema);
}
final Struct updatedValue = new Struct(updatedSchema);

updatedValue.schema().fields()
.forEach(field -> updatedValue.put(field.name(),
                              value.get(field.name()))));
return newRecord(connectRecord, updatedSchema, updatedValue);
}
```

Here, with the schema-record combo, we first need to get a `Struct`, which is very similar to the `HashMap` in the schemaless version, but it contains all the type information for the fields in the record. First, we check whether we've already created an updated schema; if not, we make it and store it in a cache. Once we've created the updated schema, there's no need to create another one since the structure for all records will be the same. Then, we iterate over the field names of our updated schema, using each one to extract the value from the old record.

You now know how to implement your custom connect `Transformation`. I've not covered all the details here, so consult the source code in `bbejeck.chapter_5` `.transformer.MultiFieldExtract` for all the details.

I want to close this chapter on Kafka Connect by saying the custom `Connector` and `Transformation` you've created here aren't for production use. They are teaching tools on how you can create custom variations of those two classes when the ones provided by Kafka Connect won't fit your needs.

Summary

- Kafka Connect is the lynchpin in moving data from an external system into Kafka and from Kafka out into an external system. This ability to move data in and out of Kafka is critical for getting existing applications outside of Kafka involved in an event streaming platform.
- Using an existing connector doesn't require any code. You upload some JSON configurations to get it up and running. There are hundreds of existing connectors, so there's usually one already developed that you can use out of the box.
- The Connect framework also provides a Single Message Transform (SMT) that can apply lightweight changes to incoming or outgoing records.
- You can implement your own connector if the existing connectors or transformations don't do what you need them to do.

Part 3

In part 3, you'll go deep into Kafka Streams. Armed with the knowledge you gained in the first two parts, you're primed to hit the ground running as you'll understand how Kafka works at this point.

You'll explore and learn the Kafka Streams DSL layer and fully appreciate what types of event-streaming applications you can create. You'll start with a simple "Hello World" application in Kafka Streams. From there, you'll quickly move on to a more practical example involving a fictional retail store. But you will continue there; from starting with the more straightforward DSL applications, you'll move on to more features, such as using state in a Kafka Streams application. You'll need to use state whenever you need to "remember" something from previous events. From there, the next stop on your learning journey is the KTable API, where a KStream is an event stream and a KTable is an update stream where records with the same key are updated to previous ones. Then, you'll move on to a concept that goes hand in hand with stateful operations—windowing. Where stateful operations on their own will continue to grow in size over time, windowing allows you to put events into discrete time "buckets," breaking the state up into more analyzable chunks by time.

But as with any well-designed abstraction, there will be times when the Kafka Streams DSL won't provide the exact thing you'll need to get the job done, and that's precisely where the Kafka Streams Processor API comes in handy. You will learn about the Processor API. While it takes more work to build an application, it gives you complete control, allowing you to meet any need in creating an event-streaming application.

Then, your learning goes beyond Kafka Streams as you get into ksqlDB, which allows you to develop event-streaming applications using familiar SQL. From ksqlDB, you get into integration with Spring. Spring is a widely used framework that is highly popular with developers as it facilitates building more modular and testable applications. Then, you'll apply what you've learned about Spring to one of Kafka Stream's most unique features—creating an interactive query service.

Finally, you'll wrap up your experience by learning how to test not only Kafka Streams applications but all of the different parts of the Kafka ecosystem you learned in this book. But there is more information in this book. Some additional topics would be helpful to you in developing event-streaming applications with Kafka and Kafka Streams. While the information is useful, it's optional when building your streaming software. So, four appendixes contain this useful but optional information:

- Appendix A contains a workshop on Schema Registry to get hands-on experience with the different schema compatibility modes.
- Appendix B presents information on using Confluent Cloud to help develop your event streaming applications.
- Appendix C is a survey of working with the different schema types Avro, Protobuf, and JSON Schema.
- Appendix D covers the architecture and internals of Kafka Streams.

Developing Kafka Streams

This chapter covers

- Introducing the Kafka Streams API
- Building our first Kafka Streams application
- Working with customer data and creating more complex applications
- Splitting, merging, and branching streams

A Kafka Streams application is a graph of processing nodes transforming event data as it streams through each node. In this chapter, you'll learn how to build a graph that makes up a stream processing application with Kafka Streams.

6.1 A look at Kafka Streams

Let's take a look at an illustration of what this means in figure 6.1. This illustration represents the generic structure of most Kafka Streams applications. There is a source node that consumes event records from a Kafka broker. There are any number of processing nodes, each performing a distinct task, and, finally, a sink node producing transformed records back out to Kafka. In chapter 4, we discussed how to use the Kafka clients to produce and consume records with Kafka. Much of what you learned in that chapter applies to Kafka Streams because, at its heart, Kafka

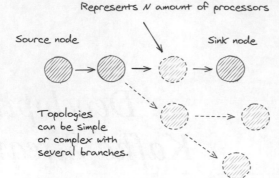

Represents *N* amount of processors

Source node

Sink node

Topologies can be simple or complex with several branches.

Figure 6.1   **Kafka Streams is a graph with a source node, any number of processing nodes, and a sink node.**

Streams is an abstraction over the producers and consumers, leaving you free to focus on your stream-processing requirements.

> **NOTE**   While Kafka Streams is the native stream processing library for Apache Kafka, it does not run inside the cluster or brokers but connects as a client application.

6.2   *Kafka Streams DSL*

The Kafka Streams DSL is the high-level API enabling you to build Kafka Streams applications quickly. This API is very well thought out, with methods to handle most stream-processing needs out of the box to create a sophisticated stream-processing program without much effort. At the heart of the high-level API is the KStream object, which represents the streaming key-value pair records.

Most of the methods in the Kafka Streams DSL return a reference to a KStream object, allowing for a fluent interface style of programming. Most KStream methods also accept single-method interfaces, allowing for the liberal use of lambda expressions, which speeds up your development. Considering these factors, you can imagine the simplicity and ease of building a Kafka Streams program.

There's also a lower-level API, the Processor API, which isn't as concise as the Kafka Streams DSL but allows for more control. We'll cover the Processor API in chapter 10. With that introduction, let's dive into the Hello World program for Kafka Streams.

6.3   *Hello World for Kafka Streams*

For the first Kafka Streams example, we'll build something fun that will get off the ground quickly so you can see how Kafka Streams works: a toy application that takes incoming messages and converts them to uppercase characters, effectively yelling at anyone who reads the message. We'll call this the Yelling app. Before diving into the code, let's take a look at the processing topology you'll assemble for this application in figure 6.2.

As you can see, it's a simple processing graph—so simple that it resembles a linked list of nodes more than the typical tree-like structure of a graph. But there's enough

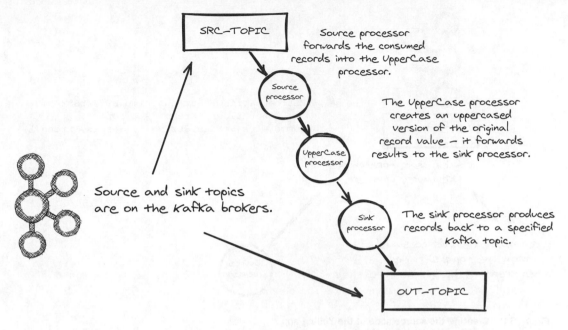

Figure 6.2 Topology of the Yelling app

here to give you substantial clues about what to expect in the code. There will be a source node, a processor node transforming incoming text to uppercase, and a sink processor writing results to a topic.

This is a trivial example, but the code shown here represents what you'll see in other Kafka Streams programs. In most of the examples, you'll see a similar pattern:

1. Defining the configuration items
2. Creating Serde instances, either custom or predefined, used in the deserialization/serialization of records
3. Building the processor topology
4. Creating and starting Kafka Streams

When we get into the more advanced examples, the principal difference will be in the complexity of the processor topology. With all this in mind, it's time to build your first application.

6.3.1 Creating the topology for the Yelling app

The first step to creating any Kafka Streams application is to create a source node, which you will do here. The source node is the root of the topology and forwards the consumed records into the application. Figure 6.3 highlights the source node in the graph.

The code in listing 6.1 creates the graph's source, or parent, node.

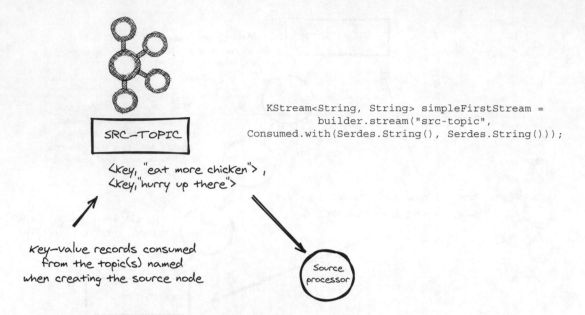

```
KStream<String, String> simpleFirstStream =
            builder.stream("src-topic",
Consumed.with(Serdes.String(), Serdes.String()));
```

Figure 6.3   Creating the source node of the Yelling app

Listing 6.1   Defining the source for the stream

```
KStream<String, String> simpleFirstStream = builder.stream("src-topic",
Consumed.with(Serdes.String(), Serdes.String()));
```

You've configured the `simpleFirstStream` instance to consume messages from the `src-topic` topic. In addition to specifying the topic name, you can add a `Consumed` object that Kafka Streams uses to configure optional parameters for a source node. In this example, you've provided Serde instances, the first for the key and the second for the value. A Serde is a wrapper object that contains a serializer and deserializer for a given type.

Based on our discussion on consumer clients in a previous chapter, the broker stores and forwards records in byte array format. For Kafka Streams to perform any work, it must deserialize the bytes into concrete objects. Here, both Serde objects are for strings since that's the key type and value type. Kafka Streams will use the Serde to deserialize the key and value into string objects. We'll explain Serdes in more detail soon. You can also use the `Consumed` class to configure a `TimestampExtractor`, the offset reset for the source node, and provide a name for the operator. We'll cover the `TimestampExtractor` in chapter 9. Since we covered offset resets in chapter 2, I won't cover them again here.

That is how to create a `KStream` to read from a Kafka topic. But a single topic is not our only choice. Let's take a quick look at some other options. Let's say that there are several topics you'd like to "yell at." In that case, you can subscribe to all of them at one time using a `Collection<String>` to specify all the topic names as shown in the following listing.

Listing 6.2 Creating the Yelling app with multiple topics as the source

```
KStream<String, String> simpleFirstStream =
  builder.stream(List.of("topicA", "topicB", "topicC"),
       Consumed.with(Serdes.String(), Serdes.String()))
```

Typically, you'd use this approach when you want to apply the same processing to multiple topics simultaneously. But what if you have a long list of similarly named topics? Do you have to write them all out? The answer is no! You can use a regular expression to subscribe to any topic that matches the pattern.

Listing 6.3 Using a regular expression to subscribe to topics in the Yelling app

```
KStream<String, String> simpleFirstStream =
  buider.source(Pattern.compile("topic[A-C]"),
      Consumed.with(Serdes.String(), Serdes.String()))
```

A regular expression for subscribing to topics is convenient when your organization uses a standard topic naming pattern related to its business function. You have to know the naming pattern, and you can subscribe to all of them concisely. Additionally, as topics are created or deleted, your subscription will automatically update to reflect the changes in the topics.

When subscribing to multiple topics, there are a few caveats to remember. The keys and values from all subscribed topics must be the same type; for example, you can't combine topics where one topic contains `Integer` keys and another has `String` keys. Also, if they contain a different number of partitions, it's up to you to repartition the data before performing any key-based operation like aggregations. We'll cover repartitioning in the next chapter. Finally, there are no ordering guarantees for the incoming records.

> **TIP** I emphasize that Kafka only guarantees order within a partition of a *single* topic. So, if you're consuming from multiple topics in your Kafka Streams application, there's no guarantee of the order of the records consumed.

You now have a source node for your application, but you must attach a processing node to use the data, as shown in figure 6.4.

Here's the code you'll use to change the incoming text to uppercase:

Listing 6.4 Mapping incoming text to uppercase

```
KStream<String, String> upperCasedStream =
  simpleFirstStream.mapValues(value -> value.toUpperCase());
```

In the introduction to this chapter, I mentioned that a Kafka Streams application is a graph of processing nodes—a directed acyclic graph (DAG), to be precise. You build

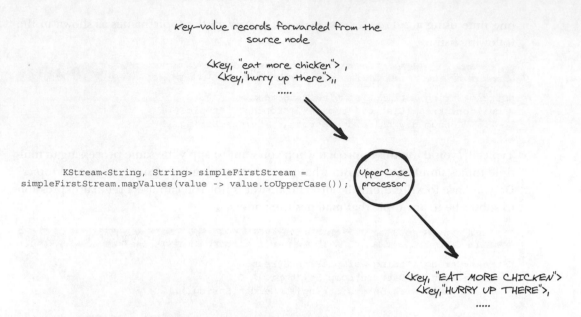

key-value records forwarded from the source node

<key, "eat more chicken"> ,
<key,"hurry up there">,,
.....

```
KStream<String, String> simpleFirstStream =
simpleFirstStream.mapValues(value -> value.toUpperCase());
```

UpperCase processor

<key, "EAT MORE CHICKEN">
<key,"HURRY UP THERE">,
.....

Figure 6.4   Adding the uppercase processor to the Yelling app

the graph one processor at a time. You establish a parent–child relationship between the graph nodes with each method call. The parent–child relationship in Kafka Streams sets the direction for the data flow; parent nodes forward records to the child node(s).

So, looking at the code example here, by executing `simpleFirstStream.mapValues`, you're creating a new processing node whose inputs are the records consumed in the source node. So the source node is the parent, and it forwards records to its child, the processing node returned from the `mapValues` operation.

> **NOTE**   As you can tell from the name, `mapValues` only affects the key-value pair's value, but the original record's key is still forwarded along.

The `mapValues()` method takes an instance of the `ValueMapper<V, V1>` interface. The `ValueMapper` interface defines only one method, `ValueMapper.apply`, making it an ideal candidate for using a lambda expression, which is what you've done here with `value -> value.toUpperCase()`.

> **NOTE**   Many tutorials are available for lambda expressions and method references. You will find good starting points in Oracle's Java documentation: "Lambda Expressions" (http://mng.bz/J0Xm) and "Method References" (http://mng.bz/BaDW).

So far, your Kafka Streams application is consuming records and transforming them to uppercase. The final step is adding a sink processor that writes the results to a topic. Figure 6.5 shows where you are in the construction of the topology.

The code in listing 6.5 adds the last processor in the graph.

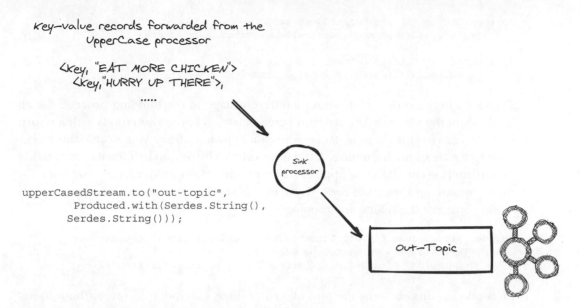

key-value records forwarded from the
        UpperCase processor

⟨key, "EAT MORE CHICKEN"⟩
⟨key, "HURRY UP THERE"⟩,
        .....

```
upperCasedStream.to("out-topic",
        Produced.with(Serdes.String(),
    Serdes.String()));
```

Sink
processor

Out-Topic

Figure 6.5   Adding a processor for writing the Yelling app results

---

Listing 6.5   Creating a sink node

```
upperCasedStream.to("out-topic",
                Produced.with(Serdes.String(), Serdes.String()));
```

The `KStream.to` method creates a processing node that writes the final transformed records to a Kafka topic. It is a child of the `upperCasedStream`, so it receives all of its inputs directly from the results of the `mapValues` operation.

   Again, you provide Serde instances, this time for serializing records written to a Kafka topic. But in this case, you use a `Produced` instance, which provides optional parameters for creating a sink node in Kafka Streams. The `Produced` configuration object also allows you to provide a custom `StreamPartitioner`. We covered the concept of custom partitioning in chapter 4, section 4.2.4, so I won't review that here. You can see an example of using a `StreamPartitioner` in chapter 9, towards the end of section 9.1.6.

> **NOTE**   You don't always have to provide Serde objects to either the `Consumed` or `Produced` objects. If you don't, the application will use the serializer/ deserializer listed in the configuration. Additionally, with the `Consumed` and `Produced` classes, you can specify a Serde for either the key or value only.

The preceding example uses three lines of code to build the topology:

```
KStream<String,String> simpleFirstStream =
    builder.stream("src-topic",
      Consumed.with(Serdes.String(), Serdes.String()));
```

```
KStream<String, String> upperCasedStream =
    simpleFirstStream.mapValues(value -> value.toUpperCase());

upperCasedStream.to("out-topic",
    Produced.with(Serdes.String(), Serdes.String()));
```

Each step in the code demonstrates a different stage of the building process. But all methods in the KStream API that don't create terminal nodes (methods with a return type of void) return a new KStream instance, which allows you to use the fluent interface style of programming. A fluent interface (https://martinfowler.com/bliki/FluentInterface.html) is an approach where you chain method calls together for more concise and readable code. To demonstrate this idea, here's another way you could construct the Yelling app topology:

```
builder.stream("src-topic", Consumed.with(Serdes.String(), Serdes.String()))
 .mapValues(value -> value.toUpperCase())
 .to("out-topic", Produced.with(Serdes.String(), Serdes.String()));
```

Method chaining shortens the program from three statements to one without losing clarity or purpose. From this point forward, you'll see all the examples using the fluent interface style unless doing so causes the clarity of the program to suffer.

> **TIP** Starting a Kafka Streams application requires a running Kafka broker. Using Docker is a very convenient way to run Kafka locally. The source code includes docker-compose.yml. To start Kafka, use this command: docker compose up -d.

You've built your first Kafka Streams topology, but we glossed over the essential configuration steps and Serde creation. We'll look at those now.

6.3.2 *Kafka Streams configuration*

Kafka Streams is highly configurable and has several properties you can adjust for your specific needs. Here, you only see the two required configuration settings, APPLICATION_ID_CONFIG and BOOTSTRAP_SERVERS_CONFIG:

```
props.put(StreamsConfig.APPLICATION_ID_CONFIG, "yelling_app_id");
props.put(StreamsConfig.BOOTSTRAP_SERVERS_CONFIG, "localhost:9092");
```

Both are required because there's no practical way to provide default values for these configurations. Attempting to start a Kafka Streams program without these two properties defined will result in a ConfigException being thrown.

The StreamsConfig.APPLICATION_ID_CONFIG property uniquely identifies your Kafka Streams application. Kafka Streams instances with the same application.id are considered one logical application. The book covers this concept in appendix D. application.id also serves as a prefix for the embedded client (KafkaConsumer and KafkaProducer) configurations. You can provide custom configurations for the embedded clients using one of the various prefix labels in the `StreamsConfig` class. However, the default client configurations in Kafka Streams are chosen to provide the best performance, so you should exercise caution when adjusting them.

The `StreamsConfig.BOOTSTRAP_SERVERS_CONFIG` property can be a single `host-name:port` pair or multiple `hostname:port` comma-separated pairs. `BOOTSTRAP_SERVERS_CONFIG` is what Kafka Streams uses to establish a connection to the Kafka cluster. We'll cover several more configuration items as we explore more examples in the book.

6.3.3 Serde creation

In Kafka Streams, the `Serdes` class provides convenience methods for creating Serde instances, as shown here:

```
Serde<String> stringSerde = Serdes.String();
```

You create the Serde instance required for serialization/deserialization using the `Serdes` class in this line. Here, you create a variable to reference the Serde for repeated use in the topology. The `Serdes` class provides default implementations for the following types: `String`, `ByteArray`, `Bytes`, `Long`, `Short`, `Integer`, `Double`, `Float`, `ByteBuffer`, `UUID`, and `Void`.

Implementations of the Serde interface are advantageous because they contain the serializer and deserializer, which keeps you from having to specify four parameters (key serializer, value serializer, key deserializer, and value deserializer) every time you need to provide a Serde in a `KStream` method. In upcoming examples, you'll use Serdes to work with Avro, Protobuf, and JSON Schema and create a Serde implementation to handle serialization/deserialization of more complex types.

Let's take a look at the whole program you just put together (some details are omitted for clarity). You can find the source in src/main/java/bbejeck/chapter_6/KafkaStreamsYellingApp.java (located at https://github.com/bbejeck/KafkaStreams InAction2ndEdition).

Listing 6.6   Hello World: The Yelling app

```java
public class KafkaStreamsYellingApp extends BaseStreamsApplication {

private static final Logger LOG =
  LoggerFactory.getLogger(KafkaStreamsYellingApp.class);

@Override
public Topology topology(Properties streamProperties) {

  Serde<String> stringSerde = Serdes.String();
  StreamsBuilder builder = new StreamsBuilder();
```

Creates the Serdes and stores in a variable used to serialize/deserialize keys and values

Creates the StreamsBuilder instance used to construct the processor topology

**Creates the actual stream with a source topic
to read from (the parent node in the graph)**

```
KStream<String, String> simpleFirstStream = builder.stream("src-topic",
        Consumed.with(stringSerde, stringSerde));
KStream<String, String> upperCasedStream =
simpleFirstStream.mapValues(value-> value.toUpperCase());

upperCasedStream.to("out-topic",
    Produced.with(stringSerde, stringSerde));

return builder.build(streamProperties);
}

public static void main(String[] args) throws Exception {
    Properties streamProperties = new Properties();
    streamProperties.put(StreamsConfig.APPLICATION_ID_CONFIG,
      "yelling_app_id");
    streamProperties.put(StreamsConfig.BOOTSTRAP_SERVERS_CONFIG,
      "localhost:9092");
    KafkaStreamsYellingApp yellingApp = new KafkaStreamsYellingApp();
    Topology topology = yellingApp.topology(streamProperties);

    try(KafkaStreams kafkaStreams =
                new KafkaStreams(topology, streamProperties)) {
      LOG.info("Hello World Yelling App Started");
      kafkaStreams.start();

      LOG.info("Shutting down the Yelling APP now");
    }
  }
}
```

**A processor
using a lambda
(the first child
node in the
graph)**

**Writes the
transformed output
to another topic
(the sink node in
the graph)**

**Kicks off the Kafka
Streams threads**

You've now constructed your first Kafka Streams application. Let's quickly review the steps involved, as it's a general pattern you'll see in most of your Kafka Streams applications:

1 Create a `Properties` instance for configurations.
2 Create a Serde object.
3 Construct a processing topology.
4 Start the Kafka Streams program.

We'll now move on to a more complex example that will allow us to explore more of the Streams DSL API.

6.4   *Masking credit card numbers and tracking purchase rewards in a retail sales setting*

Imagine you work as an infrastructure engineer for the retail giant ZMart. ZMart has adopted Kafka as its data processing backbone and is looking to capitalize on its ability to process customer data quickly, which will help ZMart do business more efficiently.

At this point, your boss tasked you with building a Kafka Streams application to work with purchase records as they come streaming in from transactions in ZMart stores. Figure 6.6 shows the requirements for the streaming program and serves as a good description of what the program will do.

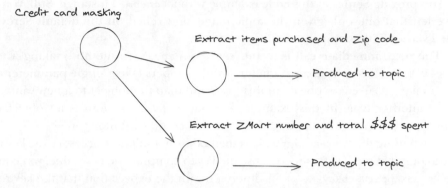

Figure 6.6   Diagram of the new requirements for the ZMart application

The requirements are as follows:

1 All Purchase objects need to have credit card numbers protected, in this case by masking the first 12 digits.
2 You must extract the items purchased and the zip code to determine regional purchase patterns and inventory control. This data will be written out to a topic.
3 You must capture the customer's ZMart member number and the amount spent and write this information to a topic. Consumers of the topic will use this data to determine rewards.

With these requirements, let's start building a streaming application that will satisfy ZMart's business requirements.

6.4.1   Building the source node and the masking processor

The first step in building the new application is to create the source node and first processor of the topology. You'll do this by chaining together two calls to the KStream API. The child processor of the source node will mask credit card numbers to protect customer privacy.

Listing 6.7   Building the source node and first processor

```
KStream<String, RetailPurchase> retailPurchaseKStream =
    streamsBuilder.stream("transactions",
    Consumed.with(stringSerde, retailPurchaseSerde))
    .mapValues(creditCardMapper);
```

You create the source node with a call to the StreamBuilder.stream method using a default String Serde, a custom Serde for RetailPurchase objects, and the name of the topic that's the source of the messages for the stream. In this case, you only specify one topic, but you could have provided a comma-separated list of names or a regular expression to match topic names instead.

You provide Serdes in this code example with a `Consumed` instance. Still, you could have left that out, only given the topic name, and relied on the default Serdes provided via configuration parameters.

The next immediate call is to the `KStream.mapValues` method, taking a `Value-Mapper<V, V1>` instance as a parameter. Value mappers take a single parameter of one type (a `RetailPurchase` object, in this case) and map that object to a new value, possibly of another type. In this example, `KStream.mapValues` returns an object of the same type (`RetailPurchase`) but with a masked credit card number.

When using the `KStream.mapValues` method, you don't have access to the key for the value computation. If you wanted to use the key to compute the new value, you could use the `ValueMapperWithKey<K, V, VR>` interface, with the expectation that the key remains the same. If you need to generate a new key along with the value, use the `KStream.map` method that takes a `KeyValueMapper<K, V, KeyValue<K1, V1>>` interface. Let's take a quick look at an example for both of these, starting with the `ValueMapperWithKey`.

Listing 6.8   `ValueMapperWithKey`

```
public class RetailValueMapperWithKey implements
  ValueMapperWithKey<String,RetailPurchase,RetailPurchase> {
@Override
public RetailPurchase apply(String customerIdKey,
                           RetailPurchase value) {        ◄─── The apply method applying the desired changes to the value only
    RetailPurchase.Builder builder = value.toBuilder();
    if(customerIdKey != null && customerIdKey.endsWith("333")){
        builder.setCreditCardNumber("0000000000");
        builder.setCustomerId("000000000");
    }
    return builder.build();
    }
}
```

`ValueMapperWithKey` operates the same way as `ValueMapper`, except you can access the key to help you with the value transformation. Here, we're assuming the key is the customer ID, and if it ends with a `"333"`, this means it was a corporate shopping "spy," and we can strip away the customer details from the purchase. The method returns the updated `RetailPurchase` object when the transformation is complete.

For a `KeyValueMapper`, you follow the same process, but you can change the key as well, as you'll see in the next listing.

Listing 6.9   `KeyValueMapper`

```
public class RetailKeyValueMapper implements
    KeyValueMapper<String, RetailPurchase,
                   KeyValue<String, RetailPurchase>> {
```

```
@Override
public KeyValue<String, RetailPurchase> apply(String key,
                                   RetailPurchase value) {
    RetailPurchase.Builder builder = RetailPurchase.newBuilder();
    if(key != null && key.endsWith("333")){            ◁───┐ The condition
        builder.setCreditCardNumber("0000000000");          │ for updating the
        builder.setCustomerId("000000000");                 │ purchase object
    }
  return KeyValue.pair(value.getStoreId(), builder.build());  ◁──┐
  }                                                 Sets the key to a │
}                                                       new value │
```

Here, `KeyValueMapper` accepts the key and value as parameters, just as `ValueMapper-WithKey` does. Still, it returns a `KeyValue` instead of just the value, meaning you can change the key to a new value. Sometimes, you may need to add a key (which still requires using the `KeyValueMapper` interface) or change the existing key when you want to perform an aggregation, which requires grouping records by key. In chapter 7, we'll cover aggregations and the impact of changing keys with stateful operations.

> **NOTE**   Keep in mind that Kafka Streams expects functions to operate without side effects, meaning the functions don't modify the original key and/or value but return new objects when making modifications.

6.4.2   Adding the purchase-patterns processor

Now, you'll build the second processor, which is responsible for extracting geographical data from the purchase, which ZMart can use to determine purchase patterns and inventory control in regions of the country. There's also an additional wrinkle with building this part of the topology. The ZMart business analysts have determined they want to see individual records for each item in a purchase and consider purchases made regionally together.

The `RetailPurchase` data model object contains all the items in a customer purchase, so you'll need to emit a new record for each one. Additionally, you'll need to add the zip code as the key to the transaction. Finally, you'll add a sink node responsible for writing the pattern data to a Kafka topic.

In the patterns processor example, you can see the `retailPurchaseKStream` processor using a `flatMap` operator. The `KStream.flatMap` method takes a `ValueMapper` or a `KeyValueMapper` that accepts a single record and returns an `Iterable` (any Java `Collection`) of new records, possibly of a different type. The `flapMap` processor "flattens" the `Iterable` into one or more records forwarded to the topology. Figure 6.7 illustrates how this works.

The process of a `flatMap` is a well-known operation from functional programming where one input creates a collection of items (the map portion of the function). But instead of returning the collection, it flattens the collection into a sequence of records. In our case here with Kafka Streams, a retail purchase of five items results in five individual `KeyValue` objects with the keys corresponding to the zip code and the values of a `PurchasedItem` object.

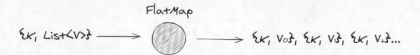

Figure 6.7   `FlatMap` **emits zero or more records from a single input record by flattening a collection returned from a** `KeyValueMapper` **or** `ValueMapper`**.**

The following listing contains the code for `KeyValueMapper`.

Listing 6.10   `KeyValueMapper` **returning a collection of** `PurchasedItem` **objects**

Extracts the zip code on the
purchase for the new key

```
KeyValueMapper<String, RetailPurchase,
 Iterable<KeyValue<String, PurchasedItem>>> retailTransactionToPurchases =
    (key, value) -> {
        String zipcode = value.getZipCode();
        return value.getPurchasedItemsList().stream()        ◁─── Uses the Java stream
                .map(purchasedItem ->                              API to create a list of
                    KeyValue.pair(zipcode, purchasedItem))         KeyValue pairs
                .collect(Collectors.toList());
}
```

The `KeyValueMapper` takes an individual transaction object and returns a list of `KeyValue` objects. The key is the zip code where the transaction occurred, and the value is an item included in the purchase. Now let's put our new `KeyValueMapper` into this section of the topology we're creating.

Listing 6.11   **Patterns processor and sink node that writes to Kafka**

Uses flatMap to create a new object
for each time in a transaction

```
KStream<String, Pattern> patternKStream = retailPurchaseKStream
                .flatMap(retailTransactionToPurchases)
                .mapValues(patternObjectMapper);        ◁─── Maps each purchase
                                                              to a pattern object
patternKStream.print(Printed.<String,Pattern>toSysOut()
                .withLabel("patterns-stream"));
patternKStream.to("patterns",
        Produced.with(stringSerde,purchasePatternSerde));    ◁───
```

Prints records to the console   *Produces each record from the purchase*
                                                to a Kafka topic called "patterns"

In this code example, you declare a variable to hold the reference of the new `KStream` instance, and you'll see why in an upcoming section. The purchase-patterns processor forwards the records it receives to a child node of its own, defined by the method call `KStream.to`, writing to the `patterns` topic. Note using a `Produced` object to provide the previously built Serde. I've also snuck in a `KStream#print` processor that prints the key-value pairs of the stream to the console; we'll talk more about viewing stream records in section 6.5.

The `KStream.to` method mirrors the `KStream.source` method. Instead of setting a source for the topology to read from, it defines a sink node used to write the data from a `KStream` instance to a Kafka topic. The `KStream.to` method also provides overloads that allow for dynamic topic selection, and we'll discuss that soon.

6.4.3 Building the rewards processor

The third processor in the topology is the customer rewards accumulator node, which will let ZMart track purchases made by members of their preferred customer club. The rewards accumulator sends data to a topic consumed by applications at ZMart HQ to determine rewards when customers complete purchases.

Listing 6.12 Third processor and a terminal node that writes to Kafka

```
KStream<String, RewardAccumulator> rewardsKStream =
        retailPurchaseKStream.mapValues(rewardObjectMapper);
rewardsKStream.to("rewards",
        Produced.with(stringSerde,rewardAccumulatorSerde));
```

You build the rewards accumulator processor using what should be, by now, a familiar pattern: creating a new `KStream` instance that maps the raw purchase data contained in the retail purchase object to a new object type. You also attach a sink node to the rewards accumulator so the rewards `KStream` results can be written to a topic and used to determine customer reward levels.

Now that you've built the application piece by piece, let's look at the entire application.

Listing 6.13 ZMart customer purchase `KStream` program

```
public class ZMartKafkaStreamsApp {

@Override
public Topology topology(Properties streamProperties) {

StreamsBuilder streamsBuilder = new StreamsBuilder();

KStream<String, RetailPurchase> retailPurchaseKStream =
        streamsBuilder.stream("transactions",
            Consumed.with(stringSerde, retailPurchaseSerde))
        .mapValues(creditCardMapper);           ◁┐ Builds the source
                                                  │ and first processor
KStream<String, Pattern> patternKStream =
    retailPurchaseKStream
        .flatMap(retailTransactionToPurchases)
        .mapValues(patternObjectMapper);        ◁─ Builds the PurchasePattern
                                                   processor
patternKStream.to("patterns",
        Produced.with(stringSerde,purchasePatternSerde));
```

```
KStream<String, RewardAccumulator> rewardsKStream =
    retailPurchaseKStream.mapValues(rewardObjectMapper);      ◄─┐

rewardsKStream.to("rewards",                                     Builds the
    Produced.with(stringSerde,rewardAccumulatorSerde));      RewardAccumulator
retailPurchaseKStream.to("purchases",                            processor
    Produced.with(stringSerde,retailPurchaseSerde));

return streamsBuilder.build(streamProperties);

}
```

> **NOTE** I've left out some details in the listing for clarity. The code examples illustrate a point and are a subset of the complete source code accompanying this book (src/main/java/bbejeck/chapter_6/ZMartKafkaStreamsApp.java).

This example is slightly more involved than the Yelling app but has a similar flow. Specifically, you still performed the following steps:

1 Create a `StreamsBuilder` instance.
2 Build one or more Serde instances.
3 Construct the processing topology.
4 Assemble all the components and start the Kafka Streams program.

You'll also notice you don't see the logic responsible for creating the various mappings from the original transaction object to new types. That is by design. First, the code for a `KeyValueMapper` or `ValueMapper` will be distinct for each use case, so the particular implementations don't matter too much.

I've mentioned using Serdes, but I haven't explained why or how you create them. Let's take some time now to discuss the role of the Serde in a Kafka Streams application.

6.4.4 *Using Serdes to encapsulate serializers and deserializers in Kafka Streams*

As you learned in previous chapters, Kafka brokers work with records in a byte array format. The client is responsible for serializing when producing records and deserializing when consuming. It's no different with Kafka Streams since it uses embedded consumers and producers. But instead of providing a specific deserializer or serializer, you configure Kafka Streams with a Serde, which contains both the serializer and deserializer.

Some Serdes are provided out of the box by the Kafka client dependency (`String`, `Long`, `Integer`, and so on), but you'll need to create custom Serdes for other objects.

In the first example, the Yelling app, you only needed a serializer/deserializer for a `String`; an implementation is provided by the `Serdes.String()` factory method. In the ZMart example, however, you need to create custom Serde instances because of the arbitrary object types. We'll look at what's involved in building a Serde for the `RetailPurchase` class. We won't cover the other Serde instances because they follow the same pattern, just with different types.

> **NOTE** I include this discussion on Serdes creation for completeness, but the source code contains a class, `SerdeUtil`, that provides a `protobufSerde` method. You'll see this in the examples, which encapsulate the steps described in this section.

Building a Serde requires implementations of the `Deserializer<T>` and `Serializer<T>` interfaces. We covered creating your own serializer and deserializer instances in section 3.7, so I won't repeat those details here. For reference, you can see the complete code for the `ProtoSerializer` and `ProtoDeserializer` in the `bbejeck.serializers` package in the source for the book.

Now, to create a `Serde<T>` object, you'll use the `Serdes.serdeFrom` factory method, taking steps like the following:

```
Deserializer<RetailPurchase> purchaseDeserializer =          Creates the Deserializer for
    new ProtoDeserializer<>();                               the RetailPurchase class
Serializer<RetailPurchase> purchaseSerializer =
    new ProtoDeserializer<>();                               Creates the Serializer for
Map<String, Class<RetailPurchase>> configs                   the RetailPurchase class
        = new HashMap<>();
    configs.put(false, RetailPurchase.class);                Configurations for
        deserializer.configure(configs,isKey);               the deserializer
Serde<RetailPurchase> purchaseSerde =
    Serdes.serdeFrom(purchaseSerializer,purchaseDeserializer);
```

Creates the Protobuf Serde
for RetailPurchase objects

As you can see, a Serde object is helpful because it serves as a container for the serializer and deserializer for a given object. You create a custom Serde for the Protobuf objects because the example does not use Schema Registry. But using Schema Registry with Kafka Streams is a perfectly valid use case. Let's take a quick pause to review how you configure your Kafka Streams application when using it with Schema Registry.

6.4.5 *Kafka Streams and Schema Registry*

In chapter 4, I discussed why you'd want to use Schema Registry with a Kafka-based application. I'll briefly describe those reasons here. The domain objects in your application represent an implicit contract between the different users of your application. For example, imagine one team of developers changing a field type from a `java.util` `.Date` to a `long` and producing those changes to Kafka; the downstream consumer applications will break due to the unexpected field type change.

So, by using a schema and Schema Resitry to store it, you make it much easier to enforce this contract by enabling better coordination and compatibility checks. Additionally, the Schema Registry project provides Schema Registry–aware (de)serializers and Serdes, alleviating the developer from writing the serialization code.

> **NOTE** Schema Registry provides both a `JSONSerde` and a `JSONSchemaSerde`, but they are not interchangeable! The `JSONSerde` is for Java objects that use

JSON to describe the object. The `JSONSchemaSerde` is for objects that use `JSONSchema` as the formal definition of the object.

So how would the `ZMartKafkaStreamsApp` change to work with Schema Registry? All that is required is to use Schema Registry–aware Serde instances. The steps for creating a Schema Registry aware Serde are simple:

1 Create an instance of one of the provided Serde instances.
2 Configure it with the URL for a Schema Registry server.

The following code provides the concrete steps you'll take:

```
KafkaProtobufSerdePurchase> protobufSerde =
 new KafkaProtobufSerde<>(Purchase.class);
String url = "https://...";
Map<String, Object> configMap = new HashMap<>();
configMap.put(
  AbstractKafkaSchemaSerDeConfig.SCHEMA_REGISTRY_URL_CONFIG,
  url);
protobufSerde.conserde(configMap, false);
```

Instantiates the KafkaProtobufSerde, providing the class type as a constructor parameter

The URL for the location of a Schema Registry instance

Puts the URL in a HashMap

Calls the KafkaProtobufSerde#configure method

So, with just a few lines of code, you've created a Schema Registry–aware Serde that you can use in your Kafka Streams application.

NOTE Since Kafka Streams contains consumer and producer clients, the same rules for schema evolution and compatibility apply.

We've covered a lot of ground so far in developing a Kafka Streams application. We still have much more to cover, but let's pause and talk about the development process and how you can make life easier while developing a Kafka Streams application.

6.5 *Interactive development*

You've built the graph to process purchase records from ZMart in a streaming fashion, and you have three processors that write out individual topics. During development, it is possible to have a console consumer running to view results. But instead of using an external tool, having your Kafka Streams application print or log from anywhere you want inside the topology would be more convenient. This visual feedback directly from the application is very efficient during development. You enable this output using the `KStream.peek()` or `KStream.print()` method.

  `KStream.peek()` allows you to perform a stateless action (via the `ForeachAction` interface) on each record flowing through the `KStream` instance. It's important to note that it's expected this will not change the incoming key and value. Instead, the `peek` operator is an opportunity to print, log, or collect information at arbitrary points in the topology. Let's take another look at the Yelling app, but now add a way to view the records before and after the application starts "yelling" (some details are omitted for clarity). The source code is found at bbejeck/chapter_6/KafkaStreamsYellingAppWithPeek.

Listing 6.14 Printing records flowing through the Yelling app

```
ForeachAction<String, String> sysout =
  (key, value) ->
   System.out.println("key " + key
   + " value " + value);


builder.stream("src-topic",
  Consumed.with(stringSerde, stringSerde))
  .peek(sysout)
  .mapValues(value -> value.toUpperCase())
  .peek(sysout)
  .to( "out-topic",
  Produced.with(stringSerde, stringSerde));
```

Prints records to the console as they enter the application

Prints the yelling events

We've strategically placed these `peek` operations that will print records to the console, both before and after the `mapValues` call.

The `KStream.print()` method is purpose-built for printing records. Some previous code snippets contained examples of using it, but we'll show it again here (some details are omitted for clarity).

Listing 6.15 Printing records using KStream.print

```
...
  KStream<...> upperCasedStream = simpleFirstStream.mapValues(...);
  upperCasedStream.print(Printed.toSysOut());
  upperCasedStream.to(...);
```

Printing the uppercased letters is an example of a terminal method in Kafka Streams.

In this case, you're printing the uppercased words immediately after transformation. What is the difference between the two approaches? You should notice with the `KStream.print()` operation, you didn't chain the method calls together like you did using `KStream.peek()` because `print` is a terminal method.

Terminal methods in Kafka Streams have a return signature of `void`; hence, you can't chain another method call afterward, as it terminates the stream. The terminal methods in `KStream` interface are `print`, `foreach`, and `to`. Aside from the `print` method we discussed, you'll use `to` when you write results back to Kafka. The `foreach` method is suitable for operating on each record when you don't need to write the results back to Kafka, such as calling a microservice. There is a deprecated `process` method that is terminal as well, but since it's deprecated we won't discuss it. The new `process` method (introduced in Apache Kafka 3.3) allows for the integration of the DSL with the Processor API, which we'll discuss in chapter 10.

While either printing method is a valid approach, I prefer the `peek` method because it makes it easy to slip a print statement into an existing stream. But this is a personal preference, so ultimately, it's up to you to decide which approach to use.

So far, we've covered some basic things we can do with a Kafka Streams application, but we've only scratched the surface. Let's continue exploring what we can do with an event stream.

6.6   *Choosing which events to process*

So far, you've seen how to apply operations to events flowing through the Kafka Streams application. But you are processing every event in the stream in the same manner. What if there are events you want to avoid handling? Or what about events with a given attribute that require you to handle them differently?

Fortunately, the API makes methods available that provide the flexibility necessary to meet those needs. The `KStream#filter` method drops records from the stream that do not match a given predicate. The `KStream#split` allows you to split the original stream into branches for different processing based on provided predicate(s) to reroute records. To make these new methods more concrete, let's update the requirements to the original ZMart application:

- The ZMart updated its rewards program and now only provides points for purchases over $10. With this change, dropping nonqualifying purchases from the rewards stream would be ideal.
- ZMart has expanded and has bought an electronics chain and a popular coffee house chain. All purchases from these new stores will flow into the streaming application you've set up. However, you'll need to separate those purchases for different treatments while still processing everything else in the application the same.

NOTE   From this point forward, all code examples are pared down to the essentials to maximize clarity. Unless we introduce something new, you can assume that the configuration and setup code remain the same. These truncated examples aren't meant to stand alone: you'll find the complete code listing for this example at src/main/java/bbejeck/chapter_6/ZMartKafkaStreamsFilteringBranchingApp.java.

6.6.1   *Filtering purchases*

The first update is to remove nonqualifying purchases from the rewards stream. To accomplish this, insert a `KStream.filter()` before the `KStream.mapValues` method. The `filter` takes a `Predicate` interface as a parameter (here we'll a lambda), and it has one method defined, `test()`, which takes two parameters—the key and the value—although, at this point, you only need to use the value.

NOTE   There is also `KStream.filterNot`, which performs filtering but in reverse. Only records that *don't* match the given predicate are processed further in the topology.

The processor topology graph changes by making these changes, as shown in listing 6.16.

Listing 6.16   Adding a filter to drop purchases not meeting rewards criteria

The KStream.filter method, which takes a
Predicate<K,V>, represented as a lambda

The original
rewards stream

```
KStream<String, RewardAccumulator> rewardsKStream =
   retailPurchaseKStream.filter((key, value) ->
                              value.getPurchasedItemsList().stream()
                              .mapToDouble((item -> item.getQuantity()
                                                 * item.getPrice())))
                        .sum() > 10.00)
          .mapValues(rewardObjectMapper);
```

Maps the purchase into a
RewardAccumulator object

You have now successfully updated the rewards stream to drop purchases that don't qualify for reward points.

6.6.2   *Splitting/branching the stream*

New events are flowing into the purchase stream, and you need to process them differently. You'll still want to mask any credit card information, but after that, the purchases from the acquired coffee and electronics chain need to get pulled out and sent to different topics. Additionally, you need to continue using the same process for the original events.

You need to split the original stream into three substreams or branches: two for handling the new events and one for processing the initial events in the topology you've already built. This splitting of streams sounds tricky, but Kafka Streams provides an elegant way to do this, as we'll see now. Figure 6.8 demonstrates the conceptual idea of what splitting a stream involves.

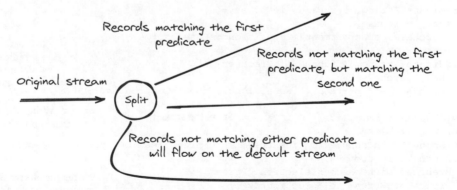

Figure 6.8   Creating branches for the two specific purchase types

The general steps you'll take to split a stream into branches are the following:

1   Use the `KStream.split()` method, which returns a `BranchedKStream` object.

2 Call `BranchedKStream.branch()` with `Predicate` and `Branched` objects as parameters. The `Predicate` contains a condition that returns `true` or `false` when tested against a record. The `Branched` object contains the logic for processing a record. Each execution of this method creates a new branch in the stream.

3 You complete the branching with a call to either `BranchedKStream.default-Branch()` or `BranchedKStream.noDefaultBranch()`. If you define a default branch, Kafka Streams will use it for records that do not match predicates. With the `noDefaultBranch` option, nonmatching records get dropped. When calling either of the branching termination methods, a `Map<String, KStream<K, V>` is returned. The `Map` may contain `KStream` objects for a new branch, depending on how you've built the `Branched` objects. We'll cover more options for branching soon.

The `Predicate` is a logical gate for its companion `Branched` object. If the condition returns `true` the "gate" opens, and the record flows into the processor logic for that branch.

NOTE When splitting a `KStream`, you can't change the types of the keys or values, as each branch has the same types as the parent or original branch.

In our case, you'll want to filter out the two purchase types into their branch and then create a default branch consisting of everything else. This default branch is the original purchase stream, which will handle all records that don't match either predicate. Now that we've reviewed the concept, let's take a look at the code you'll implement (some details are omitted for clarity). The source code is found at bbejeck/chapter_6/ZMartKafkaStreamsFilteringBranchingApp.

Listing 6.17   Splitting the stream

```
Predicate<String, Purchase> isCoffee =
  (key, purchase) ->
   purchase.getDepartment().equalsIgnoreCase("coffee");          ◁── Creates the
                                                                      predicates for
                                                                      determining
Predicate<String, Purchase> isElectronics =                          branches
  (key, purchase) ->
   purchase.getDepartment().equalsIgnoreCase("electronics");     ◁──

purchaseKStream.split()           ◁──── Splits the stream      Writes the coffee
.branch(isCoffee,                                              purchases out to a topic
  Branched.withConsumer(coffeeStream -> coffeeStream.to("coffee-topic")))  ◁──
.branch(isElectronics,
  Branched.withConsumer(electronicStream ->
  electronicStream.to("electronics"))                          Writes the electronic
                                                          ◁──  purchases out to a topic
.defaultBranch(Branched.withConsumer(retailStream ->
             retailStream.to("purchases")));              ◁──
                                                          The default branch where
                                                          nonmatching records go
```

In this example, you've split the purchase stream into two new streams, one each for coffee and electronic purchases. Branching provides an elegant way to process records differently within the same stream. While in this initial example, each is a single processor writing records to a topic, these branched streams can be as complex as you need to make them.

> **NOTE** This example sends records to several different topics. Although you can configure Kafka to create topics automatically, relying on this mechanism is not a good idea. If you use auto-creation, the topics will use the default values from the server.config properties file, which may or may not be the settings you need. You should plan for the topics you'll need, the number of partitions, and the replication factor and create them before running your Kafka Streams application.

In this branching example, you've split out discrete KStream objects, which are stand-alone and don't interact with anything else in the application, and that is an acceptable approach. But now let's consider a situation where you have an event stream you want to tease out into separate components, but you need to combine the new streams with existing ones in the application.

Consider you have Internet of Things (IoT) sensors, and early on, you combined two related sensor readings into one topic, but as time passed, newer sensors started to send results to distinct topics. The older sensors are fine as is, and it would be cost-prohibitive to go back and make the necessary changes to fit the new infrastructure. So, you'll need an application to split the legacy stream into two streams and combine or merge them with the newer streams of a single reading type. Another factor is that older proximity readings use feet, but the new ones are in meters. So, in addition to extracting the proximity reading into a separate stream, you need to convert the reading values into meters.

Now let's walk through an example of how you'll do splitting and merging, starting with the splitting (some details are omitted for clarity).

Listing 6.18 Splitting the stream in a way you have access to new streams

Splits the stream and provides the base name for the map keys

```
KStream<String, Sensor> legacySensorStream =
    builder.stream("combined-sensors", sensorConsumed);

Map<String, KStream<String, Sensor>> sensorMap =
        legacySensorStream.split(Named.as("sensor-"))
        .branch(isTemperatureSensor, Branched.as("temperature"))
        .branch(isProximitySensor,
            Branched.withFunction(
                ps -> ps.mapValues(feetToMetersMapper), "proximity"))
        .noDefaultBranch();
```

Creates the temperature reading branch and names the key

Creates the proximity sensor branch with a ValueMapper function

Specifies no default branch because we know all records fall into only two categories

Overall, each branch call results in a `Map` entry. The key is the concatenation of the name passed into the `KStream.split()` method and string provided in the `Branched` object. The value is a `KStream` instance resulting from each `branch` call.

In the first branching example, the split and subsequent branching calls also return a `Map`, but it would have been empty. The reason is that when you pass in a `Branched.withConsumer` (a `java.util.Consumer` interface), it's a void method, so the `Branched` operator places no entry in the map. But the `Branched.withFunction` (a `java.util.Function` interface) accepts a `KStream<K, V>` object as a parameter and returns a `KStream<K, V>` instance, so it goes into the map as an entry. The function takes the branched `KStream` object. It executes a `MapValues` to convert the proximity sensor reading values from feet to meters since the sensor records in the updated stream are in meters.

I want to point out some subtlety with this example. The `branch` call does not provide a function but still ends up in the resulting `Map`. How is that so? When you only provide a `Branched` parameter with a name, it's treated the same as if you had used a `java.util.Function` that returns the provided `KStream` object, also known as an *identity function*. So what's the determining factor to use either `Branched.withConsumer` or `Branched.withFunction`? I can answer that question best by going over the next block of code in our example.

Listing 6.19   Splitting the stream and gaining access to the newly created streams

```
KStream<String, Sensor> temperatureSensorStream =        ◁──  The stream with the new
    builder.stream("temperature-sensors", sensorConsumed);     temperature IoT sensors

KStream<String, Sensor> proximitySensorStream =          ◁──  The stream with the
    builder.stream("proximity-sensors", sensorConsumed);       updated proximity
                                                               IoT sensors

temperatureSensorStream.merge(sensorMap.get("sensor-temperature"))
    .to("temp-reading", Produced.with(stringSerde, sensorSerde));      ◁──

proximitySensorStream.merge(sensorMap.get("sensor-proximity"))
    .to("proximity-reading", Produced.with(stringSerde, sensorSerde));
```

Merges the updated to meters proximity stream with the new proximity stream

Merges the legacy temperature readings with the new ones

The requirements for splitting the stream were to extract the different IoT sensor results by type, place them in the same stream as the new updated IoT results, and convert any proximity readings into meters. You accomplish this task by extracting the `KStream` from the map with the corresponding keys created in the branching code in the previous code block.

To combine the branched legacy stream with the new one, you use a DSL operator `KStream.merge`, the functional analog of `KStream.split`. It merges different `KStream` objects into one. With `KStream.merge`, there are no ordering guarantees between records of the different streams, but the relative order of each stream remains. In other words, the processing order between the legacy stream and the updated one is not guaranteed, but the order within each stream is preserved.

It should now be clear why you use `Branched.withConsumer` or `Branched.with-Function`. In the latter case, you need to get a handle on the branched `KStream` to integrate into the outer application, while with the former, you don't need access to the branched stream.

That wraps up our discussion on branching and merging, so let's cover naming topology nodes in the DSL.

6.6.3   *Naming topology nodes*

When you build a topology in the DSL, Kafka Streams creates a graph of processor nodes, giving each one a unique name. Kafka Streams generates the node names by taking the processor function name and appending a globally incremented number. To view this topology description, you must get the `TopologyDescription` object and print it to the console or a log entry.

Listing 6.20   Getting a description of the topology and printing it out

```
TopologyDescription topologyDescription =
  streamsBuilder.build().describe();
System.out.println(topologyDescription.toString());
```

Running this code yields the following output on the console.

Listing 6.21   Full topology description of the `KafkaStreamsYellingApplication`

```
Topologies:
   Sub-topology: 0
     Source: KSTREAM-SOURCE-0000000000 (topics: [src-topic])
        --> KSTREAM-MAPVALUES-0000000001
     Processor: KSTREAM-MAPVALUES-0000000001 (stores: [])
        --> KSTREAM-SINK-0000000002
        <-- KSTREAM-SOURCE-0000000000
     Sink: KSTREAM-SINK-0000000002 (topic: out-topic)
        <-- KSTREAM-MAPVALUES-0000000001
```

The source node name

The processor that the source node sends records to

The name of the map values processor

The processor that provided input to the map values processor

The name of the sink node

From looking at the names, you can see the first node ends in `0`, the second node `KSTREAM-MAPVALUES` ending in `1`, etc. The `Sub-topology` listing indicates a portion of the topology containing a distinct source node; every processor downstream of the source node is a member of the given `Sub-topology`. If you were to define a second stream with a new source, it would show up as `Sub-topology: 1`. We'll see more about sub-topologies later in the book when we cover repartitioning.

The arrows pointing to the right (`-→`) show the records flow in the topology. The arrows pointing left (`←-`) indicate the lineage of the record flow, where the current processor received their input. Note that a processor could forward records to more than one node, and a single node could get input from multiple nodes.

The topology description gives you a good sense of the structure of the Kafka Streams application. However, once you build more complex applications, the generic names with the numbers become hard to follow. For this reason, Kafka Streams provides a way to name the processing nodes in the DSL.

All of the methods in the Streams DSL have an overload that takes a `Named` object where you can specify the name used for the node in the topology. Providing the name is crucial as you can make it relate to the processing node *role* in your application, not just what the processor *does*. Configuration objects like `Consumed` and `Produced` have a `withName` method for giving a name to the operator. Let's revisit the `KafkaStreamsYellingApplication`, but this time, we'll add a name for each processor.

Listing 6.22 Updated `KafkaStreamsYellingApplication` with names

```
builder.stream("src-topic",
            Consumed.with(stringSerde, stringSerde)
                 .withName("Application Input"))        <-- Names the source node
        .mapValues((key, value) -> value.toUpperCase(),
               Named.as("Convert to Yelling"))          <-- Gives a name to the mapValues processor
        .to("out-topic",
            Produced.with(stringSerde, stringSerde)
                 .withName("Application Output"))        <-- Names the sink node
```

The description from the updated topology with names will now look like the following code listing.

Listing 6.23 Full topology description with provided names

```
Topologies:
   Sub-topology: 0
    Source: Application-Input (topics: [src-topic])
      --> Convert-to-Yelling
    Processor: Convert-to-Yelling (stores: [])
      --> Application-Output
      <-- Application-Input
    Sink: Application-Output (topic: out-topic)
      <-- Convert-to-Yelling
```

Now, you can view the topology description and get a sense of the role of each processor in the overall application instead of just what the processor itself does. Naming the processor nodes becomes critical for your application when state is involved, but we'll get to that later.

Next, we'll look at how you can use dynamic routing for your Kafka Streams application.

6.6.4 Dynamic routing of messages

Say you need to differentiate which store department the purchase comes from—for example, housewares or shoes. You can use dynamic routing to accomplish this task on a per-record basis. The `KStream.to()` method has an overload that takes a `Topic-NameExtractor`, which will dynamically determine the correct Kafka topic name to use. Note that the topics must exist ahead of time; by default, Kafka Streams will not automatically create extracted topic names.

So, let's go back to the branching example. Each object has a `department` field, so instead of creating a branch, we will process these events with everything else and use the `TopicNameExtractor` to determine the topic to which we route the events.

The `TopicNameExtractor` has one method, `extract`, which you implement to provide the logic for determining the topic name. What you're going to do here is check whether the department of the purchase matches one of the unique conditions for routing the purchase events to a different topic. If it does match, it returns the department's name for the topic name (you created it beforehand). Otherwise, it returns the name of the topic for all other purchase events.

Listing 6.24 Implementing the extract method to determine the topic name

```
@Override
public String extract(String key,
                      Purchase value,
                      RecordContext recordContext) {
    String department = value.getDepartment();
    if (department.equals("coffee")
            || department.equals("electronics")) {      // Checks whether the
        return department;                               // department matches
    } else {                                             // one of the special
        return "purchases";      // The default case for  // cases
    }                            // the topic name
}
```

NOTE The `TopicNameExtractor` interface only has one method to implement; I've chosen to use a concrete class because you can then write a test for it.

Although the code example here uses the value to determine the topic to use, it could very well use the key or a combination of the key and the value. But the third parameter to the `TopicNameExtractor#extract` method is a `RecordContext` object. The `RecordContext` is associated with a record in Kafka Streams.

The context contains metadata about the record—the original timestamp, offset, topic, partition, and the `Headers`. We discussed headers in chapter 4; we won't repeat the coverage here. One of the primary use cases for headers is routing information, and Kafka Streams exposes them via the `ProcessorContext`.

Let's look at one possible example for retrieving the topic name via a `Header`. In this example, you'll extract the `Headers` from the `RecordContext`. First, you need to check that the `Headers` are not null, and then you drill down to get the specific routing information.

> **TIP** Kafka Streams only exposes the `RecordContext` to the `TopicNameExtractor` interface. If you need access to the headers of a record in Kafka Streams, you'll want to use the Processor API or the `process` method of the DSL and then use the `Record.headers()` method to access the headers.

From there, you return the topic's name to use based on the value stored in the `Header`. Since `Headers` are optional and may not exist or contain the specific "routing" header, you've defined a default value in the `TopicNameExtractor`.

Listing 6.25   Using information in a `Header` to determine the topic name

```
public String extract(String key,
                      Purchase value,
                      RecordContext recordContext) {

    Headers headers = recordContext.headers();        ⟵  Retrieves the headers
    if (headers != null) {                                from the RecordContext
    Iterator<Header> routingHeaderIterator =
      headers.headers("routing").iterator();

        if (routingHeaderIterator.hasNext()) {        Extracts the
            Header routing = routingHeaderIterator.next();   ⟵  specific routing
                                                                Header
            return new String(routing.value(),
                              StandardCharsets.UTF_8);   ⟵  Returns the name
        }                                                    of the topic to use
    }                                                        from the Header
    return defaultTopicName;      ⟵                          value
}
              If no routing information is found,
              returns a default topic name
```

Now, you've learned about using the Kafka Streams DSL API.

Summary

- Kafka Streams is a graph of processing nodes called a topology. Each node in the topology is responsible for performing some operation on the key-value records flowing through it. A Kafka Streams application is minimally composed of a source node that consumes records from a topic and a sink node that produces results back to a Kafka topic. You minimally configure a Kafka Streams application with the `application.id` and bootstrap servers configuration. Multiple Kafka Streams applications with the same `application.id` are logically considered one application.
- You can use the `KStream.mapValues` function to map incoming record values to new values, possibly of a different type. These mapping changes shouldn't modify

the original objects. Another method, `KStream.map`, performs the same action but can be used to map both the key and the value to something new.

- To selectively process records, you can use the `KStream.filter` operation, where records that don't match a predicate get dropped. A predicate is a statement that accepts an object as a parameter and returns `true` or `false` depending on whether that object matches a given condition. The `KStream.filterNot` method does the opposite: it only forwards key-value pairs that don't match the predicate.

- The `KStream.branch` method uses predicates to split records into new streams when a record matches a given predicate. The processor assigns a record to a stream on the first match and drops unmatched records. Branching is an elegant way of splitting a stream into multiple streams where each stream can operate independently. For the opposite action, `KStream.merge` merges two `KStream` objects into one stream.

- You can modify an existing key or create a new one using the `KStream.selectKey` method.

- For viewing records in the topology, you can use `KStream.print` or `KStream.peek` (by providing a `ForeachAction` that does the actual printing). `KStream.print` is a terminal operation, meaning you can't chain methods after calling it. `KStream.peek` returns a `KStream` instance, making it easier to embed before and after `KStream` methods.

- You can view a Kafka Streams application's generated graph using the `Topology.describe` method. All graph nodes in Kafka Streams have autogenerated names by default, making the graph hard to understand when the application grows in complexity. You can avoid this situation by providing names to each `KStream` method so that when you print the graph, you have names describing the role of each node.

- You can route records to different topics by passing a `TopicNameExtractor` parameter to the `KStream.to` method. The `TopicNameExtractor` can inspect the key, value, or headers to determine the correct topic name for producing records back to Kafka. You need to create the topics ahead of time.

Streams and state

7

This chapter covers

- Adding stateful operations to Kafka Streams
- Using state stores in Kafka Streams
- Enriching event streams with joins
- Learning how timestamps drive Kafka Streams

In the last chapter, we dove headfirst into the Kafka Streams DSL and built a processing topology to handle streaming requirements from purchase activity. Although you created a nontrivial processing topology, it was one-dimensional in that all transformations and operations were stateless. You considered each transaction in isolation, without regard to other events coinciding or within certain time boundaries, either before or after the transaction. Also, you only dealt with individual streams, ignoring any possibility of gaining additional insight by joining streams together.

In this chapter, you'll extract the maximum amount of information from the Kafka Streams application. To get this level of information, you'll need to use state. State is nothing more than the ability to recall information you've seen before and connect it to current information. You can utilize state in different ways. We'll look at one example when we explore the stateful operations provided by the Kafka Streams DSL, such as the accumulation of values.

188

We'll get to another example of using state when we discuss the joining of streams. Joining streams are closely related to the joins performed in database operations, such as joining records from the employee and department tables to generate a report on who staffs which departments in a company.

We'll also define what the state needs to look like and the requirements for using state when we discuss state stores in Kafka Streams. Finally, we'll weigh the importance of timestamps and look at how they can help you work with stateful operations, such as ensuring you only work with events occurring within a given time frame or allowing you to work with data arriving out of order.

7.1 *Stateful vs. stateless*

Before giving examples, let's discuss the difference between stateless and stateful. In a stateless operation, the current event contains enough information to complete the desired action. Stateful operations are more complex because they involve retrieving data from previous events. A basic example of a stateful operation is an aggregation. For example, consider the following code listing.

Listing 7.1 Stateless function example

```
public boolean numberIsOnePredicate (Widget widget) {

    return widget.number == 1;
}
```

Here, the `Widget` object contains all the information needed to execute the predicate; there's no need to look up or store data. Now, let's take a look at an example of a stateful function.

Listing 7.2 Stateful function example

```
public int count(Widget widget) {

  int widgetCount = hashMap.compute(widget.id,
    (key, value) -> (value == null) ? 1 : value + 1)

  return widgetCount;
}
```

Here, in the `count` function, we are computing the total of widgets with the same ID. To perform the count, we first must look up the current number by ID, increment it, and then store the new number. If the number doesn't exist, we provide an initial value of 1.

While this is a trivial example of using state, the principals involved are what matters here. We use a common identifier across different objects, called a key, to store and retrieve some value type to track a given state we want to observe. Additionally, we use an initializing function to create an initial value for the first calculation. We will explore and use these core steps in this chapter, although it will be far more robust than using the humble `HashMap`!

7.2   *Adding stateful operations to Kafka Streams*

So the next question is, why do you need to use state when processing an event stream? The answer is any time you need to track information or progress across related events. For example, consider a Kafka Streams application monitoring players' progress in an online poker game. Participants play in rounds, and their score from each round is transmitted to a server and then reset to zero for the start of the next round. The game server produces the player's score to a topic.

A stateless event stream will allow you to work with the current score from the latest round. But for tracking the player's total, you'll need to keep the state of all their previous scores.

This scenario leads us to our first example of a stateful operation in Kafka Streams. For this example, we're going to use a `reduce` operation. A `reduce` or, more generically, a `fold` operation takes multiple values and reduces or merges them into a single result. Let's look at figure 7.1 to help understand how this process works.

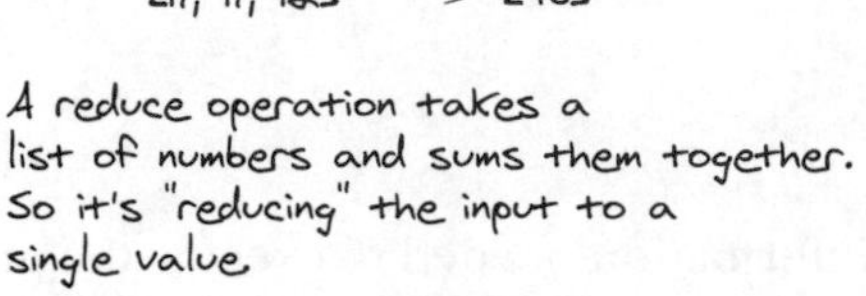

[17, 17, 12] ⟶ [46]

A reduce operation takes a
list of numbers and sums them together.
So it's "reducing" the input to a
single value.

Figure 7.1 A reduce takes several inputs and merges them into a single result of the same type.

As you can see in the illustration, the `reduce` operation takes three numbers and "reduces" them to a single result by summing the numbers together. So Kafka Streams takes an unbounded stream of scores and continues to add them per player. At this point, we've described the `reduce` operation itself, but there's some additional information we need to cover regarding setup.

When describing our online poker game scenario, I mentioned that there are individual players, so it stands to reason that we want to calculate the total scores for each individual. But we aren't guaranteed the order of the incoming player scores, so we need the ability to group them. Remember, Kafka works with key-value pairs, so we'll assume the incoming records take the form of `playerId-score` for the key-value pair.

So, if the key is the `player-id`, all Kafka Streams needs to do is bucket or group the scores by the ID, and you'll end up with the summed scores per player. It will probably be helpful for us to view an illustration of the concept in figure 7.2.

By grouping the scores by `player-id`, you are guaranteed only to sum the scores for each player. This group-by functionality in Kafka Streams is similar to the SQL group-by when performing aggregation operations on a database table.

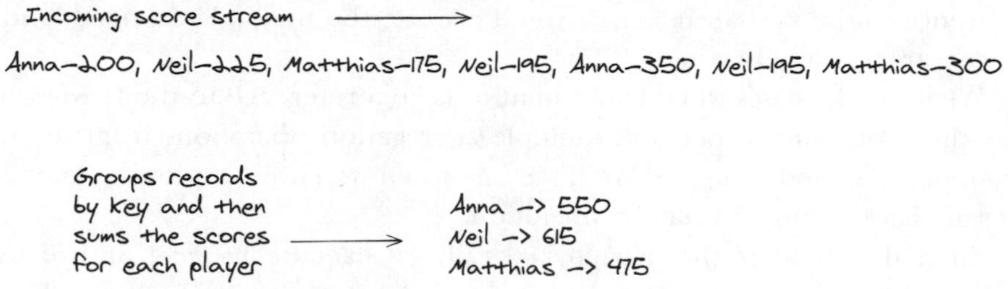

Incoming score stream ⟶

Anna—200, Neil—225, Matthias—175, Neil—195, Anna—350, Neil—195, Matthias—300

Groups records
by key and then   Anna —> 550
sums the scores ⟶   Neil —> 615
for each player   Matthias —> 475

Figure 7.2   Grouping the scores by `player-id` ensures we only sum the scores for the individual players.

NOTE   From now on, I'm not showing the basic setup code needed (i.e., creating the `StreamBuilder` instance and Serdes for the record types). You've learned in the previous chapter how these components fit into an application, so you can refer back if you need to refresh your memory.

Now let's see the `reduce` in action with Kafka Streams.

Listing 7.3   Performing a `reduce` in Kafka Streams

```
KStream<String, Double> pokerScoreStream = builder.stream("poker-game",
        Consumed.with(Serdes.String(), Serdes.Double()));

pokerScoreStream                              Groups by key so that scores are
        .groupByKey()                          calculated by individual keys
        .reduce(Double::sum,
                Materialized.with(Serdes.String(), Serdes.Double()))
        .toStream()
        .to("total-scores",
                Produced.with(Serdes.String(), Serdes.Double()));
```

Reducer as a method reference · `.reduce(Double::sum,`

`.toStream()` — Converts the KTable to a stream

Writes the results out to a topic

This Kafka Streams application results in key-value pairs like `Neil`, `650`, and it's a continual stream of summed scores, continually updated.

Looking over the code, you can see you first perform a `groupByKey` call. It's important to note that grouping by key is a prerequisite for stateful aggregations in Kafka Streams. So what do you do when it doesn't exist or you need to choose a new one? For the case of selecting a different key, the `KStream` interface provides a `groupBy` method that accepts a `KeyValueMapper` parameter that you use to choose a new key. We'll see an example of selecting a new key in section 7.2.3.

7.2.1   Group-by details

We should take a quick detour to briefly discuss the return type of the group-by call, which is a `KGroupedStream`. `KGroupedStream` is an intermediate object that provides methods `aggregate`, `count`, and `reduce`. In most cases, you won't need to keep a reference to the KGroupedStream; you'll execute the method you need, and its existence is transparent to you.

What are the cases when you'd want to keep a reference to the KGroupedStream? Any time you want to perform multiple aggregation operations from the same key grouping is a good example. We'll see one when we cover windowing later on. Now, let's get back to our first stateful operation.

Immediately after the groupByKey call, we execute reduce; as I've explained before, the KGroupedStream object is transparent to us in this case. The reduce method has overloads taking anywhere from one to three parameters; in this case, we're using the two-parameter version, which accepts a Reducer interface and a Materialized configuration object as parameters. For the Reducer, you're using a method reference to the static method Double.sum, which sums the previous total score with the newest score from the game.

The Materialized object provides the Serdes the state store uses for (de)serializing keys and values. Under the covers, Kafka Streams uses local storage to support stateful operations. The stores store key-value pairs as byte arrays, so you need to provide the Serdes to serialize records on input and deserialize them on retrieval. We'll get into the details of state stores in an upcoming section.

After reduce, you call toStream because the result of all aggregation operations in Kafka Streams is a KTable object (which we haven't covered yet, but we will in the next chapter), and to forward the aggregation results to downstream operators, we need to convert it to a KStream.

Then, we can send the aggregation results to an output topic via a sink node represented by the to operator. But stateful processors don't have the same forwarding behavior as stateless ones, so we'll take a minute here to describe that difference.

Kafka Streams provides a caching mechanism for the results of stateful operations. Only when Kafka Streams flushes the cache are stateful results forwarded to downstream nodes in the topology. There are two scenarios when Kafka Streams will flush the cache. The first is when the cache is full, which, by default, is 10 MB; the second is when Kafka Streams commits (every 30 seconds with default settings). Figure 7.3 will help to cement your understanding of how the caching works in Kafka Streams.

From the illustration, you see that the cache sits in front of forwarding records so that you won't observe several of the intermediate results. You will always see the latest updates during a cache flush. Caching also has the effect of limiting writes to the state store and its associated changelog topic. Changelog topics are internal topics created by Kafka Streams for fault tolerance of the state stores. We'll cover changelog topics in section 7.4.1.

> **TIP** If you want to observe every result of a stateful operation, you can turn off the cache by setting StreamsConfig.CACHE_MAX_BYTES_BUFFERING_CON-FIG to 0. However, using this configuration affects every state store in the topology. For an individual store, you apply no caching via the Materialized

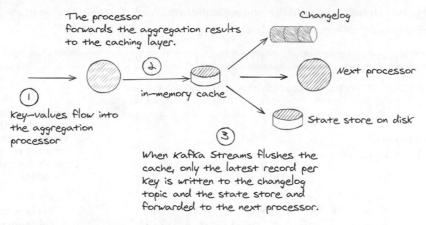

The processor forwards the aggregation results to the caching layer.

Changelog

② in-memory cache

Next processor

State store on disk

① key-values flow into the aggregation processor

③ When Kafka Streams flushes the cache, only the latest record per key is written to the changelog topic and the state store and forwarded to the next processor.

Figure 7.3 Caching intermediate results of an aggregation operation

config object via the `Materialized.withCachingDisabled()` function. Setting the cache level to zero is best applied in a development environment.

7.2.2 Aggregation vs. reducing

You've learned about one stateful operator, but we have another option for stateful operations. If you notice, a `reduce` operation returns the same type. But sometimes, you'll want to build a different result type, and for that, you'll want to use the `aggregate` operation. The concept behind an aggregation is similar, but you can return a type different from the record value. Let's look at an example to answer why you would use `aggregate` over `reduce`.

Imagine you work for ETrade. You need to create an application that tracks stock transactions of individual customers, not large institutional traders. You want to keep a running tally of the total volume of shares bought and sold, the dollar volume of sales and purchases, and the highest and lowest price seen at any point.

You'll need to create a custom data object to provide this information. Needing a custom data object is where the `aggregate` comes into play because it allows for a different return type from the incoming value. In this case, the incoming record type is a singular stock transaction object. The aggregation result will be another type containing the required information described in the previous paragraph.

Since we'll need to put this custom object in a state store that requires serialization, we'll create a Protobuf schema to generate it and use utility methods for creating a Protobuf Serde. Since this application has detailed aggregation requirements, we'll implement the `Aggregator<K, V, VR>` interface as a concrete class, allowing us to test it independently.

Let's take a look at part of the aggregator implementation. Since this class contains some logic unrelated to learning Kafka Streams, I will only show partial information.

To view full details, consult the source code and look for the `bbejeck.chapter_7` `.aggregator.StockAggregator` class.

Listing 7.4   Aggregator implementation used for creating stock transaction summaries

```
public class StockAggregator implements Aggregator<String,
                                      Transaction,
                                      Aggregate> {

    @Override
    public Aggregate apply(String key,
                    Transaction transaction,
                    Aggregate aggregate) {

    Aggregate.Builder currAggregate =
                        aggregate.toBuilder();

    double transactionDollars =
        transaction.getNumberShares()
      * transaction.getSharePrice();

    if (transaction.getIsPurchase()) {
        long currentPurchaseVolume =
            currAggregate.getPurchaseShareVolume();
        currAggregate.setPurchaseShareVolume(
                currentPurchaseVolume
              + transaction.getNumberShares());

        double currentPurchaseDollars =
            currAggregate.getPurchaseDollarAmount();

        currAggregate.setPurchaseDollarAmount(
                currentPurchaseDollars
              + transactionDollars);
    }
```

> **Implementation of the apply method: the second parameter is the incoming record, and the third parameter is the current aggregate.**

> **You need to use a builder to update the Protobuf object.**

> **Gets the total dollars of the transaction**

> **If the transaction is a purchase, updates the purchase-related details**

I will not go into much detail about the `Aggregator` instance here since the main point of this section is how to build a Kafka Streams aggregation application; the particulars of how you implement the aggregation will vary from case to case. But this code shows how we're building up the transactional data for a given stock. Let's look at how we'll plug this `Aggregator` implementation into a Kafka Streams application to capture the information. The source code for this example can be found in bbejeck.chapter_7.StreamsStockTransactionAggregations.

NOTE   There are some details I will leave out of the source code as presented in the book, such as printing records to the console. From now on, our Kafka Streams applications will get more complex, and learning the essential part of the lesson will be easier if I only show the necessary details. Rest assured, the source code is complete.

Listing 7.5 Kafka Streams aggregation

```
KStream<String, Transaction> transactionKStream =
    builder.stream("stock-transactions",
                   Consumed.with(stringSerde, txnSerde));
```
Creates the KStream instance

```
transactionKStream.groupBy((key, value) -> value.getSymbol(),
    Grouped.with(Serdes.String(), txnSerde))
  .aggregate(() -> initialAggregate,
             new StockAggregator(),
             Materialized.with(stringSerde, aggregateSerde))
  .toStream()
  .peek((key, value) -> LOG.info("Aggregation result {}", value))
  .to("stock-aggregations", Produced.with(stringSerde, aggregateSerde));
```
Groups by key and provides a function to select the key

Calls the aggregate function

Converts the resulting aggregation KTable to a KStream

Writes the aggregation results out to a topic

This application starts in familiar territory, creating the KStream instance by subscribing to a topic and providing the Serdes for deserialization. You've seen a groupByKey in the reduce example, but in this example, the incoming records use client ID for a key, but we need to group records by the stock ticker or symbol. So, to change the key, you use GroupBy, which takes a KeyValueMapper, a lambda function in our code example. In this case, the lambda returns the ticker symbol in the record to enable proper grouping.

Since the topology changes the key, Kafka Streams must repartition the data. In the next section, I'll discuss repartitioning in more detail, but it's enough to know now that Kafka Streams takes care of it for you.

Listing 7.6 Kafka Streams aggregation

```
transactionKStream.groupBy((key, value) -> value.getSymbol(),
    Grouped.with(Serdes.String(), txnSerde))
  .aggregate(() -> initialAggregate,
             new StockAggregator(),
             Materialized.with(stringSerde, aggregateSerde))
  .toStream()
  .peek((key, value) -> LOG.info("Aggregation result {}", value))
  .to("stock-aggregations", Produced.with(stringSerde, aggregateSerde));
```
Calls the aggregate function

Converts the resulting aggregation KTable to a KStream

Writes the aggregation results out to a topic

By the end of the code, we get to the crux of our example—applying the aggregation operation. Aggregations differ slightly from the reduce operation because aggregations require an initial value for the first calculation. With reduce, the first input serves as the initial value.

Since there's no way for Kafka Streams to know what the aggregation will create, you need to give it an initial value to seed it. In our case, it's an instantiated `Aggregate` object with uninitialized fields.

The second parameter you provide is the `Aggregator` implementation, which contains your logic to build up the aggregation for each record it encounters. The optional third parameter is a `Materialized` object, which you're using here to supply the Serdes required by the state store.

The final parts of the application convert the `KTable` resulting from the aggregation to a `KStream` so that you can forward the aggregation results to a topic. Here, you're also using a `peek` operation before the sink processor to view results directly from the application flow. Using a `peek` operator this way is typically for development or debugging purposes only.

> **NOTE**   Remember when running the examples that Kafka Streams uses caching, so you won't immediately observe results until the cache gets flushed.

So, at this point, you've learned about the primary tools for stateful operations in the Kafka Streams DSL: `reduce` and aggregation. There's another stateful operation that deserves mention here, and that is the `count` operation. The `count` operation is "syntactic sugar" for incrementing a counter aggregation. You'd use the `count` when you need a running tally of a total—say, the number of times a user has logged into your site or the total number of readings from an IoT sensor. I won't show an example here, but you can see one in the source code at bbejeck/chapter_7/StreamsCounting Application.

In this previous example where we built stock transaction aggregates, I mentioned that changing the key for an aggregation requires repartitioning the data. Let's discuss this in more detail in the next section.

7.2.3   *Repartitioning the data*

In the aggregation example, we saw how changing the key required a repartition. Let's have a more detailed conversation on why Kafka Streams repartitions the data and how it works. Let's talk about the why first.

We learned in a previous chapter that the key of a Kafka record determines the partition. When you modify or change the key, there's a strong probability it belongs on another partition. So, if you've changed the key and you have a processor that depends on it, an aggregation, for example, Kafka Streams, will repartition the data to place records with the new key on the correct partition. Let's look at figure 7.4, which demonstrates this process in action.

As you can see here, repartitioning is nothing more than producing records to a topic and then immediately consuming them again. When the Kafka Streams–embedded producer writes the records to the broker, it uses the updated key to select the

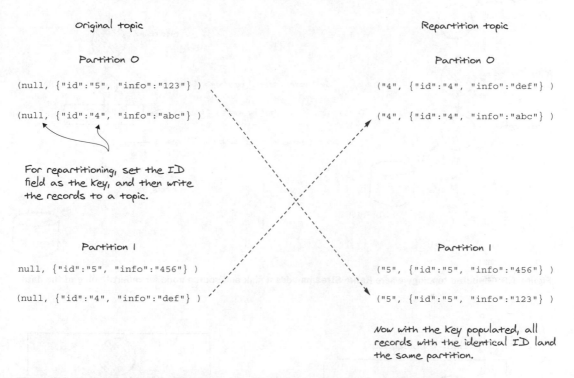

The keys are originally null, so distribution is done round-robin, resulting in records with the same ID across different partitions.

Original topic

Repartition topic

Partition 0

```
(null, {"id":"5", "info":"123"} )

(null, {"id":"4", "info":"abc"} )
```

Partition 0

```
("4", {"id":"4", "info":"def"} )

("4", {"id":"4", "info":"abc"} )
```

For repartitioning, set the ID field as the key, and then write the records to a topic.

Partition 1

```
null, {"id":"5", "info":"456"} )

(null, {"id":"4", "info":"def"} )
```

Partition 1

```
("5", {"id":"5", "info":"456"} )

("5", {"id":"5", "info":"123"} )
```

Now with the key populated, all records with the identical ID land the same partition.

Figure 7.4 Repartitioning: Changing the original key to move records to a different partition

new partition. Under the covers, Kafka Streams inserts a new sink node for producing the records and a new source node for consuming them. Figure 7.5 shows the before and after state where Kafka Streams updated the topology.

The newly added source node creates a new sub-topology in the overall topology for the application. A sub-topology is a portion of a topology that shares a common source node. Figure 7.6 shows the updated version of the repartitioned topology demonstrating the sub-topology structures. So, any processors after the new source node are part of the new sub-topology.

What is the determining factor that causes Kafka Streams to repartition? The determining factor is whether you have a key changing operation *and* a downstream operation relies on the key, such as a groupByKey, aggregation or join (we'll get to joins soon). Otherwise, if no downstream operations are dependent on the key, Kafka Streams will leave the topology as is. Let's look at a couple of examples in listings 7.7 and 7.8 to help clarify this point.

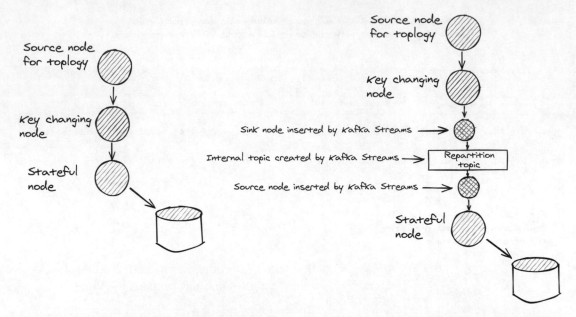

Figure 7.5   Updated topology where Kafka Streams adds a sink and source node for repartitioning of the data

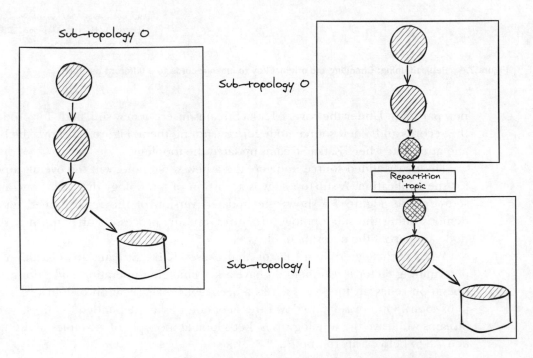

Figure 7.6   Adding a sink and source node for repartitioning creates a new sub-topology.

Listing 7.7   Examples of when repartitioning is needed

```
myStream.groupBy(...).reduce(...)...
```
◄─┐ **Uses groupBy followed by a reduce**

```
myStream.map(...).groupByKey().reduce(...)...
```
◄─┐ **Executes a map followed
by a groupByKey**

```
filteredStream = myStream.selectKey(...).filter(...);
....
filteredStreaam.groupByKey().aggregate(...)...
```
◄

**Uses a selectKey to choose a new key; the
resulting KStream later calls groupByKey.**

These code examples demonstrate that when you execute an operation where you could change the key, Kafka Streams sets an internal Boolean flag, `repartition-Required`, to `true`. Since Kafka Streams can't possibly know whether you changed the key, it will automatically repartition the data when it finds an operation dependent on the key and the internal flag evaluates to `true`. On the other hand, even if you change the key but don't do an aggregation or join, the topology remains the same.

Listing 7.8   Examples of when repartitioning is not needed

```
myStream.map(...).peek(...).to(...);
```
◄─┐ **Uses a map, but no downstream
operation depends on the key.**

```
myStream.selectKey(...).filter(...).to(...);
```
◄

**Uses a selectKey, but no downstream
operations rely on the key.**

In these examples, even if you updated the key, it doesn't affect the results of the downstream operators. For example, filtering a record depends on whether the predicate evaluates to `true` or not. Additionally, since these `KStream` instances produce to a topic, the records with updated keys will end up on the correct partition.

So the bottom line is only to use key-changing operations (`map`, `flatMap`, `trans-form`) when you need to change the key. Otherwise, it's best to use processors that only work on values (i.e., `mapValues`, `flatMapValues`, etc.). This way, Kafka Streams won't needlessly repartition the data. There are overloads to xValues methods that provide access to the key when updating a value, but in this case, changing the key will lead to undefined behavior.

> **NOTE** When grouping records before an aggregation, only use `groupBy` when you need to change the key; otherwise, favor `groupByKey`.

Before we wrap up coverage of repartitioning, we should talk about an additional subject—inadvertently creating redundant repartition nodes and ways to prevent it. Let's say you have an application with two input streams. You need to perform an aggregation on the first stream and join it with the second stream. Your code would look something like the following listing (some details are omitted for clarity).

Listing 7.9 Changing the key and then aggregate and join

```
KStream<String, String> originalStreamOne = builder.stream(...);

KStream<String, String> inputStreamOne =
    originalStreamOne.selectKey(...);

KStream<String, String> inputStreamTwo = builder.stream(...);

inputStreamOne.groupByKey().count().toStream().to(...);

KStream<String, String> joinedStream =
    inputStreamTwo.join(inputStreamOne,
            (v1, v2)-> v1+":"+v2,
            JoinWindows.ofTimeDifferenceWithNoGrace(...),
            StreamJoined.with(...));
....
```

Changes the key of the original stream setting the "needsRepartition" flag

The second stream

Performs a group-by-key, triggering a repartition

Performs a join between inputStreamOne and inputStreamTwo, triggering another repartition

This code example is simple enough. You take the `originalStreamOne` and change the key since you'll need to do an aggregation and a join with it. So you use a `select-Key` operation, which sets the `repartitionRequired` flag for the returned `KStream`. Then you perform a `count()` and a `join` with `inputStreamOne`. What is not apparent here is that Kafka Streams will automatically create two repartition topics, one for the `groupByKey` operator and the other for the `join` when you only need one repartition.

It will help to fully understand what's going on here by looking at the topology for this example in figure 7.7. Notice there are two repartitions, but you only need the first one where the key is changed.

When you use the key-changing operation on `originalStreamOne`, the resulting `KStream`, `inputStreamOne`, now carries the `repartitionRequired = true` setting. So any `KStream` resulting from `inputStreamOne` that uses a processor involving the key will trigger a repartition.

What can you do to prevent this from happening? There are two choices here. The first option is to manually repartition earlier, which sets the repartition flag to `false` so any subsequent streams won't trigger a repartition. The other option is to let Kafka Streams handle it by enabling optimizations. Let's talk about using the manual approach first.

NOTE While repartition topics take up disk space, Kafka Streams actively purges records from them, so you don't need to be concerned with the size on disk. However, since repartitions increase processing latency due to increased network round trips, avoiding redundant repartitions is always a good idea.

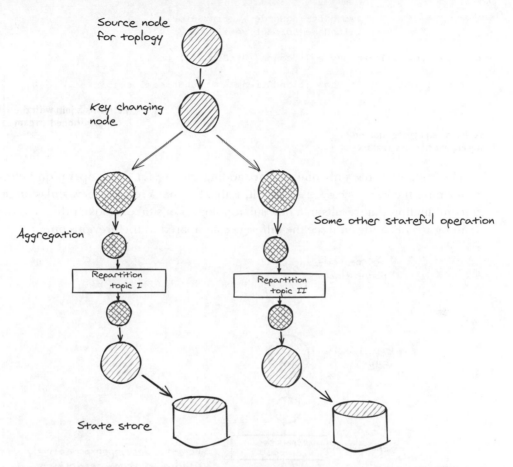

Figure 7.7   Redundant repartition nodes due to a key changing operation occurring previously in the topology

7.2.4   Proactive repartitioning

For the times when you might need to repartition the data yourself, the KStream API provides the `repartition` method. The following listing shows you how to manually repartition after changing a key (some details are omitted for clarity).

Listing 7.10   Changing the key, repartitioning, and performing an aggregation and a join

```
KStream<String, String> originalStreamOne = builder.stream(...);
KStream<String, String> inputStreamOne =
    originalStreamOne.selectKey(...);

KStream<String, String> inputStreamTwo = builder.stream(...);

KStream<String, String> repartitioned =
    inputStreamOne.repartition(Repartitioned
```

◁─┤ **Changes the key setting the "needs repartition" flag**

◁─┤ **Calls the repartition method and providing key-value Serdes and a name for the repartition topic**

```
            .with(stringSerde, stringSerde)
            .withName("proactive"));

repartitioned.groupByKey().count().toStream().to(...);

KStream<String, String> joinedStream = inputStreamTwo.join(...)

.....
```

**Performs a join with the
repartitioned stream**

**Performs an aggregation on
the repartitioned stream**

The code here has only one change, adding the `repartition` operation before performing the `groupByKey`. As a result, Kafka Streams creates a new sink-source node combination that results in a new sub-topology. Let's take a look at the topology now in figure 7.8, and you'll see the difference compared to the previous one.

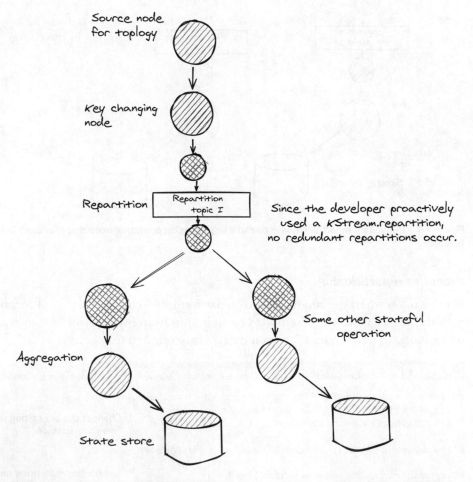

**Figure 7.8   Now, only one repartition node due to a proactive repartition, which allows for
more stateful operations without repartitioning**

This new sub-topology ensures that the new keys end up on the correct partition, and equally as important, the returned `KStream` object has the `needsRepartition` flag set to `false`. As a result, all downstream stateful operations that are descendants of this `KStream` object don't trigger any further repartitions (unless you change the key again).

The `KStream.repartition` method accepts one parameter, the `Repartitioned` configuration object. `Repartitioned` allows you to specify:

1 The Serdes for the key and value
2 The base name for the topic
3 The number of partitions to use for the topic
4 A `StreamPartitioner` instance, should you need to customize the distribution of records to partitions

Let's pause on our current discussion and review some of these options.

Providing a base name for the repartition topic is always a good idea. I'm using the term base name because Kafka Streams takes the name you provide and adds a prefix of `<application-id>-`, which comes from the value you supplied in the configs, and a suffix of `-repartition`.

Given an application ID of "streams-financial" and a repartition name of "stock-aggregation," results in a repartition topic named "streams-financial-stock-aggregation-repartition." The reason it's a good idea always to provide a name is twofold. First, having a meaningful topic name is always helpful to understand its role when you list the topics on your Kafka cluster. Second, and more important, the name you provide remains fixed even if you change your topology upstream of the repartition. Remember, when you don't provide names for processors, Kafka Streams generates names for them, and part of the name includes a zero-padded number generated by a global counter.

So, if you add or remove operators upstream of your repartition operation and you haven't explicitly named them, their names will change due to changes in the global counter. This name shift can be problematic when redeploying an existing application. In section 7.4.5, I'll talk more about the importance of naming stateful components of a Kafka Streams application.

> **NOTE** Although there are four parameters for the `Repartitioned` object, you don't have to supply all of them. You can use any combination of the parameters that suit your needs.

Specifying the number of partitions for the repartition topic is particularly useful in two cases: co-partitioning for joins and increasing the number of tasks to enable higher throughput. Let's discuss the co-partitioning requirement first. When performing joins, both sides must have the same number of partitions (we'll discuss why this is when covering joins starting in section 7.3). So, by using the `repartition` operation, you can change the number of partitions to enable a join without changing the source topic, keeping the changes internal to the application.

7.2.5 *Repartitioning to increase the number of tasks*

The number of partitions drives the number of tasks, ultimately determining the number of active threads an application can have. One way to increase the processing power is to increase the number of partitions since that leads to more tasks and, ultimately, more threads that can process records. Kafka Streams attempts to assign tasks to all applications with the same ID evenly, so this approach to increase throughput is particularly useful in an environment where you can elastically expand the number of running instances.

While you could increase the number of partitions for the source topic, this might not always be possible. The source topic(s) of a Kafka Streams application is typically public, meaning other developers and applications use that topic. In most organizations, changes to shared infrastructure resources can be difficult.

Let's look at an example of performing a repartition to increase the number of tasks (found in bbejeck.chapter_7.RepartitionForThroughput).

Listing 7.11   Increasing the number of partitions for a higher task count

```
KStream<String, String> repartitioned =

initialStream.repartition(Repartitioned
        .with(stringSerde, stringSerde)
        .withName("multiple-aggregation")
        .withNumberOfPartitions(10));
```

Increases the number of partitions

Now, this application will have 10 tasks, meaning there can be up to 10 stream threads processing records driven by the increase in the number of partitions. Note that the example of using 10 partitions here is arbitrary. The example only demonstrates how to use the `KStream.repartition` method and does not imply the exact value to use in a production system.

However, you need to remember that adding partitions for increased throughput will work best when there is a relatively even distribution of keys. For example, if 70 percent of your key space lands on one partition, increasing the number of partitions will only move those keys to a new partition. But since the keys' overall distribution is relatively unchanged, you won't see any gains in throughput since one partition, hence one task, is shouldering most of the processing burden.

So far, we've covered how to repartition when changing the key proactively. But this requires you to know when to repartition and always remember to do so. But there's a better approach using Kafka Stream's optimizations. Using optimizations, Kafka Streams will automatically handle redundant repartition topics for you.

7.2.6 *Using Kafka Streams optimizations*

While you're busy creating a topology with various methods, Kafka Streams builds a graph or internal representation under the covers. You can also consider the graph a logical representation of your Kafka Streams application. In your code, when you execute the `StreamBuilder#build` method, Kafka Streams traverses the graph and builds the final or physical representation of the application. At a high level, it works like this: as you apply each method, Kafka Streams adds a node to the graph, as depicted in figure 7.9.

```
KStream<String, String> myStream = builder.stream("topic")
 myStream.filter(...).map(...).to("output")
```

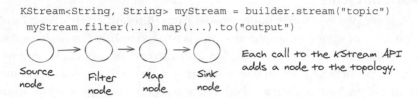

Source node   Filter node   Map node   Sink node

Each call to the KStream API adds a node to the topology.

Figure 7.9   Each call to a KStream method adds a node to the graph.

When you make an additional method call, the previous node becomes the parent of the current one. This process continues until you finish building your application.

Along the way, Kafka Streams will record metadata about the graph it's building. Specifically, it records if it has encountered a repartition node. Then, when using the `StreamsBuilder#build` method to create the final topology, Kafka Streams will examine the graph for redundant repartition nodes. If found, it will rewrite your topology to have only one! Optimizations are opt-in behavior for Kafka Streams, so to get this feature working, you'll need to enable optimizations by doing the following.

Listing 7.12   Enabling optimizations in Kafka Streams

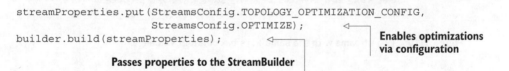

```
streamProperties.put(StreamsConfig.TOPOLOGY_OPTIMIZATION_CONFIG,
                    StreamsConfig.OPTIMIZE);
builder.build(streamProperties);
```

Enables optimizations via configuration

Passes properties to the StreamBuilder

So, to enable optimizations, first, you need to set the proper configuration because they are disabled by default. The second step is to pass the properties object to the `StreamBuilder#build` method. Then Kafka Streams will optimize your repartition nodes when building the topology.

> **NOTE**   If you have more than one key-changing operation with a stateful one further downstream, the optimization will not remove that repartition. It only takes away redundant repartitions for a single key-changing processor.

But when you enable optimizations, Kafka Streams automatically updates the topology by removing the three repartition nodes preceding the aggregation and inserts a new single repartition node immediately after the key-changing operation, which results in a topology that looks like the illustration in figure 7.8.

So, with a configuration setting and passing the properties to the `StreamBuilder`, you can automatically remove any unnecessary repartitions! The decision on which one to use comes down to personal preference, but enabling optimizations guards against you overlooking where you may need it.

Now we've covered repartitioning, let's move on to our next stateful operation, joins.

7.3   Stream-stream joins

Sometimes, you may need to combine records from different event streams to "complete the picture" of what your application wants to accomplish. Say we have a stream of purchases with the customer ID as the key and a stream of user clicks, and we want to join them to connect pages visited with purchases. To do this in Kafka Streams, you use a join operation. You may already be familiar with the concept of a join from SQL and the relational database world, and the idea is the same in Kafka Streams. Let's look at an illustration to demonstrate the concept of joins in Kafka Streams in figure 7.10.

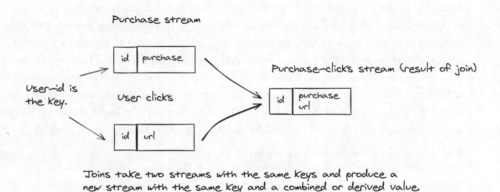

Figure 7.10   Two streams with the same keys but different values

Figure 7.10 shows that two event streams use the same item for the key, a customer ID, but the values differ. In one stream, the values are purchases, and in the other, the values are links to pages the user clicked visiting the site.

> **NOTE**   Since joins depend on identical keys from different topics residing on the same partition, the same rules apply when using a key-changing operation. If a `KStream` instance contains `repartitionRequired=true`, Kafka Streams will partition it before the join operation to ensure co-partitioning. So, all the information in this chapter's repartitioning section also applies to joins.

In this section, you'll combine events from two streams with the same key to form a new event. The best way to learn about joining streams is to look at a concrete example so we'll return to the world of retail. Consider a big box retailer that sells just about anything you can imagine. To lure more customers into the store, the retailer partners with a national coffee house and embeds a cafe in each store.

To encourage customers to come into the store, the retailer has started a special promotion where if you are a member of the customer club and you buy a coffee drink from the embedded cafe and purchase anything in the store (in either order), you'll automatically earn loyalty points after your second purchase. The customers can redeem those points for items from either store. The rules state that purchases must made within 30 minutes of each other to qualify for the promotion.

Since the main store and the cafe run on separate computing infrastructure, the purchase records are in two event streams, but that's not an issue as they both use the customer ID from the club membership for the key. Hence, it's a case of using a stream-stream join to complete the task.

7.3.1 Implementing a stream-stream join

The next step is to perform the actual join. So, let's show the code for the join (some details are omitted for clarity) in the following listing. Since there are a couple of components that make up the join, I'll explain them in a section following the code example. You'll find the source code for this example in src/main/java/bbejeck/chapter_7/ KafkaStreamsJoinsApp.java.

Listing 7.13   Using the `join()` method to combine two streams into one

```
KStream<String, CoffeePurchase>
                coffeePurchaseKStream = builder.stream(...)    ⟵  The
                                                                   streams
KStream<String, RetailPurchase>                                    you will
                retailPurchaseKStream = builder.stream(...)   ⟵   join

ValueJoiner<CoffeePurchase,                      ValueJoiner instance that
            RetailPurchase,                   produces the joined result object
            Promotion> purchaseJoiner =
                          new PromotionJoiner();    ⟵

JoinWindows thirtyMinuteWindow =
     JoinWindows.ofTimeDifferenceWithNoGrace(Duration.minutes(30));   ⟵

KStream<String, Promotion> joinedKStream =                          JoinWindow
    coffeePurchaseKStream.join(retailPurchaseKStream,               specifying the
                       purchaseJoiner,                               max time
                       thirtyMinuteWindow,                           difference
                       StreamJoined.with(stringSerde,                between
                                      coffeeSerde,                   records to
                                      storeSerde)                    participate
                       .withName("purchase-join")                    in join
                       .withStoreName("join-stores"));
```

Constructs the join — StreamJoined configuration object

You supply four parameters to the `KStream.join` method:

- `retailPurchaseKStream`—The stream of purchases from to join with.
- `purchaseJoiner`—An implementation of the `ValueJoiner<V1, V2, R>` interface. `ValueJoiner` accepts two values (not necessarily of the same type). The

ValueJoiner.apply takes both values for the key from the joined streams, performs some implementation-specific logic, and returns a (possibly new) object of type R. In this example, purchaseJoiner will extract some information from both Purchase objects and return a PromotionProto object.

- thirtyMinuteWindow—A JoinWindows instance. The JoinWindows.ofTime-DifferenceWithNoGrace method specifies a maximum time difference between the two values to include them in the join. Specifically, the timestamp on the secondary stream, retailPurchaseKStream, can only be a maximum of 30 minutes before or after the timestamp of a record from the coffeePurchaseK-Stream with the same key.

- A StreamJoined instance—Provides optional parameters for performing joins. In this case, it's the key, the value Serde for the calling stream, and the value Serde for the secondary stream. When joining records, you only have one key, Serde; keys must be the same type. The withName method provides the name for the node in the topology and the base name for a repartition topic (if required). The withStoreName is the base name for the state stores used for the join. I'll cover join state stores usage in section 7.4.

NOTE   Serde objects are required for joins because Kafka Streams materializes join participants in windowed state stores. You provide only one Serde for the key because both sides of the join must have a key of the same type.

Joins in Kafka Streams are one of the most powerful operations you can perform, and they're also one of the more complex ones to understand. Let's take a minute to dive into the internals of how joins work.

7.3.2   *Join internals*

Under the covers, the KStream DSL API does a lot of heavy lifting to make joins operational. But it will be helpful for you to understand how joins operate under the covers. Kafka Streams creates a join processor with a state store for each side of the join. Figure 7.11 shows how this looks conceptually.

When building the processor for the join for each side, Kafka Streams includes the name of the state store for the reciprocal side of the join—the left side gets the name of the right side store, and the right side processor contains the left store name. Why does each side have the name of the opposite side store? The answer gets at the heart of how joins work in Kafka Streams. Let's look at another illustration in figure 7.12 to demonstrate.

When a new record comes in (we're using the left-side processor for the coffee-PurchaseKStream), the processor puts the record in its store but then looks for a match by retrieving the right-side store (for the retailPurchaseKStream) by name. The processor retrieves records with the same key and within the time range specified by the JoinWindows.

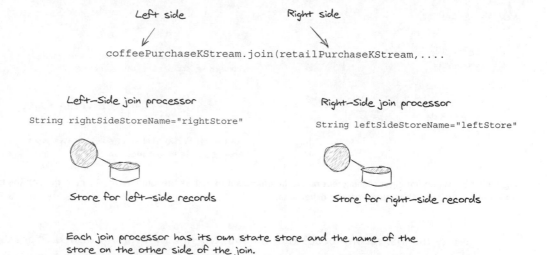

Figure 7.11   In a stream-stream join, both sides of the join have a processor and state store.

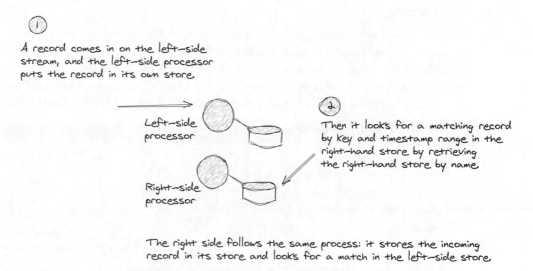

Figure 7.12   Joining processors and looking in the other side's state store for matches when a new record arrives

Now, the final part to consider is whether a match occurs. Let's look at one more illustration in figure 7.13 to help us see what's going on.

So now, after an incoming record finds a match by looking in the store from the other join side, the join processor (the `coffeePurchaseKStream` in our illustration) takes the key and the value from its incoming record and the value for each record it has retrieved from the store and executes the `ValueJoiner.apply` method, which

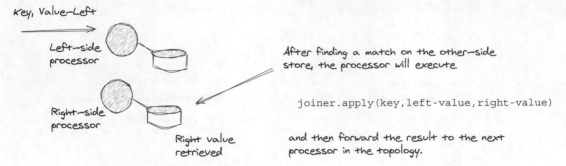

key, Value-Left

Left-side processor

Right-side processor

Right value retrieved

After finding a match on the other-side store, the processor will execute

```
joiner.apply(key,left-value,right-value)
```

and then forward the result to the next processor in the topology.

Figure 7.13   When finding matching record(s), the processor executes the joiner's `apply` method with the key, its record value, and the value from the other side.

creates the join record specified by the implementation you've provided. From there, the join processor forwards the key and join result to any downstream processors.

Now that we've discussed how joins operate internally, let's discuss some of the join parameters in more detail.

7.3.3   *ValueJoiner*

To create the joined result, you create an instance of a `ValueJoiner<V1, V2, R>`. The `ValueJoiner` takes two objects, which may or may not be of the same type, and it returns a single object, possibly of a third type. In this case, `ValueJoiner` takes a `Coffee-Purchase` and a `RetailPurchase` and returns a `Promotion` object. Let's take a look at the code (found in src/main/java/bbejeck/chapter_7/joiner/PromotionJoiner.java).

Listing 7.14   `ValueJoiner` implementation

```
public class PromotionJoiner
    implements ValueJoiner<CoffeePurchase,
                           RetailPurchase,
                           Promotion> {

    @Override
    public Promotion apply(                                       ← Extracts how much
            CoffeePurchase coffeePurchase,                          was spent on coffee
            RetailPurchase retailPurchase) {

    double coffeeSpend = coffeePurchase.getPrice();    ←
    double storeSpend = retailPurchase.getPurchasedItemsList()  ←   Sums the total
         .stream()                                                  of purchased
         .mapToDouble(pi -> pi.getPrice() * pi.getQuantity()).sum();  items
    double promotionPoints = coffeeSpend + storeSpend;  ←
    if (storeSpend > 50.00) {                                       Calculates the
        promotionPoints += 50.00;                                   promotion points
    }
    return Promotion.newBuilder()                       ←          Builds and returns the
         .setCustomerId(retailPurchase.getCustomerId())            new Promotion object
         .setDrink(coffeePurchase.getDrink())
```

```
                        .setItemsPurchased(retailPurchase.getPurchasedItemsCount())
                        .setPoints(promotionPoints).build();
}
```

To create the `Promotion` object, you extract the amount spent from both sides of the join and calculate the total points to reward the customer. The `ValueJoiner` interface only has one method, `apply`, so you could use a lambda to represent the joiner. But in this case, you create a concrete implementation because you can write a separate unit test for the `ValueJoiner`. We'll come back to this approach in chapter 14.

> **NOTE** Kafka Streams also provides a `ValueJoinerWithKey` interface, which allows access to the key for calculating the value of the join result. However, the key is considered read-only, and changing it in the joiner implementation will lead to undefined behavior.

7.3.4 JoinWindows

The `JoinWindows` configuration object plays a critical role in the join process; it specifies the difference between the timestamps of records from both streams to produce a join result. Let's refer to the illustration in figure 7.14 to understand the `JoinWindows` role.

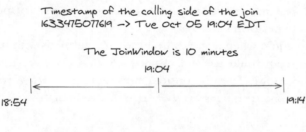

Figure 7.14 The `JoinWindows` configuration specifies the max difference (before or after) from the timestamp of the calling side the secondary side can have to create a join result.

More precisely, the `JoinWindows` setting is the maximum difference before or after the secondary (other) side's timestamp can be from the primary side timestamp to create a join result. From the example, the join window has a setting of 30 minutes. So let's say a record from the `coffeeStream` has a timestamp of 12:00 p.m.; for a corresponding record in the `storeStream` to complete the join, it will need a timestamp between 11:30 a.m. and 12:30 p.m.

Two additional `JoinWindows()` methods are available, `after` and `before`, which you can use to specify the timing and possibly the order of events for the join. Let's say you set the opening window of the join at 30 minutes, but you want the closing window to be shorter, say 5 minutes. For example, if a given record has a timestamp of 12:00 p.m., you're OK with joining another record with a timestamp of at least 11:30

a.m. but no later than 12:05 p.m. To do this, you'd use the `JoinWindows.after` method (still using the example in listing 7.13).

```
coffeeStream.join(storeStream,...,
    thirtyMinuteWindow.after(Duration.ofMinutes(5))....
```

Here, the opening window stays the same; the `storeStream` record can have a timestamp of at least 11:30 a.m., but the closing window for the join is shorter; the latest it can be is now 12:05 p.m..

The `JoinWindows.before` method works similarly in the opposite direction. Let's say you want to shorten the opening window, so you'll now use the following code.

Listing 7.16   The `JoinWindows.before` method changes the opening side

```
coffeeStream.join(storeStream,...,
    thirtyMinuteWindow.before(Duration.ofMinutes(5))....
```

Now you've changed things so the timestamp of the `storeStream` record can be at most 5 minutes before the timestamp of a `coffeeStream` record. So the acceptable timestamps for a join (`storeStream` records) now start at 11:55 a.m. and end at 12:30 p.m.. You can also use `JoinWindows.before` and `JoinWindows.after` to specify the order of arrival of records to perform a join.

For example, to set up a join when a store purchase only happens within 30 minutes *after* a cafe purchase, you would use `JoinWindows.of(Duration.ofMinutes(0)` `.after(Duration.ofMinutes(30)`. To only consider store purchases *before*, you would use `JoinWindows.of(Duration.ofMinutes(0).before(Duration.ofMinutes(30))`.

7.3.5   *Co-partitioning*

To perform a join in Kafka Streams, you must ensure that all join participants are co-partitioned, meaning they have the same number of partitions and the keys are the same type. Co-partitioning requires all Kafka producers to use the same partitioning class when producing to Kafka Streams source topics. Likewise, you need to use the identical `StreamPartitioner` for any operations writing Kafka Streams sink topics via the `KStream.to()` method. If you stick with the default partitioning strategies, you won't need to worry about partitioning strategies.

As you can see, the `JoinWindows` class gives you plenty of options to control joining two streams. It's important to remember that the timestamps on the records drive the join behavior. The timestamps can be set by Kafka (broker or producer) or embedded in the record payload. To use a timestamp embedded in the record, you'll need to provide a custom `TimestampExtractor`, and I'll cover that and timestamp semantics in chapter 9.

7.3.6   StreamJoined

The final parameter to discuss is the `StreamJoined` configuration object. With `Stream-Joined`, you can provide the Serdes for the key and the values involved in the join. Note that when using Schema Registry, you'll supply the appropriate Schema Registry–aware Serdes to the `StreamJoined` object. Providing the Serdes for the join records is always a good idea because you may have different types than the application-level configurations. You can also name the join processor and the state stores used for storing record lookups to complete the join. I'll discuss naming state stores in the section 7.4.5.

Before moving on from joins, let's discuss other join options available.

7.3.7   Other join options

The join in the listing for the current example is an *inner join*. With an inner join, if either record isn't present, the join doesn't occur, and you don't emit a `Promotion` object. But other join options don't require both records. Kafka Streams will emit a result if the other side of the join isn't present. These are useful if you need information even when the desired record for joining isn't available.

7.3.8   Outer joins

Outer joins always output a record, but the result may not include both sides of the join. You'd use an outer join when you wanted to see a result, regardless of whether it was a successful join or not. If you need to use an outer join for the join example, you'd do so like this:

```
coffeePurchaseKStream.outerJoin(retailPurchaseKStream,..)
```

An outer join sends a result that contains records from either side or both. For example, the join result could be `left+right`, `left+null`, or `null+right`, depending on what's present. Figure 7.15 demonstrates the three possible outcomes of the outer join.

7.3.9   Left-outer join

A left-outer join always produces a result. But the difference from the outer-join is the left or calling side of the join is always present in the outcome—`left+right` or `left+null`, for example. You'd use a left-outer join when considering the left or calling side stream records essential for your business logic. If you wanted to use a left-outer join in listing 7.13, you'd do so like this:

```
coffeePurchaseKStream.leftJoin(retailPurchaseKStream..)
```

Figure 7.16 shows the outcomes of the left-outer join.

You've learned the different join types, so what are the cases when you need to use them? Let's start with the current join example. An inner join makes sense since you

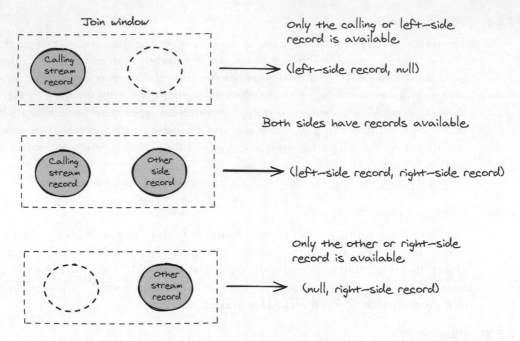

Figure 7.15   Three outcomes are possible with outer joins: only the calling stream's event, both events, and only the other stream's event.

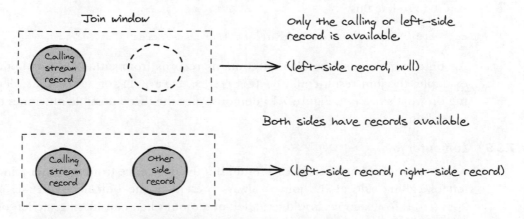

Figure 7.16   Two outcomes are possible with a left-outer join: left and right side or left and null.

are determining a promotional reward based on purchasing two items, each in their stream. If there is no corresponding purchase on the other side, you don't have an actionable result, so to emit nothing is desired.

For cases where one side of the join is critical and the other is beneficial but not essential, a left-side join is a good choice where you'd use the critical stream on the left or calling side. I'll cover an example when we get to stream-table joins in chapter 8.

Finally, for a case where you have two streams where both sides enhance each other, but each one is important on its own, an outer join fits the bill. Consider IoT, where you have two related sensor streams. Combining the sensor information provides a more complete picture, but you want information from either side if it's available.

In the next section, we'll go into the details of the workhorse of stateful operations, the state store.

7.4 State stores in Kafka Streams

So far, we've discussed the stateful operations in the Kafka Streams DSL API but glossed over those operations' underlying storage mechanisms. In this section, we'll look at the essentials of using state stores in Kafka Streams and the critical factors related to using state in streaming applications in general. This will enable you to make practical choices when using state in your Kafka Streams applications.

Before I go into any specifics, let's cover some general information. At a high level, the state stores in Kafka Streams are key-value stores, and they fall into two categories: persistent and in-memory. Both types are durable because Kafka Streams uses changelog topics to back the stores. I'll talk more about changelog topics here and in section 7.4.1.

Persistent stores store their records on a local disk, maintaining their contents over restarts. The in-memory stores place records in memory, so they need to be restored after a restart. All state stores use the changelog topic to restore their contents. To understand how a state store uses a changelog topic for restoration, let's look at how Kafka Streams implements them.

In the DSL, when you apply a stateful operation to the topology, Kafka Streams creates a state store for the processor (persistent is the default type). Along with the store, Kafka Streams also makes a changelog topic backing the store. As Kafka Streams writes records to the store, it also writes them to the changelog. Figure 7.17 demonstrates this process.

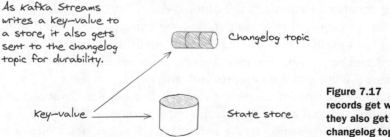

As Kafka Streams writes a key-value to a store, it also gets sent to the changelog topic for durability.

Changelog topic

key-value ⟶ State store

Figure 7.17 As the key-value records get written to the store, they also get written to the changelog topic for data durability.

So, as Kafka Streams places a record into a state store, it also sends it to a Kafka topic that backs the state store. Earlier in the chapter, I mentioned that you don't see every update with an aggregation as Kafka Streams initially uses a cache to hold the results. Only when Kafka Streams flushes the cache, either at a commit or when it's full, do records from the aggregation go to downstream processors. At this point, Kafka Streams will produce records to the changelog topic.

> **NOTE**   If you've turned off the cache, every record gets sent to the state store, meaning every record goes to the changelog topic.

7.4.1   *Changelog topics restoring state stores*

So, how does the Kafka Stream use the changelog topic? Let's first consider the case of an in-memory state store. Since an in-memory store doesn't maintain its contents across restarts, when starting up, any in-memory stores will rebuild their contents from the head record of the changelog topic. So even though the in-memory store loses all its contents on application shutdown, it picks up where it left off when restarted.

Persistent stores usually need to do a full restore only after losing all local states or detecting data corruption. For persistent stores, Kafka Streams maintains a checkpoint file, and it will use the offset in the file as a starting point to restore, instead of restoring from scratch. If the offset is no longer valid, Kafka Streams will remove the checkpoint file and restore it from the beginning of the topic. Kafka Streams has a slightly different use of checkpoint files in `exactly_once` and `exactly_once_v2` (EOS mode). Kafka Streams deletes the checkpoint file after opening a state store in EOS mode. And after a clean shutdown, Kafka Streams will re-generate it. Following this process in EOS mode ensures that complete restoration occurs only after shutting down due to an error.

This difference in restoration patterns brings an interesting twist to the discussion of the tradeoffs of persistent or in-memory stores. While an in-memory store should yield faster lookups as it doesn't need to go to disk for retrieval, under "happy path" conditions, the topology with persistent stores will generally resume processing faster as it will not have as many records to restore.

Another situation to consider is the makeup of running Kafka Streams applications. If you recall from our discussion on task assignments, you can dynamically change the number of running applications, either by expansion or contraction. Kafka Streams will automatically assign tasks from existing applications to new members or add tasks to those still running from an application that has dropped out of the group. A task responsible for a stateful operation will have a state store as part of its assignment (I'll talk about state stores and tasks next).

Let's consider the case of a Kafka Streams application that loses one of its members; remember, you can run Kafka Streams applications on different machines, and those with the same application ID are considered all part of one logical application. Kafka Streams will issue a rebalance and reassign the tasks from the defunct application. For any reassigned stateful operations, since Kafka Streams creates a new *empty* store for the newly assigned task, it will need to restore from the beginning of the changelog topic before it resumes processing. Figure 7.18 demonstrates this situation.

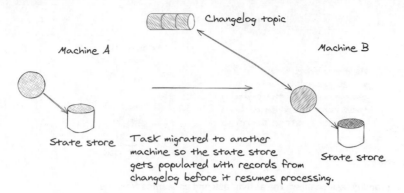

Figure 7.18 When a stateful task gets moved to a new machine, Kafka Streams rebuilds the state store from the beginning of the changelog topic.

So, by using changelog topics, you can be assured your applications will have a high degree of data durability even in the face of application loss, but processing is delayed until the store is entirely online. Fortunately, Kafka Streams offers a remedy for this situation, the standby task.

7.4.2 Standby tasks

Kafka Streams provides the standby task to enable fast failover from an application instance dropping out of the group. A standby task "shadows" an active task by consuming from the changelog topic into a state store local to the standby. Then, should the active task drop out of the group, the standby becomes the new active task. But since it's been consuming from the changelog topic, the new active task will come online with minimum recovery time.

> **NOTE** To enable standby tasks, you need to set the `num.standby.replicas` configuration to a value greater than 0, and you need to deploy an *N*+ 1 number of Kafka Streams instances (with *N* being equal to the number of desired replicas). Ideally, you'll also deploy those Kafka Streams instances on separate machines.

While the concept is straightforward, let's review the standby process by walking through the illustration in figure 7.19.

So, following along with the illustration, a standby task consumes records from the changelog topic and puts them in its local state store. A standby task does not process any records. Its only job is to keep the state store in sync with the state store of the active task. Like any standard producer and consumer application, there's no coordination between the active and standby tasks.

With this process, since the standby stays fully caught up to the active task or, at a minimum, it will be only a handful of records behind, when Kafka Streams reassigns the task, the standby becomes the active task, and processing resumes with minimal

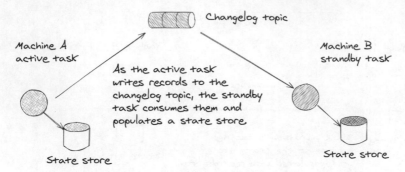

Figure 7.19 A standby task shadows the active task and consumes from the changelog topic, keeping a local state store in sync with the store of the active task.

latency. As with anything, there is a tradeoff to consider with standby tasks. By using standby, you end up duplicating data, but with the benefit of near-immediate failover, it's worth consideration.

> **NOTE** Significant work went into improving the scaling out the performance of Kafka Streams with Kafka KIP-441 (http://mng.bz/0Gml). When you enable standby tasks and the standby instance becomes the active one, if at a later time, Kafka Streams determines a more favorable assignment is possible, that stateful task may get migrated to another instance.

So far, we've covered how state stores enable stateful operations and how the stores are robust due to changelog topics and using standby tasks to enable quick failover. But we still have some more ground to cover. First, we'll go over state store assignments; from there, you'll learn how to configure state stores by specifying a store type, including an in-memory store, and configuring changelog topics if needed.

7.4.3 *Assigning state stores in Kafka Streams*

In the previous chapter, we discussed the role of tasks in Kafka Streams. I want to reiterate that tasks operate in a shared-nothing architecture and only in a single thread. While a Kafka Streams application can have multiple threads and each thread can have multiple tasks, they share nothing. Again, I emphasize this shared-nothing architecture because when a task is stateful, only the owning task will access its state store; there are no locking or concurrency issues.

Going back to the stock aggregation example in section 7.2.2, let's say the source topic has two partitions, meaning it has two tasks. Let's look at figure 7.20, which is an updated illustration of the tasks assignment with state stores for that example.

By looking at this illustration, you can see that the task associated with the state store is the only task that will ever access it. Now, let's discuss how Kafka Streams places state stores in the filesystem.

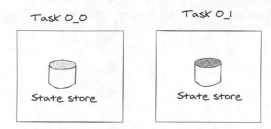

Each task is the sole owner of the
assigned store and is the only one
to read and write to it.

**Figure 7.20   Stateful tasks have
a state store assigned to them.**

7.4.4   State stores' location on the filesystem

In a stateful application, when Kafka Streams first starts up, it creates a root directory
for all state stores from the `StreamsConfig.STATE_DIR_CONFIG` configuration. If not
set, the `STATE_DIR_CONFIG` defaults to the temporary directory for the Java virtual
machine (JVM), followed by the system-dependent separator and then "kafka-streams."
For example, on my MacOS, the default root directory for state stores is /var/folders/
lk/d_9__qr558zd6ghbqwty0zc80000gn/T/kafka-streams.

> **NOTE**   The value of the `STATE_DIR_CONFIG` configuration must be unique for
> each Kafka Streams instance that shares the same filesystem.

Next, Kafka Streams appends the application ID, which you have to provide in the
configurations, to the path. Again, on my laptop, the path is /var/folders/lk/d_9__
qr558zd6ghbqwty0zc80000gn/T/kafka-streams/test-application/.

> **TIP**   To view the system-dependent temporary directory on your machine,
> you can start a Java shell from a terminal window by running the `jshell` com-
> mand. Then type in `System.getProperty("java.io.tmpdir")` and press the
> Return key, and it will be displayed on the screen.

The directory structure branches out to unique directories for each task. Kafka
Streams creates a directory for each stateful task using the sub-topology ID and partition (separated by an underscore) for the directory name. For example, a stateful
task from the first sub-topology assigned to partition 0 would use `0_0` for the directory name.

The next directory is named for the store's implementation, `rocksdb`. So, at this
point, the path would look like /var/folders/lk/d_9__qr558zd6ghbqwty0zc80000gn/
T/kafka-streams/test-application/0_0/rocksdb. It is under this directory there is the
final directory from the processor (unless provided by a `Materialized` object, which
I'll cover that soon). Let's look at the following code listing of a stateful Kafka Streams
application and the generated topology names to understand how the final directory
gets its name.

Listing 7.17   A simple Kafka Streams stateful application

```
builder.stream("input")
    .groupByKey()
    .count()
    .toStream()
    .to("output")
```

This application has a topology named accordingly.

```
Topologies:
  Sub-topology: 0
    Source: KSTREAM-SOURCE-0000000000 (topics: [input])
      --> KSTREAM-AGGREGATE-0000000002
    Processor: KSTREAM-AGGREGATE-0000000002
     (stores: [KSTREAM-AGGREGATE-STATE-STORE-0000000001])
      --> KTABLE-TOSTREAM-0000000003
      <-- KSTREAM-SOURCE-0000000000
    Processor: KTABLE-TOSTREAM-0000000003 (stores: [])
      --> KSTREAM-SINK-0000000004
      <-- KSTREAM-AGGREGATE-0000000002
    Sink: KSTREAM-SINK-0000000004 (topic: output)
      <-- KTABLE-TOSTREAM-0000000003
```

The name of the aggregate processor

The name of the store assigned to the processor

From the topology here, Kafka Streams generates the name `KSTREAM-AGGREGATE-0000000002` for the `count()` method. Notice its associated with the store named `KSTREAM-AGGREGATE-STATE-STORE-0000000001`. So Kafka Streams takes the base name of the stateful processor and appends a `STATE-STORE` and the number generated from the global counter. Now let's take a look at the full path you would find this state store: `/var/folders/lk/d_9__qr558zd6ghbqwty0zc80000gn/T/kafka-streams/test-application/0_0/rocksdb/KSTREAM-AGGREGATE-STATE-STORE-0000000001`.

So it's the final directory `KSTREAM-AGGREGATE-STATE-STORE-0000000001` in the path that contains the RocksDB files for that store. Now, if you were to check the topics on the broker after starting the Kafka Streams application, you'd see this name in the list `test-application-KSTREAM-AGGREGATE-STATE-STORE-0000000001-changelog`. This topic is the changelog for the state store; notice that Kafka Streams uses a naming convention of <application-id>-<state store name>-changelog for the topic.

7.4.5   *Naming stateful operations*

This naming raises an interesting question: What happens if we add an operation before the `count()`? Let's say you want to add a filter to exclude certain records from the counting. You'd update the topology like in the following listing.

Listing 7.18   Updated topology with a filter

```
builder.stream("input")
    .filter((key, value) -> !key.equals("bad"))
    .groupByKey()
    .count()
```

```
.toStream()
.to("output")
```

Remember, Kafka Streams uses a global counter for naming the processor nodes, so since you've added an operation, every processor downstream of it will have a new name since the number will be greater by 1. The new topology will look like the following listing.

Listing 7.19 Updated topology names

```
Topologies:
   Sub-topology: 0
    Source: KSTREAM-SOURCE-0000000000 (topics: [input])
      --> KSTREAM-FILTER-0000000001
    Processor: KSTREAM-FILTER-0000000001 (stores: [])
      --> KSTREAM-AGGREGATE-0000000003
      <-- KSTREAM-SOURCE-0000000000
    Processor: KSTREAM-AGGREGATE-0000000003
      (stores: [KSTREAM-AGGREGATE-STATE-STORE-0000000002])
      --> KTABLE-TOSTREAM-0000000004
      <-- KSTREAM-FILTER-0000000001
    Processor: KTABLE-TOSTREAM-0000000004 (stores: [])
      --> KSTREAM-SINK-0000000005
      <-- KSTREAM-AGGREGATE-0000000003
    Sink: KSTREAM-SINK-0000000005 (topic: output)
      <-- KTABLE-TOSTREAM-0000000004
```

The new name for the aggregation operation

The new name for the state store

Notice how the state store name has changed, which means there is a new directory named `KSTREAM-AGGREGATE-STATE-STORE-0000000002`, and the corresponding changelog topic is now called `test-application-KSTREAM-AGGREGATE-STATE-STORE-0000000002-changelog`.

> **NOTE** Any changes before a stateful operation could result in the generated name shift; that is, removing operators will have the same shifting effect.

What does this mean to you? When you redeploy this Kafka Streams application, the directory will only contain some basic RocksDB files but not your original contents. They are in the previous state store directory. Usually, an empty state store directory does not present a problem, as Kafka Streams will restore it from the changelog topic. Except in this case, the changelog topic is also new, so it's also empty. So, while your data is still safe in Kafka, the Kafka Streams application will start over with an empty state store due to the name changes.

While it's possible to reset the offsets and process data again, a better approach is to avoid a name-shifting situation altogether by providing a name for the state store instead of relying on the generated one. In the previous chapter, I covered naming processor nodes to provide a better understanding of what the topology does. But in this case, it goes beyond a better understanding of its role in the topology: it makes your application robust in the face of a changing topology.

Returning to the simple `count()` example in this section, you'll update the application by passing the `Materialized` object to the `count()` operation.

Listing 7.20 Naming the state store using a `Materialized` object

```
builder.stream("input")
       .groupByKey()
       .count(Materialized.as("counting-store"))    ◁──┐ Explicitly names
       .toStream()                                      │ the state store
       .to("output")
```

By providing the name of the state store, Kafka Streams will name the directory on disk `counting-store`, and the changelog topic becomes `test-application-counting-store-changelog`, and both of these names are "frozen." They will stay the same regardless of any updates you make to the topology. It's important to note that the names of state stores within a topology must be unique. Otherwise, you'll get a `TopologyException`.

> **NOTE** Only stateful operations are affected by name shifting. But since stateless operations don't keep any state, changes in processor names from topology updates will have no effect.

The bottom line is to *always* name state stores and repartition topics using the appropriate configuration object. By naming the stateful parts of your applications, you can ensure that topology updates don't break compatibility. Table 7.1 summarizes which configuration object to use and the operation(s) it applies to.

Table 7.1 Kafka Streams configuration objects for naming state stores and repartition topics

Configuration object	What's named	Where used
Materialized	State store, changelog topic	Aggregations
Repartitioned	Repartition topic	Repartition (manual by user)
Grouped	Repartition topic	groupBy (automatic repartitioning)
StreamJoined	State store, changelog topic, repartition topic	Joins (automatic repartitioning)

Naming state stores provides the added benefit of being able to query them while your Kafka Streams application is running, providing live, materialized views of the streams. I'll cover interactive queries in the next chapter.

So far, you've learned how Kafka Streams uses state stores to support stateful operations. You also learned that the default is for Kafka Streams to use persistent stores, and in-memory store implementations are available. In the next section, I will cover how you can specify different store types and configuration options for the changelog topics.

7.4.6 *Specifying a store type*

All the examples in this chapter use persistent state stores, but I've stated that you can also use in-memory stores. So the question is, how do you use an in-memory store? So far, you've used the `Materialized` configuration object to specify Serdes and the name for a store, but you can use it to provide a custom `StateStore` instance. Kafka Streams makes it easy to use an in-memory version of the available store types (so far, I've only covered "vanilla" key-value stores, but I'll get to session, windowed, and time-stamped stores in the next chapter).

The best way to learn how to use a different store type is to change one of our existing examples. Let's revisit the first stateful example used to keep track of scores in an online poker game.

Listing 7.21 Performing a reduce in Kafka Streams updated to use in-memory stores

```
KStream<String, Double> pokerScoreStream = builder.stream("poker-game",
        Consumed.with(Serdes.String(), Serdes.Double()));

pokerScoreStream
        .groupByKey()                                             Passing a StoreSupplier
        .reduce(Double::sum,                                      to specify an in-memory
                Materialized.<String, Double>as(                  store
            Stores.inMemoryKeyValueStore("memory-poker-score-store"))    <--
                    .withKeySerde(Serdes.String())
                    .withValueSerde(Serdes.Double()))       <--  Specifying the
        .toStream()                                               Serdes for the value
        .to("total-scores",
                Produced.with(Serdes.String(), Serdes.Double()));
```

Specifying the Serdes for the key

So, by using the overloaded `Materialized.as` method, you provide a `StoreSupplier` using one of the factory methods available from the `Stores` class. Notice that you still pass the Serde instances needed for the store. And that's all it takes to switch the store type from persistent to in-memory.

> **NOTE** Switching to a different store type is straightforward, so I only have one example here.

Why would you want to use an in-memory store? An in-memory store will give you faster access since it doesn't need to go to disk to retrieve values. So, a topology using in-memory stores should have a higher throughput than one using persistent ones. But there are tradeoffs you should consider.

First, an in-memory store has limited storage space, and once it reaches its memory limit, it could cause a crash from an `OutOfMemoryError`. Note that you can avoid the memory issue by using the `Stores.lruMap` method, which will evict entries when reaching its maximum configured size. The second consideration is when you stop and restart a Kafka Streams application under happy-path conditions, the one with persistent stores will start processing faster because it will have all its state already, but the in-memory stores will always need to restore from the changelog topic.

Kafka Streams provides a factory class `Stores` that provides methods for creating either `StoreSuppliers` or `StoreBuilders`. The choice of which one to use depends on the Kafka Streams API. When using the DSL, you'll use `StoreSuppliers` with a `Materialized` object. You'll use a `StoreBuilder` in the Processor API and directly add it to the topology. We'll cover the Processor API in chapter 10.

> **TIP** To see all the different store types, you can create a view of the JavaDoc for the `Stores` class (http://mng.bz/eEez).

Now that you've learned how to specify a different store type, let's move on to one more topic to cover with state stores: how to configure the changelog topic.

7.4.7 Configuring changelog topics

There's nothing special about changelog topics. You can use any configuration parameters available for topics. But the default settings suffice for the most part, so you should only consider changing the configurations when necessary.

> **NOTE** State store changelogs are compacted topics discussed in chapter 2. As you may recall, the delete semantics require a `null` value for a key, so if you want to remove a record from a state store permanently, you'll need to do a `put(key, null)` operation.

Let's revisit the previous example, where you provided a custom name for the state store. The data processed by this application also has a large key space. The changelogs in Kafka Streams are compacted topics. Compacted topics use a different approach to cleaning up older records.

Instead of deleting log segments by size or time, log segments are *compacted* by keeping only the latest record for each key—older records with the same key are deleted. However, since the key space is large, compaction may not be enough, as the size of the log segment will keep growing. In that case, the solution is simple: you can specify a cleanup policy of `delete` and `compact`.

Listing 7.22 Setting a cleanup policy using `Materialized`

```
Map<String, String> changeLogConfigs = new HashMap<>();
changeLogConfigs.put("cleanup.policy", "compact,delete");

builder.stream("input")
    .groupByKey()
    .count(Materialized.as("counting-store")
        .withLoggingEnabled(changeLogConfigs))      <-- Uses the withLoggingEnabled method to set a configuration
    .toStream()
    .to("output")
```

You can adjust the configurations for this specific changelog topic. Earlier, I mentioned that to turn off the caching that Kafka Streams uses for stateful operations, you'd set the `StreamsConfig.CACHE_MAX_BYTES_BUFFERING_CONFIG` to 0. But since it's in the configuration, it is globally applied to all stateful operations. If you only wanted to turn off the cache for a specific one, you can disable it by calling the `Materialized` `.withCachingDisabled()` method when passing in the `Materialized` object.

> **WARNING** The `Materialized` object also provides a method to turn off logging. Doing so will cause the state store not to have a changelog topic. Hence, it is subject to getting in a state where it can't restore its previous contents. Only use this method if absolutely necessary. While working with Kafka Streams, I've never encountered a good reason for using this method.

Summary

- Stream processing needs state. Stateless processing is acceptable in many cases, but you'll need to use stateful operations to make more complex decisions.
- Kafka Streams provide stateful operations that reduce, aggregate, and join. The state store is created automatically for you; by default, they use persistent stores.
- You can use in-memory stores for any stateful operation by passing a `Store-Supplier` from the `Stores` factory class to the `Materialized` configuration object.
- To perform stateful operations, your records need to have valid keys. If your records don't have a key or you'd like to group or join records by a different key, you can change it, and Kafka Streams will automatically repartition the data for you.
- It's important always to provide a name for state stores and repartition topics to keep your application resilient from breaking when you make topology changes.

The KTable API 8

This chapter covers

- Changelog streams, the `KTable`, and the `GlobalKTable`
- Aggregating records with a `KTable`
- Enriching event streams with joins
- Joining a `KTable` with another `KTable`

This chapter will introduce you to a new API in Kafka Streams, the `Ktable`. The `KTable` is an update or changelog stream. You've already used a `KTable` as any aggregation operations in Kafka Streams result in a `KTable`. The `KTable` is an essential abstraction for working with records with the same key. In a `KStream`, records with the same key are independent events. But in the `KTable`, a record updates the previous one with the same key.

Why is learning about an update stream necessary? Sometimes, you'll only care about the latest entry for a given piece of data. For example, consider a user profile. When someone updates their profile, only the newest entry is correct. All previous versions of the profile don't matter. Compared to a relational database, the event stream (a `KStream`) could be considered a series of inserts where the primary key is an auto-incrementing number. Each insert of a new record has no relationship

226

to previous ones. But with a `KTable`, the key in the key-value pair is the primary key. So, instead of inserting a new row, an update to the row results.

You'll learn how to aggregate with a `KTable`. Aggregations in the `KTable` work differently from the `Kstream` because you don't want to group by primary key. You'll only ever have one record that way. Instead, you'll need to consider how you want to group the records to calculate the aggregate.

You can use the `KTable` as a lookup table, enriching event stream records by joining them with records in the table for additional details. You can also join two tables together, even using a foreign key. You'll also learn about a unique construct called the `GlobalKTable`. The `KTable` is sharded by partition, meaning instance each only contains the data for a single partition. But the `GlobalKTable` includes all records from its underlying source topic across all application instances.

8.1   KTable: The update stream

To fully understand the concept of an update stream, it will be helpful to compare it with an event stream to see the differences between the two. Let's use a concrete example of tracking stock price updates (figure 8.1).

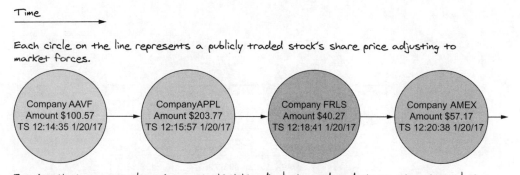

Figure 8.1   A diagram for an unbounded stream of stock quotes

You can see that each stock price quote is a discrete event, and they aren't related. Even if the same company accounts for many price quotes, you only look at them one at a time. This view of events is how the `KStream` works—a stream of records.

Now, let's see how this concept ties into database tables. Each record is an insert into the table, but the primary key is a number increment for each insert, depicted in a simple stock quote table in figure 8.2.

Next, let's take another look at the record stream. Because each record stands independently, the stream represents inserts into a table. Figure 8.3 shows these two concepts in action.

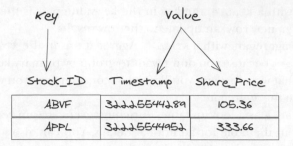

The rows from the table can be recast as key-value pairs.
For example, the first row can be represented by this key-value pair:

```
{
    key: { stock_id:ABVF },
    value: { ts: 32225544289, price: 105.36 }
}
```

Figure 8.2   A simple database table represents stock prices for companies. There's a key column, and the other columns contain values. You can consider this a key-value pair if you lump the other columns into a "value" container.

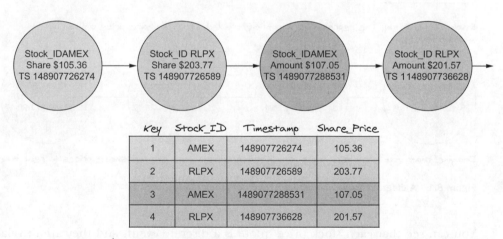

This shows the relationship between events and inserts into a database. Even though it's stock prices for two companies, it counts as four events because you consider each item on the stream as a singular event.

As a result, each event is an insert, and you increment the key by one for each insert into the table.

With that in mind, each event is a new, independent record or insert into a database table

Figure 8.3   A stream of individual events compared to inserts into a database table. You could similarly imagine streaming each row from the table.

What's important here is that you can view a stream of events in the same light as inserts into a table, which can help give you a deeper understanding of using streams for working with events. The next step is to consider the case where events in the stream *are* related.

8.1.1 Updates to records or the changelog

Let's say you want to track customer purchase behavior, so you take the same stream of customer transactions but now track activity over time. If you add a key, the customer ID, the purchase events can be related to each other, and you'll have an update stream instead of an event stream.

If you consider the stream of events as a log, you can view this stream of updates as a changelog. Figure 8.4 demonstrates this concept.

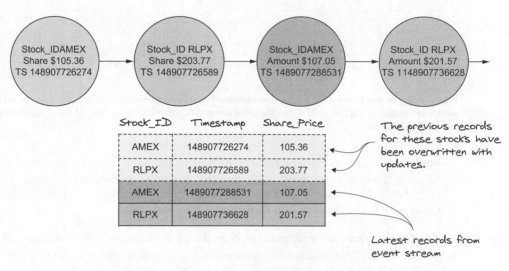

If you use the stock ID as a primary key, subsequent events with the same key are updates in a changelog. In this case, you only have two records, one per company. Although more records can arrive for the same companies, the records won't accumulate.

Figure 8.4 In a changelog, each incoming record overwrites the previous one with the same key. You'd have four events with a record stream, but for an update stream, you have only two.

Here, you can see the relationship between a stream of updates and a database table. A log and a changelog represent incoming records appended to the end of a file. In a log, you see all the records; in a changelog, you only keep the latest record for any given key.

> **NOTE** With a log and a changelog, records are appended to the end of the file as they come in. The distinction between the two is that in a log, you want to see *all* records, but in a changelog, you only want the *latest* one for each key.

To trim a log while maintaining the latest records per key, you can use log compaction, which we discussed in chapter 2. You can see the effect of compacting a log in figure 8.5. Because you only care about the latest values, you can remove older key-value pairs.

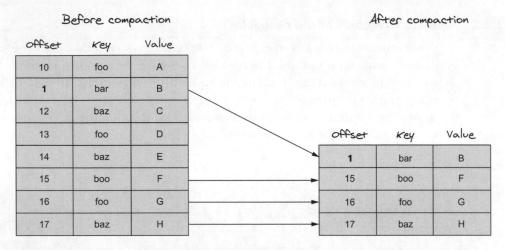

Figure 8.5   On the left is a log before compaction. You'll notice duplicate keys with different values, which are updates. On the right is the log after compaction. You keep the latest value for each key, but the log is smaller.

> **NOTE**   This section derived information from Jay Kreps's "Introducing Kafka Streams: Stream Processing Made Simple" (http://mng.bz/49HO) and "The Log: What Every Software Engineer Should Know About Real-time Data's Unifying Abstraction" (http://mng.bz/eE3w).

You're already familiar with event streams from working with the KStream. Now that we've established the relationship between streams and tables, the next step is to compare an event stream to an update stream. We'll use an abstraction known as the KTable for a changelog or stream of updates.

8.1.2   *KStream and KTable API in action*

Let's compare the KStream and the KTable. We'll do this by running a simple stock ticker application. A KStream and a KTable will read and write the records to the console via the print() method. The stock ticker will produce three iterations of stock quotes totaling nine records.

> **NOTE**   The KTable does not have methods like print() or peek() in its API, so to do any printing of records, you'll need to convert the KTable from an update stream to an event stream by using the toStream() method first.

The following code listing is an example program for printing stock ticker results to the console (found in src/main/java/bbejeck/chapter_8/KStreamVsKTableExample.java).

You can find the source code at https://github.com/bbejeck/KafkaStreamsInAction 2ndEdition).

Listing 8.1 `KTable` and `KStream` printing to the console

```
KTable<String, StockTickerData> stockTickerTable =
builder.table(STOCK_TICKER_TABLE_TOPIC);                    ← Creates the
                                                              KTable instance
KStream<String, StockTickerData> stockTickerStream =
builder.stream(STOCK_TICKER_STREAM_TOPIC);                 ← Creates the
                                                            KStream instance

stockTickerTable.toStream()
  .print(Printed.<String, StockTickerData>toSysOut()      ← KTable prints results
  .withLabel("Stocks-KTable"));                              to the console.

stockTickerStream
  .print(Printed.<String, StockTickerData>toSysOut()      ← KStream prints results
  .withLabel("Stocks-KStream"));                             to the console.
```

Using default Serdes

In creating the `KTable` and `KStream`, you didn't specify any Serdes to use. The same applies to both calls to the `print()` method. You could do this because you registered a default Serdes in the configuration like so:

```
props.put(StreamsConfig.DEFAULT_KEY_SERDE_CLASS_CONFIG,
    Serdes.String().getClass().getName());
props.put(StreamsConfig.DEFAULT_VALUE_SERDE_CLASS_CONFIG,
    StreamsSerdes.StockTickerSerde().getClass().getName());
```

If you used different types, you'd need to provide Serdes in the overloaded methods for reading or writing records.

Figure 8.6 shows the results of running the application. As you can see, the `KStream` printed all nine records. We'd expect the `KStream` to behave this way because it views each record individually. In contrast, the `KTable` printed only three records because it views records as updates to previous ones.

From the `KTable`'s point of view, it didn't receive nine individual records. The `KTable` received three original records and two rounds of updates, and it only printed the last round of updates. Notice that the `KTable` records are the same as the last three records published by the `KStream`. We'll discuss the mechanisms of how the `KTable` emits only the updates in the next section.

The takeaway is that records in a stream with duplicate keys are updates, not new records. A stream of updates is the central concept behind the `KTable`, the backbone of stateful operations in Kafka Streams.

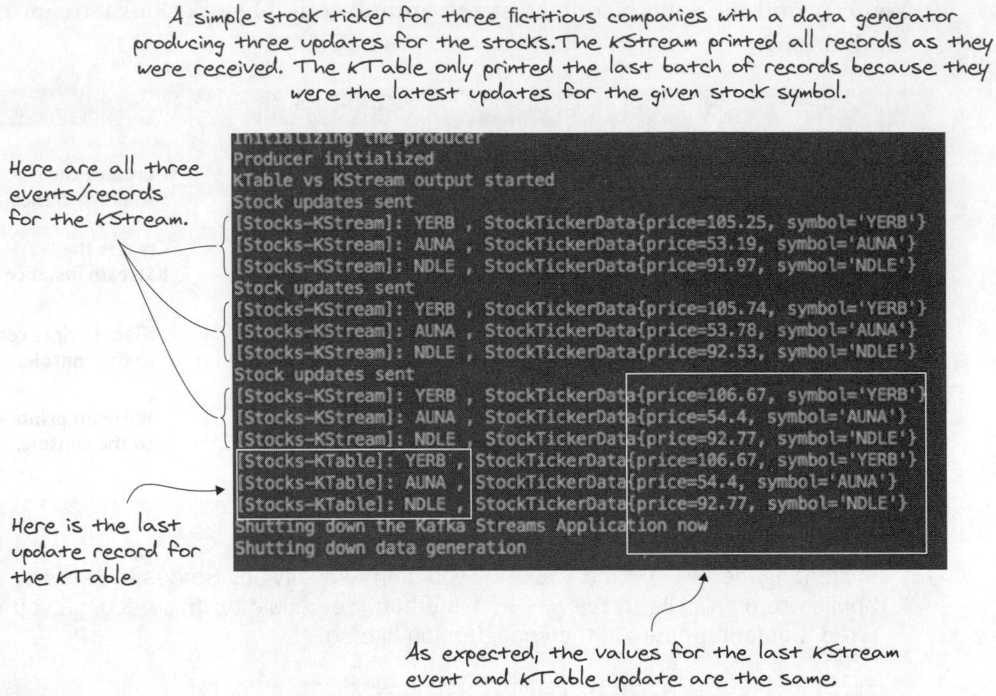

A simple stock ticker for three fictitious companies with a data generator
producing three updates for the stocks. The kStream printed all records as they
were received. The kTable only printed the last batch of records because they
were the latest updates for the given stock symbol.

Here are all three
events/records
for the kStream.

```
Initializing the producer
Producer initialized
KTable vs KStream output started
Stock updates sent
[Stocks-KStream]: YERB , StockTickerData{price=105.25, symbol='YERB'}
[Stocks-KStream]: AUNA , StockTickerData{price=53.19, symbol='AUNA'}
[Stocks-KStream]: NDLE , StockTickerData{price=91.97, symbol='NDLE'}
Stock updates sent
[Stocks-KStream]: YERB , StockTickerData{price=105.74, symbol='YERB'}
[Stocks-KStream]: AUNA , StockTickerData{price=53.78, symbol='AUNA'}
[Stocks-KStream]: NDLE , StockTickerData{price=92.53, symbol='NDLE'}
Stock updates sent
[Stocks-KStream]: YERB , StockTickerData{price=106.67, symbol='YERB'}
[Stocks-KStream]: AUNA , StockTickerData{price=54.4, symbol='AUNA'}
[Stocks-KStream]: NDLE , StockTickerData{price=92.77, symbol='NDLE'}
[Stocks-KTable]: YERB , StockTickerData{price=106.67, symbol='YERB'}
[Stocks-KTable]: AUNA , StockTickerData{price=54.4, symbol='AUNA'}
[Stocks-KTable]: NDLE , StockTickerData{price=92.77, symbol='NDLE'}
Shutting down the Kafka Streams Application now
Shutting down data generation
```

Here is the last
update record for
the kTable

As expected, the values for the last kStream
event and kTable update are the same.

Figure 8.6   `KTable` vs. `KStream` printing messages with the same keys

8.2   *KTables are stateful*

In the previous example, when you created the table with the `builder.table()` statement, Kafka Streams also creates a `StateStore` for tracking the state, and by default,
it's a persistent store. Since state stores only work with byte arrays for the keys and values, you'll need to provide the Serde instances so the store can (de)serialize the keys
and values. Just as you can provide specific Serdes to an event stream with a `Consumed`
configuration object, you can do the same when creating a `KTable`:

```
builder.table(STOCK_TICKER_TABLE_TOPIC,
           Consumed.with(Serdes.String(),
                   StockTradeSerde()));
```

Now the Serdes you've provided with the `Consumed` object get passed along to the state
store. An overloaded version of `StreamsBuilder.table` also accepts a `Materialized`
instance, allowing you to customize the type of store and provide a name to make it
available for querying. We'll discuss interactive queries later in chapter 13.

   Creating a `KTable` directly using the `KStream.toTable` method is also possible.
Using this method changes the interpretation of the records from events to updates.
You can also use the `KTable.toStream` method to convert the update stream into an
event stream. We'll discuss this conversion from an update stream to an event stream

when we discuss the KTable API in the next section. The main point is you create a KTable directly from a topic, and Kafka Streams establishes a state store supporting the KTable.

So far, I've talked about how the KTable handles inserts and updates, but what about when you need to delete a record? To remove a record from a KTable, you send a key-value pair with the value set to null. A null value acts as a tombstone marker, ultimately getting deleted from the state store and the backing changelog topic, hence deleted from the table.

Like the KStream, the KTable is spread over tasks determined by the number of partitions in the underlying source topic. This distribution by task means that the records for the table can reside on separate application instances.

8.3 The KTable API

The KTable API offers similar methods to what you'd see with the KStream—filter, filterNot, mapValues, and transformValues. Executing these methods also follows the fluent pattern; they return a new KTable instance.

While the functionality of these methods is very similar to the ones in the KStream API, there are some differences in how they operate. The differences come into play because a key-value pair where the value is null has delete semantics.

The delete semantics have the following effects on how the KTable operates:

1 If the incoming key-value pair contains a null value, the processor doesn't evaluate the record and forwards it to the new table as a tombstone marker.
2 In the case of the filter and filterNot methods, records that don't match the predicate result in a tombstone, which is forwarded to the new table.

For example, see the KTableFilterExample in the bbejeck.chapter_8 package. It runs a simple KTable.filter example where some of the incoming values are null and filters out some non-null values. But since we've discussed filtering previously, I won't review the example here, and I'll leave it up to you to do this exercise on your own.

Let's now discuss aggregations and joins with a KTable.

8.4 KTable aggregations

Aggregations in the KTable operate differently than those in the KStream. Let's illustrate this difference with an example. Imagine you build an application to track stock prices. You're only interested in any symbol's latest price, so using a KTable makes sense. Additionally, you'd like to track how different market segments perform. For example, you'd group the stocks of Google, Apple, and Confluent into the tech market segment. So, you'll need to complete an aggregation and group different stocks by their market segment. Your KTable aggregation would look like the following code listing.

Listing 8.2 Aggregates with a KTable

```
KTable<String, StockAlert> stockTable =                        Creates the
                builder.table("stock-alert",                   original KTable
                  Consumed.with(stringSerde, stockAlertSerde));

stockTable.groupBy((key, value) ->
                      KeyValue.pair(value.getMarketSegment(), value),
                    Grouped.with(stringSerde, stockAlertSerde))
          .aggregate(segmentInitializer,
                adderAggregator,
                subtractorAggregator,
                Materialized.with(stringSerde, segmentSerde))
          .toStream()

          .to("stock-alert-aggregate",
                Produced.with(stringSerde, segmentSerde));
```

Annotations: Creates the aggregate; Provides the adder Aggregator; Provides the subtractor Aggregator; Groups by the market segment and provides Serdes for the repartition via a Grouped configuration object

You create the KTable, perform a groupBy, and update the key to be the market segment, which will force a repartition. Since the original key is the stock symbol, all stocks from a given market segment may not reside on the same partition.

But this requirement somewhat hides the fact that with a KTable aggregation, you'll *always* need to perform a groupBy operation. Why? Remember that with a KTable, the incoming key is considered a primary key. Like in a relational database, grouping by the primary key always results in a single record. So, you'll need to group records by another field because combining the primary key and the grouped field(s) will yield results suitable for aggregation. Like the KStream API, calling the KTable.groupBy method returns an intermediate table—KGroupedTable, which you'll use to execute the aggregate method.

The second difference also occurs. With the KTable aggregations, just like with the KStream, the first parameter you provide is an Initializer instance to give the default value for the first aggregation. However, you then supply two aggregators, one that adds the new value and another one that subtracts the old value from the aggregation for the previous entry with the same key. Let's look at figure 8.7 to help make this process clear.

```
(key, newValue, aggr) -> {
  aggr.add(newValue);
  return aggr;               The adder adds
}                            the new value for the key
                             into the aggregation.


(key, previousValue, aggr) -> {
  aggr.subtract(previousValue);
  return aggr;               The subtractor
}                            removes the previous value for the
                             key from the aggregation.
```

Figure 8.7 KTable aggregations use an Adder aggregator and a Subtractor aggregator.

Here's another way to think about it. If you were to perform the same thing on a relational table, summing the values in the rows created by a grouping, you'd only get the latest, single value per row. For example, the SQL equivalent of this `KTable` aggregation could look something like the following listing.

> **Listing 8.3 SQL of `KTable` aggregation**

```
SELECT market_segment,
      sum(share_volume) as total_shares,
      sum(share_price * share_volume) as dollar_volume
      FROM stock_alerts
      GROUP BY market_segment;
```

When a new record arrives, the first step is to update the alerts table. Then, run the aggregation query to get the updated information. This precise process is taken by the `KTable`. The new incoming record updates the table for the `stock_alerts`, and it's forwarded to the aggregation. Since you can only have one entry per stock symbol in the rollup, add the new record into the aggregation and then remove the previous value for the given stock ticker.

This process can be challenging to understand fully, so let's clarify things by following along with some illustrations. The current state of the table is that some records have arrived, triggering the calculation of some aggregations (figure 8.8).

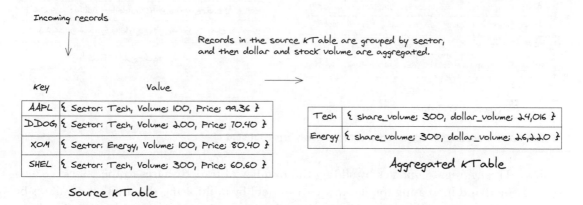

Figure 8.8 Records arrive at `KTable`, which calculates an aggregation for each one

The aggregation is summing the number of shares in each transaction and the dollar volume of the trade, which is calculated by multiplying the share price by the number of shares. Next, a new stock trade occurs (figure 8.9).

The stock trade involving CFLT arrives in the source topic and flows to the source `KTable`; since there's no previous entry for CFLT, it's an insert-only operation into the source table. Because the incoming trade involves the tech sector, that aggregation needs updating (figure 8.10).

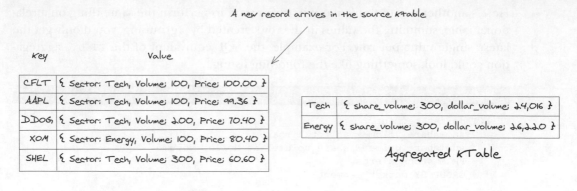

Figure 8.9 **A new trade arrives and updates the source** `KTable`

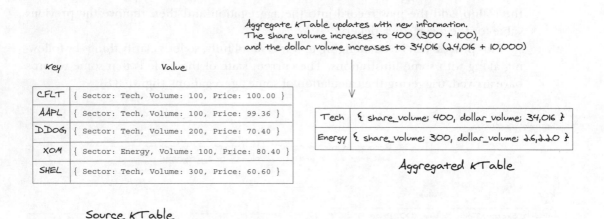

Figure 8.10 **Updating the aggregation with the newly arrived record**

The aggregate updates by adding the number of traded shares to the `share_volume` field and increasing the `dollar_volume` field by multiplying the number of shares by the price per share field. Things get more interesting when another trade involving CFLT arrives. Let's take a look at what happens in figure 8.11.

Since there's an existing record for CFLT in the `KTable`, the natural thought is to update the CFLT key with a new value. But Kafka Streams must first get the previous entry from the state store and save it to a variable; then, it can add the new value to the table. We must keep the last value because it's needed to update the downstream aggregation.

Each of the different sector aggregations contains exactly one entry for each stock it has. When a new entry arrives for the group, there's a two-step update process. First, the previous record is subtracted, and the new one is added to the aggregation. Addi-

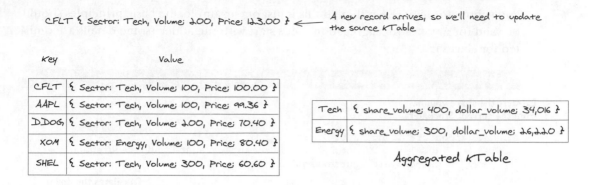

Figure 8.11 Another trade for CFLT arrives and triggers a series of updating events.

tionally, since the key returned from the `groupBy` function may have changed, Kafka Streams will forward the old and new values separately. Figure 8.12 summarizes the entire KTable aggregation process.

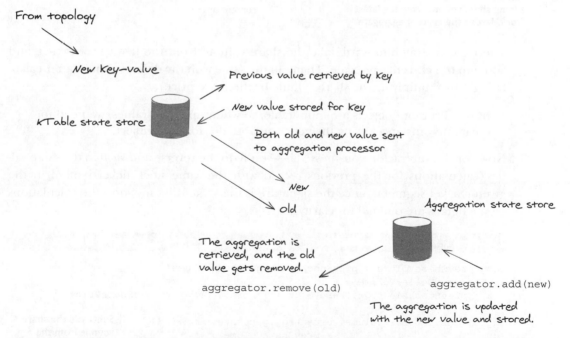

Figure 8.12 Summary of the entire KTable aggregation update process

Now that we've covered how the `KTable` aggregation works, let's look at the `Aggregator` instances. But since we've covered them in chapter 7, let's only consider the adder and subtractor logic. Even though this is just one example, the basic principles should be valid for any KTable aggregation. Let's start with the adder (some details are omitted for clarity).

Listing 8.4 KTable adder Aggregator

Extracts the share volume from the current StockAlert

```
final Aggregator<String, StockAlert,
                                   SegmentAggregate> adderAggregator =
   (key, newStockAlert, currentAgg) -> {

 long currentShareVolume =
   newStockAlert.getShareVolume();
 double currentDollarVolume =
   newStockAlert.getShareVolume() * newStockAlert.getSharePrice();

   aggBuilder.setShareVolume(currentAgg.getShareVolume() +
     currentShareVolume);
   aggBuilder.setDollarVolume(currentAgg.getDollarVolume() +
     currentDollarVolume);
 }
```

Calculates the dollar volume for the current StockAlert

Sets the total dollar volume by adding the calculated dollar volume to the current aggregate

Sets the total share volume by adding share volume from the latest StockAlert to the current aggregate

The logic is straightforward: take the share volume from the latest StockAlert and add it to the current aggregate. Then, do the same with the dollar volume (after calculating it by multiplying the share volume by the share price).

NOTE Protobuf objects are immutable, so when updating values, we need to create new instances using the generated builder for each object.

Now, for the subtractor, you guessed it—you'll do the reverse and *subtract* the same values/calculations for the previous record with the same stock ticker symbol in the given market segment. Since the signature is the same, I'll only show the calculations (some details are omitted for clarity).

Listing 8.5 KTable subtractor Aggregator

```
long prevShareVolume = prevStockAlert.getShareVolume();
double prevDollarVolume =
     prevStockAlert.getShareVolume() * prevStockAlert.getSharePrice();

aggBuilder.setShareVolume(currentAgg.getShareVolume()
                  - prevShareVolume);
aggBuilder.setDollarVolume(currentAgg.getDollarVolume()
                  - prevDollarVolume);
```

Subtracts the share volume from the previous StockAlert

Subtracts the dollar volume from the previous StockAlert

The logic is straightforward; you subtract the values from the `StockAlert` that Kafka Streams replaced in the aggregate.

In summary, the `KTable` aggregation keeps only the latest value for each unique combination of the original `KTable` key and the key used to execute the grouping. It's worth noting here that the `KTable` API also provides `reduce` and `count` methods, for which you'll take similar steps. You first perform a `groupBy` and, for the `reduce`, provide an adder and subtractor `Reducer` implementation. I won't cover them here as it's repetitive, but there are examples of both `reduce` and `count` in the source code for the book.

At this point, we've wrapped up our coverage of the `KTable` API. Still, before we move on to more advanced operations with the `KTable`, I'd like to review another table abstraction, the `GlobalKTable`.

8.5 GlobalKTable

I alluded to the `GlobalKTable` earlier in the chapter when we discussed that the `KTable` is partitioned; hence, it's distributed out among Kafka Streams application instances (with the same application ID, of course). In other words, a `KTable` only contains records from a single partition from a topic. What makes the `GlobalKTable` unique is that it entirely consumes all data of the underlying source topic. Completely consuming the topic means a full copy of all records is in the table for all application instances. Let's look at figure 8.13 to help make this clear.

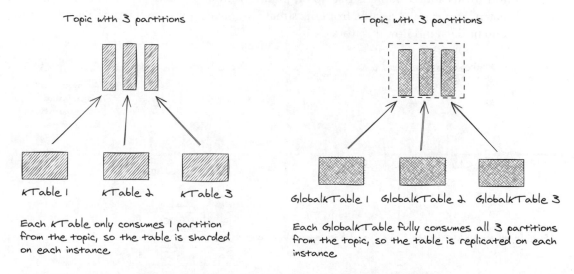

Figure 8.13 `GlobalKTable` **contains all records in a topic on each application instance.**

As you can see, the source topic for the `KTable` has three partitions, and with three application instances, each `KTable` is responsible for one data partition. But the `GlobalKTable` has the full copy of its three-partition source topic on each instance. Kafka Streams materializes the `GlobalKTable` on local disk in a `KeyValueStore`, but it does not create a changelog topic as the source topic serves as the backup for recovery. The following code listing shows you how to create a `GlobalKTable` in your application.

```
StreamsBuilder builder = new StreamsBuilder();
GlobalKTable<String, String> globalTable =
  builder.globalTable("topic",
                      Consumed.with(Serdes.String(),
                                    Serdes.String()));
```

The `GlobalKTable` does not offer an API. So the natural question is, when should you use `GlobalKTable` versus a `KTable`? The `GlobalKTable` is especially useful for broadcasting information to every Kafka Streams instance for use in joins. For example, consider you have a stream of purchases with a user ID. You can extract limited details with a sequence of characters and numbers representing the person behind the transaction.

But if you can add a name, address, age, occupation, etc., you will gain more insight into these events. Since user information doesn't change frequently (i.e., people don't change jobs or addresses weekly), a `GlobalKTable` is well suited for this reasonably static data. Since each table has a full copy of the data, it shines when used to enrich a stream of events.

Another advantage coming from the `GlobalKTable` due to its consuming all partitions of its source topic is that when joining with a `KStream`, the keys don't have to match. You can use a value from the stream to make a join. Look at the figure 8.14 to help understand how this works.

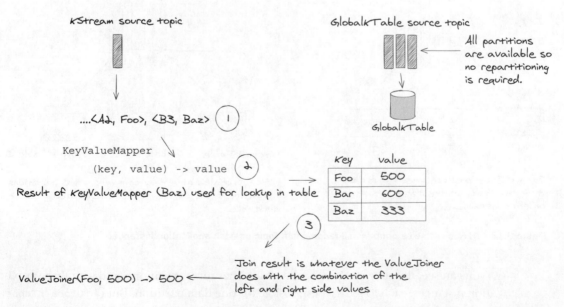

Figure 8.14   Since a `GlobalKTable` **materializes all partitions of its source topic, you can use a value from a stream for a join.**

Since the stream contains data matching the table's key in its value and the table sources all partitions, you can extract the information needed to join the table from its value. You'll see how to put this in action and enrich a stream in an upcoming section when we cover the `KStream–GlobalKTable` join in section 8.6.3.

8.6 Table joins

In the previous chapter, you learned about performing joins with two `KStream` objects, but you can also perform `KStream–KTable`, `KStream–GlobalKTable`, and `KTable–KTable` joins. Why would you want to join a stream and a table? Stream–table joins represent an excellent opportunity to create an enriched event with additional information. Both sides must be co-partitioned for the stream–table and table–table joins, meaning the underlying source topics must have the same number of partitions. Figure 8.15 gives you a view of what co-partitioning looks like at the topic level.

```
KStream<String, ClickEvent> clickStream = builder.stream("click-events");
```

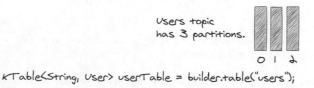

```
KTable<String, User> userTable = builder.table("users");
```

Figure 8.15 Co-partitioned topics have the same number of partitions, so `Kstream` and `KTable` instances will have tasks working with the same partition.

As you can see, the concept is nothing more than different topics having the same number of partitions. Now, let's look at a couple of illustrations to help understand why it's so important to perform a join by first looking at an example of the positive case. Let's start with figure 8.16.

From what you can see here, both keys are identical. As a result, both will land on the 0 partition, so a join is possible in this case. Next, let's look at the negative case in figure 8.17.

So even though the keys are identical, because the number of partitions differs, they will end up on different partitions, meaning a join won't work. However, this is not to say that you can't do something to enable joins. If you have `KStream` and `KTable` instances you want to join but aren't co-partitioned, you'll need to do a `repartition` operation to

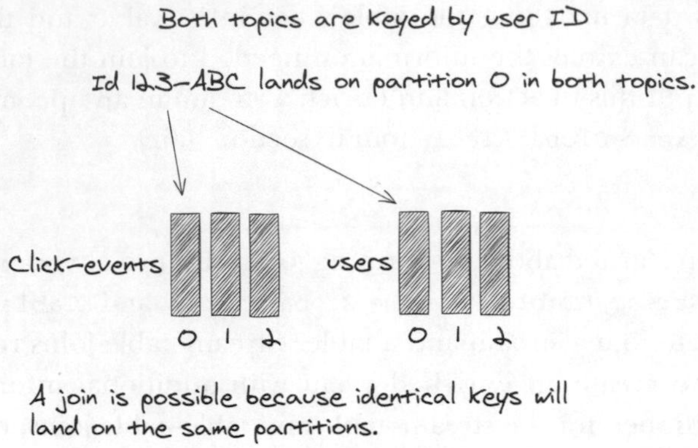

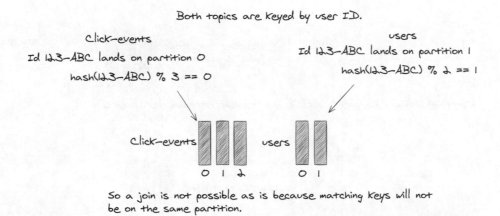

Both topics are keyed by user ID

Id 123-ABC lands on partition 0 in both topics.

Click-events   users

0 1 2   0 1 2

A join is possible because identical keys will land on the same partitions.

Figure 8.16   Co-partitioned topics can join because the primary keys share the same partition.

Both topics are keyed by user ID.

Click-events
Id 123-ABC lands on partition 0
hash(123-ABC) % 3 == 0

users
Id 123-ABC lands on partition 1
hash(123-ABC) % 2 == 1

Click-events   users

0 1 2   0 1

So a join is not possible as is because matching keys will not be on the same partition.

Figure 8.17   Topics with a different number of partitions place the same keys on different partitions.

make it possible. We'll look at an example of how to do this in section 8.6.1. Note that since the GlobalKTable has a full copy of the records, there isn't a co-partitioning requirement for stream–global table joins.

Having covered the partitioning requirements, let's give a concrete example of why you'd want to join a stream with a table. Let's say you have an event stream of user activity on a website, a clickstream, but you also maintain a table of current users logged into the system. The clickstream event object only contains a user ID and the link to the visited page, but you'd like more information. Well, you can join the clickstream against the user table, and you will have much more helpful information about the usage patterns of your site in real time. The following listing is an example to work through.

Listing 8.7	Stream–table join to enrich the event stream

```
KStream<String, ClickEvent> clickEventKStream =
            builder.stream("click-events",
                    Consumed.with(stringSerde, clickEventSerde));

KTable<String, User> userTable =
        builder.table("users",
                Consumed.with(stringSerde, userSerde));

clickEventKStream.join(userTable, clickEventJoiner)
        .peek(printKV("stream-table-join"))
        .to("stream-table-join",
                Produced.with(stringSerde, stringSerde));
```

Looking at the code in this example, you create the click-event stream and a table of logged-in users. In this case, we'll assume the stream has the user ID for the key, and the user tables' primary key is the user ID, so we can perform a join between them as is. From there, you call the `join` method of the stream passing in the table as a parameter.

8.6.1 Stream–table join details

At this point, I'd like to cover a few differences between the stream–table joins and the stream–stream join you learned about in a previous chapter. Stream–table joins aren't reciprocal: the stream is always on the left or calling side, and the table is always on the right.

Stream–table joins aren't windowed. When the stream side has a newly arriving record, Kafka Streams does a key lookup on the right side table. There's no evaluation of the timestamps involved for either side unless you're using versioned state stores, which we'll over in the next section.

To capture the join result, you provide a `ValueJoiner` object that accepts the value from both sides and produces a new value that can be the same type on either side or a new one altogether. You can perform an inner (equi) (demonstrated here) join or a left-outer join with stream-table joins.

Only newly arriving records on the stream trigger a join; new records to the table update the value for the key in the table but don't emit a new join result. Let's look at a couple of illustrations to help clarify what this means. First, figure 8.18 illustrates the `KStream` receiving an update.

When the new record arrives, Kafka Streams will do a key lookup on the `KTable` and apply the `ValueJoiner` logic, producing a join result. Now let's look at the other case, where the `KTable` receives an update (figure 8.19).

Here, you can see that when the `KTable` receives a new record, the entry for the associated key is updated, but no join action results. With this example of stream–table joins, timing is unimportant as the `KTable` has relatively static user data. But if you're working with a scenario where timing or temporal semantics are essential, you need to consider that.

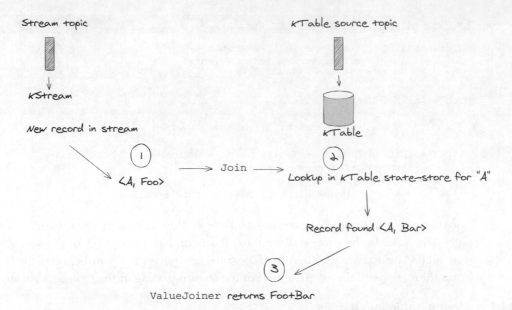

Figure 8.18 Stream–table joins produce a join result when the stream side has an update.

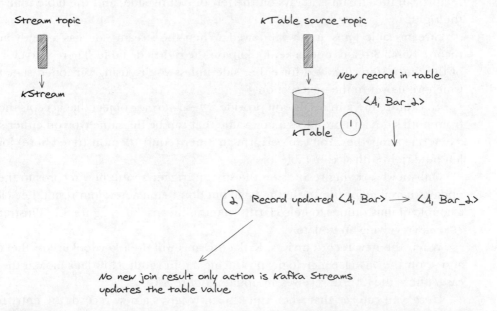

Figure 8.19 When the table updates in a stream–table join, no updated join result occurs; only the table gets updated.

8.6.2   *Versioned KTables*

Figure 8.20 demonstrates what the scenario of timing joins looks like.

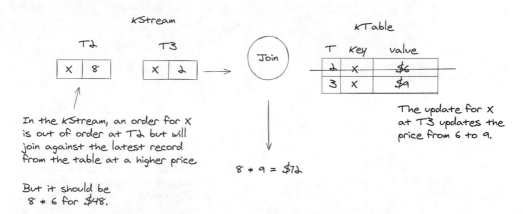

Figure 8.20   **An update to the `KTable` occurs, and a join against an out-of-order `Kstream` record applies an incorrect join result.**

The illustration shows a `KStream` with an order or a dynamically priced commodity, and the table contains the prices. When a user placed an order at time `T2`, the price was set at $6; at time `T3`, the price was updated to $9. But the `KStream` record placed at `T2` arrives out of order. As a result, a join occurs with the `T2` time order against the `T3` time price, which means the customer pays 8 * $9 = 72 instead of the expected 8 * $6 = 48.

This outcome is because when a new record arrives in a `KTable`, Kafka Streams automatically applies the update to the table. Any join with a `KStream` record will use the current corresponding record in the table. But we need to prevent the out-of-order pricing problem, which is the ability to incorporate time semantics into the join so that the stream side record's timestamp can be considered and the `KTable` will have an update from a previous timestamp. Figure 8.21 demonstrates this concept.

This illustration shows that even though the customer bid is out of order, the join uses the correct price contained in the table at time `T2`, producing the expected and correct join result. How can you put this temporal correct `KTable` into your Kafka Streams application? It's as simple as using a *versioned* state store to back the `KTable`. To use a versioned state store, you'll first create a versioned `StoreSupplier` with the following code.

Listing 8.8   Creating a versioned state store

```
KeyValueBytesStoreSupplier versionedStoreSupplier =
        Stores.persistentVersionedKeyValueStore(
                "user-details-table",
                Duration.ofSeconds(30));
```

The name of the state store ⟶

The history retention period that older records are available for joins ⟵

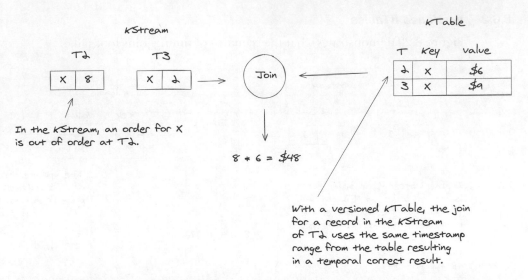

In the KStream, an order for X is out of order at T2.

$8 * 6 = \$48$

With a versioned KTable, the join for a record in the KStream of T2 uses the same timestamp range from the table resulting in a temporal correct result.

Figure 8.21 A `KTable` with historical records by timestamp can perform temporal correct joins for an out-of-order `KStream` record.

The critical point here is that the second parameter is a `Duration` object, which specifies how long you want to make older records available. There is an overload of the method that accepts a `Duration` object to determine the size of segments for storing older records.

Next, you'll need to plug the `StoreSupplier` into your `KTable` definition, as in the following code listing.

Listing 8.9 Plugging the `StoreSupplier` into the `KTable` definition

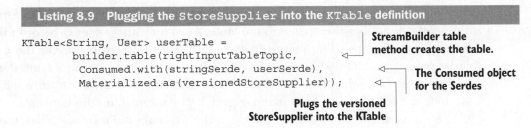

```
KTable<String, User> userTable =
      builder.table(rightInputTableTopic,
         Consumed.with(stringSerde, userSerde),
         Materialized.as(versionedStoreSupplier));
```

StreamBuilder table method creates the table.

The Consumed object for the Serdes

Plugs the versioned StoreSupplier into the KTable

With these two steps, you have enabled versioned state stores in your `KTable`, allowing for temporal correct joins.

8.6.3 *Stream–global table join details*

One main point about the example in the previous section, clickstream, is that the `KStream` had keys. But remember, it's completely valid to have a topic with `null` keys; hence, a `KStream` would also have `null` keys. In those cases, how can you perform a join against a table? Sure, you could repartition as we have seen in previous chapters, but is there a more direct approach? For the solution to that problem, let's continue to see how we can use `Stream-GlobalTable` joins.

First, it's the second join (KTable foreign key joins are the other) in Kafka Streams that does not require co-partitioning. Remember that the GlobalKTable is not sharded like the KTable but instead contains all the data of its source topic. As a result, global tables support joins with streams where the stream's key does not match the key of the global table or it doesn't exist at all. Now, let's look at how to use this information with a practical application.

Let's say you work for a manufacturer that uses IoT sensors to monitor temperature and proximity in the manufacturing process. The sensor information gets collected and produced to a Kafka topic. The data engineers didn't use any keys during the initial service setup. Later, they created an additional topic containing metadata for the different sensors. This time, the keys for this topic are the sensor ID, and the value contains the metadata.

You need to include the metadata for each sensor reading processed by your Kafka Streams app. Given the sensor records have no keys set, this situation is tailor-made for a stream–global table join. First, let's look at setting up the stream–global table join in Kafka Streams, and then we'll go into the details of each component.

Listing 8.10   `KstreamGlobalTable` **join example**

```
sensorKStream.join(sensorInfoGlobalKTable,        ◁────┐   The GlobalTable
                   sensorIdExtractor,             ◁────┤   to join against
                   sensorValueJoiner);            ◁────┘

            The ValueJoiner instance        A key selector to
            to compute the result           perform the join
```

With the KStream-GlobalKTable join, the second parameter is a KeyValueMapper that takes the key and value of the stream and creates the key used to join against the global table (in this way, it is similar to the KTable foreign-key join). The join result will have the stream's key (which could be a null value) regardless of the GlobalTable key or what the supplied function returns. Figure 8.22 illustrates the key extractor to help explain its role.

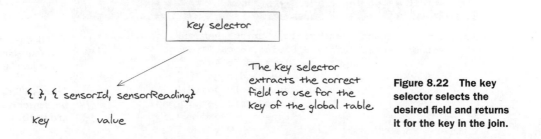

key selector

{ }, { sensorId, sensorReading}

key   value

The key selector
extracts the correct
field to use for the
key of the global table

Figure 8.22   The key selector selects the desired field and returns it for the key in the join.

You can see from this illustration the key selector knows which field from the value to return for use as the key to complete the join. The key selector can use all information available to create the key; you're not limited to using a single field on the value. Now, let's take a look at the code.

Listing 8.11   `KeyValueMapper` interface

```
KeyValueMapper<String, Sensor, String> sensorIdExtractor =
    (key, value) -> value.getId();
```

The implementation returns the ID of the sensor.

Declares the KeyValueMapper as a Java functional interface

To refresh your memory, a `KeyValueMapper` accepts two parameters, a key and a value, and returns a single object, in this case, a `String` representing the sensor ID. Kafka Streams uses this ID to search the global table for a matching record.

Now let's look into the `ValueJoiner`. First, take a look at figure 8.23, which demonstrates the concept of what it does.

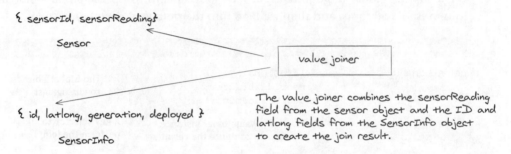

Figure 8.23   `ValueJoiner` uses the left- and right-side objects of the join to create a new result object.

This illustration shows that `ValueJoiner` accepts both records with the same key and combines them into another object—the result of the join. The following code for the `ValueJoiner` implementation shows the join result contents, which contains information about the sensor.

Listing 8.12   The `ValueJoiner` implementation

```
ValueJoiner<Sensor, SensorInfo, String> sensorValueJoiner =
    (sensor, sensorInfo) -> String.format("Sensor %s
                  located at %s
                  had reading %s",
                  sensorInfo.getId(),
                  sensorInfo.getLatlong(),
                  sensor.getReading());
```

Declares the ValueJoiner as a Java lambda

Creates a String with the sensor reading and its ID and location

---

Here, you can see the role of the `ValueJoiner`: it combines the sensor's ID, location, and the reading it recorded. You now have successfully joined a keyless `KStream` with a `GlobalKTable`, and as a result, you've added the required information to augment the sensor result.

The semantics of global table joins are different. Kafka Streams process incoming `KTable` records along with every other incoming record by its timestamp, so timestamps align the records with a stream–table join. But with a `GlobalKTable`, Kafka Streams applies updates when records are available (figure 8.24).

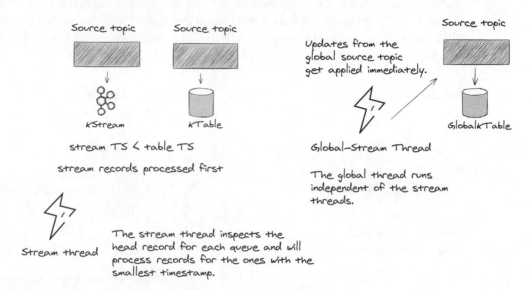

Figure 8.24   Kafka Streams has a separate thread dedicated to updating `GlobalKTables`. These updates occur outside of the normal processing of incoming records.

The update of a `GlobalKTable` is done separately from the other components of the Kafka Streams application. Kafka Streams uses a separate thread for updating any global stores or tables. So, incoming records to a `GlobalKTable` are immediately applied without considering the timestamps of those records.

Of course, every decision involves some tradeoff. A `GlobalKTable` means using more local disk space and a more significant load on the broker since it does not shard the data. The stream–global table join is not reciprocal; the `KStream` is always on the calling or left side of the join. Additionally, only updates on the stream produce a join result; a new record for the `GlobalKTable` only updates the internal state of the table. Finally, either inner or left-outer joins are available.

So what's best to use when joining with a `KStream`, `KTable`, or `GlobalKTable`? That's a tricky question, as there are flexible guidelines. But a good rule of thumb is to use a `GlobalKTable` for cases where you have relatively static lookup data you want to join with a stream. If the data in your table is extensive, strongly consider using a `KTable` since Kafka Streams will distribute it across multiple instances.

8.6.4   *Table–table join details*

Next, let's talk about table–table joins. Joins between the two tables are similar to what you've seen with the join functionality. You provide a `ValueJoiner` instance that calculates the join results and can return an arbitrary type. Also, the constraint that the source topic for both sides has the same number of partitions applies here. Joins between two tables are similar to stream–stream joins, except there is no windowing, but updates to either side will trigger a join result. So when the left-side table receives a new record, the action looks something like figure 8.25.

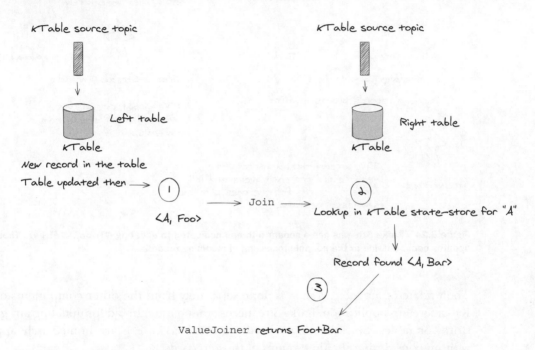

Figure 8.25   Left-side table in a table–table join is updated, and a new join result is triggered.

A new record arriving on the left table causes the left table to be updated and potentially triggers a join. The exact process occurs should the right-side table receive an update (figure 8.26). As you saw with the left-side table update process, the same steps occur when the right-side table involved in a join gets updated with a new record.

But there's a limitation with table–table joins, so let's explore that now by working through another example. Imagine now you work for a commodities trading firm and use a `KTable` that tracks the latest contract proposal for a given client. Another `KTable` has the latest prices for a given set of commodities. You need to join these two tables to have a record of the client contract and the current price of the commodity.

But there's a problem here: the client contracts table uses the client's ID as the primary key, and the commodity table uses the commodity code as its primary key. So, with different primary keys, joining is out of the question (figure 8.27).

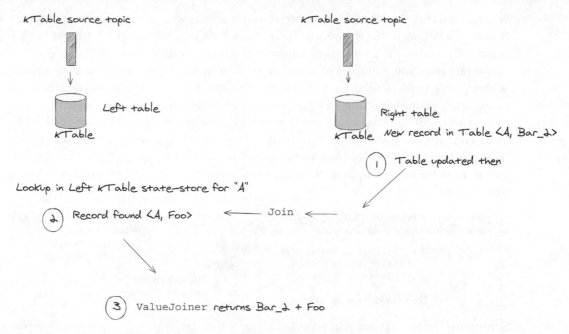

kTable source topic

kTable source topic

Left table

kTable

Right table

kTable New record in Table <A, Bar_2>

1 Table updated then

Lookup in Left kTable state-store for "A"

2 Record found <A, Foo> ← Join ←

3 ValueJoiner returns Bar_2 + Foo

Figure 8.26 Right-side table in a table–table also potentially triggers a join.

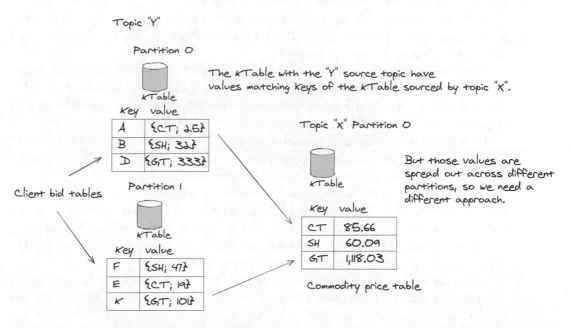

Topic "Y"

Partition 0

kTable

The kTable with the "Y" source topic have values matching keys of the kTable sourced by topic "X".

Key	value
A	{CT; 25}
B	{SH; 32}
D	{GT; 333}

Client bid tables

Partition 1

kTable

Key	value
F	{SH; 47}
E	{CT; 19}
K	{GT; 10}

Topic "X" Partition 0

kTable

But those values are spread out across different partitions, so we need a different approach.

Key	value
CT	85.66
SH	60.09
GT	1,118.03

Commodity price table

Figure 8.27 Client contracts table uses the client ID for keys, but the value contains a code that matches the key on another table containing commodity information.

You can see that the value of one table contains a field with the key of another table. But the tricky part is that the contracts are spread out on different partitions while the commodity codes, the primary key for the table, map to exactly one partition.

So far, all of the examples of joins you've seen have involved tables with the same primary key. A join is still possible if one table's value contains a field that matches the primary key of another table; in other words, you can use a foreign key join.

In this particular case, you have this precise scenario, as the client contract table value contains the code for the commodity. The `KTable` API offers a foreign key join option to join the client contract and commodity table. Implementing the foreign key join is straightforward: you use the signature of the `KTable.join` method that looks like the following listing.

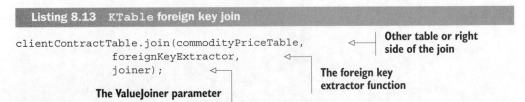

Listing 8.13   `KTable` foreign key join

```
clientContractTable.join(commodityPriceTable,                   ← Other table or right
        foreignKeyExtractor,                                      side of the join
        joiner);        ←                            The foreign key
                                                      extractor function
        The ValueJoiner parameter
```

Setting up the foreign key join is done like any other table–table join except that you provide an additional parameter, a `java.util.Function` object, that extracts the key needed to complete the join. Specifically, the function extracts the key from the left-side value to correspond with the key of the right-side table.

If the function returns `null`, no join occurs. An additional note about foreign key joins: since you're joining against the table's value on the right side, the tables involved don't need to be co-partitioned.

Inner and left-outer joins support using a foreign key. As with primary-key table joins, an update on either side will trigger a potential join result. But there are additional details we should discuss to understand how a foreign-key join works, so let's step through a series of illustrations to help you know what's going on.

In addition to extracting a key to complete the join, the contract action is fast and furious, and sometimes there are updates to the contract table, so we'll also need a mechanism to ensure we don't join with inaccurate results.

Finally, we must account for the many-to-one relationship nature of the right-side table to the left side. The value of a left table entry maps to exactly one entry on the right side, but the right-side key will map to many values on the left since it's mapping to values and not a primary key. Let's take a look at the first step—extracting the key from the value of the table on the left side (figure 8.28).

The join method extracts the key from the value in the left table using the supplied function. Next, it computes a hash of the value, which you'll see soon where this comes into play.

Then a record is produced to a repartition topic where the result of your provided function is the key, and the value contains the left-side key and the hashed value. This repartition topic is co-partitioned with the right table.

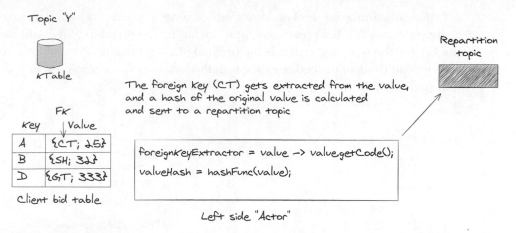

Figure 8.28 Extracting the foreign key from the value, calculating a hash of the original value, and sending both to a repartition topic

Next, let's take a look at what happens after the left side produces to the repartition topic (figure 8.29).

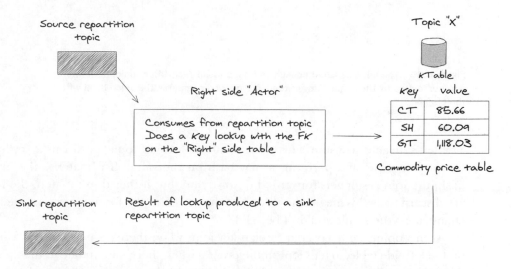

Figure 8.29 The right side of the join consumes the extracted key and performs a key lookup on the right-side table, sending results back to a result repartition topic.

We can think of the code handling the right side of the join as an actor or agent. This "agent" consumes from the "incoming" repartition topic and uses the result to perform a lookup by key on the right-side table. It also materializes the key and hash in a state store for use when there is an update to the right-side table, which you'll learn about in a subsequent step.

After obtaining the lookup result (discarding any `null` results), the right-side agent publishes another repartition topic, including the original value hash. Note that the key for the outgoing results is the original one for the left table. As with the right side, we can think of the code dealing with the left side as an agent for that part of the join (figure 8.30).

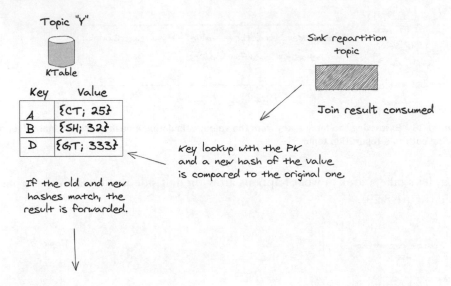

Figure 8.30 The left-side agent consumes from a result repartition topic, performs another lookup by the original key of the left side, and compares the value hash with the original.

The left-side actor consumes from the result repartition topic, performs a key lookup on the left table, and performs a new hash on the value. If it matches the original hash, the join result gets forwarded downstream, but if they don't match, the results are discarded as this means the value has changed and the foreign key has potentially changed, so the result wouldn't be valid.

At this point, we've covered foreign key joins when there's a new record or update on the left-side table, so let's look at the process when there's an update to the right-side table. The steps taken are different because we need to account for the fact that a single record in the right-side table could map to many records on the left (figure 8.31).

When the right-side table receives a new record, it will perform a prefix (consisting of only the right-side key) scan on the materialized table, matching any previously sent foreign keys from a left-side table update. Then, the right-side actor will produce all the results to the result repartition topic, and the left-side actor consumes them following the same process for handling single join results from the right-side table.

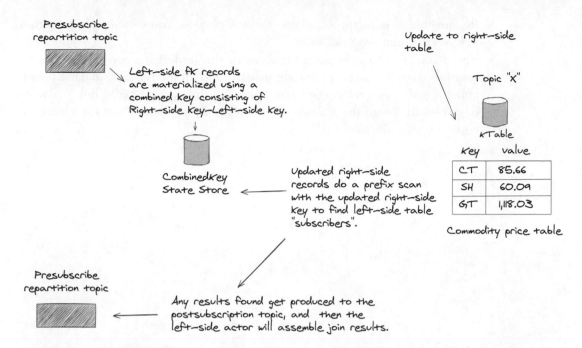

Figure 8.31 **After a right-side update, the actor scans the composite key state store and produces all matches to the result repartition topic.**

NOTE There aren't corresponding explicit foreign key joins available in the KStream API, and that is intentional. The KStream API offers the map and selectKey methods, so you can easily change the key of a stream to facilitate a join.

At this point, we've covered the different joins available on the KTable and GlobalK-Table. There's more to cover with tables, viewing the contents of the tables with interactive queries and suppressing output from a table (KTable only) to achieve a single final result. We'll cover interactive queries in chapter 13. But we'll get to suppression in our next chapter when we discuss windowing.

Summary

- The KTable is an update stream and models a database table where the primary key is the key from the key-value pair in the stream. Records with the same key are considered updates to previous ones with the same key. Aggregations with the KTable are analogous to running a Select . . . GroupBy SQL query against a relational database table.
- Performing joins with a KStream against a KTable is a great way to enrich an event stream. The KStream contains the event data, and the KTable contains the facts or dimension data.

- It's possible to perform joins between two KTables. You can also do a foreign key join between two KTables.

- The GlobalKTable contains all records of the underlying topic as it doesn't shard its data. So, each application instance has all the records, making it suitable for acting as a reference table. Joins with the GlobalKTable don't require co-partitioning with the KStream. You can supply a function that calculates the correct key for the join.

Windowing *9*
and timestamps

9

> **This chapter covers**
> - Understanding the role of windows and the different types
> - Handling out-of-order data
> - Suppressing intermediate results
> - Groking the importance of timestamps

In previous chapters, you learned how to perform aggregations with `KStream` and `KTable`. This chapter will build on that knowledge and allow you to apply it to get more precise answers to problems involving aggregations. The tool you'll use for this is windows. Using windows or windowing is putting aggregated data into discrete time buckets. This chapter teaches you how to apply windowing to your specific use cases.

Windowing is critical to apply because, otherwise, aggregations will continue to grow over time, and retrieving helpful information becomes difficult if all you have is a giant ball of facts without much context. As a high-level example, consider you're responsible for staffing a pizza shop in the student union at your university (figure 9.1). The shop is open from 11 a.m. to 5 p.m. Total sales usually amount to 20 pizzas (this is your aggregation).

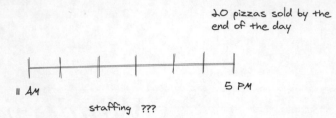

Figure 9.1   Just looking at a large aggregation doesn't give the full picture.

Determining how best to staff the shop is only possible with additional information, as you only know the total sold for the day. So you are left to guess that 20 pizzas over 6 hours amounts to roughly 3 pizzas per hour (figure 9.2), easily handled by two student workers, but is that the best choice?

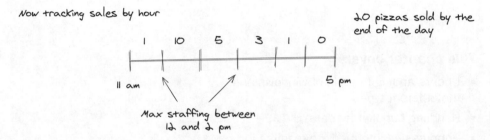

Figure 9.2   Analyzing sales by the hour (manual windowing) gives better insight into decisions.

From a course-grained aggregate, you can tell that half to three-quarters of those sales, 10 to 15 pizzas, come between 12 p.m. and 1 p.m., the classic lunch rush. By looking at your sales by the hour (a form of windowing!), your staffing needs become more evident. While this is a fictitious example, you can see how applying windowing can help get a better picture of the behavior driving an aggregation.

Now, before we move on, there are some details I should mention upfront about windowing. While both the KStream and KTable have aggregations, windowing is only available on the KStream API. This is due to the nature of how both abstractions work.

Remember, KStream is an event stream where records with the same key are considered unrelated events. Figure 9.3 gives you an idea of aggregations with an event stream.

Looking at this picture, when aggregating KStream incoming records with the same key, Kafka Streams will continue adding them, building up a larger whole. It's a grouping over a timeline. This grouping lends itself naturally to windowing; from this illustration, you can take an aggregation of 30 minutes and break it down into three 10-minute windows, and you'll get a finer-grained count of activity for every 10 minutes.

But with the KTable aggregations work a bit differently. Follow along with figure 9.4 to help you understand.

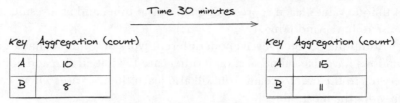

KStream aggregation are calculations by key over time — each incoming key increases the count it easily lends itself to windowing.

Figure 9.3 KStream **aggregations are groupings over time, naturally lending themselves to windowing.**

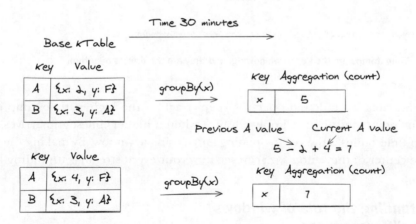

Each aggregation is a "point in time," not a grouping over time.

Figure 9.4 KTable **aggregations are point-in-time groupings and don't fit into windowing.**

Remember, with a KTable, you first do a groupBy with a field other than the primary key. But as new records arrive at the base table, Kafka Streams updates the previous entry (with the same key) with the new one and removes the old entry from the aggregation while adding the new one. So, each aggregation in a KTable is a combination of unique values, each representing a point in time, and as a result, it does not make sense to have windowing.

Additionally, you'll learn about different types of windows that are available. Some windows have fixed time sizes, and others are flexible in size and adjust based on the records in the event stream. You will also learn the use cases where each window type provides the most value.

Finally, you'll also learn about the importance of timestamps in this chapter, as they are the engine that moves windowing. Kafka Streams is an event-driven application, meaning the record's timestamps or event time drives the action in a window. Let's look at figure 9.5 to show what I'm talking about.

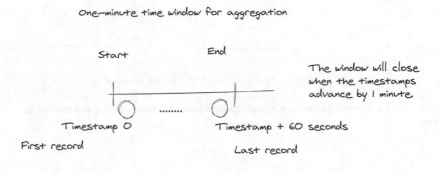

One-minute time window for aggregation

Start   End

The window will close when the timestamps advance by 1 minute.

Timestamp 0   Timestamp + 60 seconds

First record   Last record

Timestamps from the events determine the window movement.

Figure 9.5   Timestamps are the key to windowing, and they are the drivers of action.

In Kafka Streams, a windowed operator keeps track of the highest timestamp it has seen at any point, and time will only advance when a higher timestamp arrives. This moving of timestamps is the key to opening and closing a window. By making the timestamp the center of the action, you can see how your event stream is unfolding as it's happening.

9.1 Understanding the role of windows and the different types

In this section, you will work through an example to help you understand how to apply windowing to gain more actionable information. Let's say you work for a manufacturing firm producing the flux capacitor. Demand has been high since Dr. Brown successfully created a time-traveling car. As a result, you're under pressure to make as many as possible, but there's a catch.

If the temperature gets too high, the production line shuts down, causing delays. So your goal is to monitor the temperature, and if it gets too high, slow down the process, avoiding those costly production stoppages. To that end, you've installed some

IoT sensors for temperature monitoring and will begin producing their results to a Kafka topic. Next, you want to develop a Kafka Streams application to process those IoT events, with an aggregation to count when the temperature gets too high. Figure 9.6 demonstrates the concept of what you're trying to achieve.

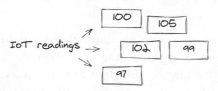

You need an application to track the temperature readings
In this case, there are 3 over 100 degrees.

IoT readings →
100
105
102
99
97

Figure 9.6 Counting the number of high-temperature events to inform you when to slow production down

So the concept is simple: create an aggregation that averages the temperature readings at their locations and records the highest. We've previously covered aggregations in KStreams in chapter 7, but it's worth a quick review here.

Listing 9.1 Creating an aggregation to track IoT temperature sensor readings

```
KStream<String,Double> iotHeatSensorStream =
     builder.stream("heat-sensor-input",
               Consumed.with(stringSerde, doubleSerde));
iotHeatSensorStream.groupByKey()
        .aggregate(() -> new IotSensorAggregation(tempThreshold),
               aggregator,
               Materialized.with(stringSerde, aggregationSerde))
        .toStream()
        .to("sensor-agg-output", Produced.with(
               stringSerde, aggregationSerde));
```

The aggregation ←

Grouping by key required for aggregating

Output of aggregation to a topic ←

This simple Kafka Stream topology does your aggregating, and you get insight into the temperatures during the manufacturing process of the flux capacitor. Although you've seen aggregators before, let's take a quick look at how this one works.

Listing 9.2 Aggregator for IoT heat sensors

```
public class IotStreamingAggregator implements
               Aggregator<String, Double, IotSensorAggregation> {
    @Override
    public IotSensorAggregation apply(String key,
                              Double reading,
                              IotSensorAggregation aggregate) {

    aggregate.temperatureSum += reading;
    aggregate.numberReadings += 1;
```

```
        if (aggregate.highestSeen < reading) {
            aggregate.highestSeen = reading;
        }
        if (reading >= aggregate.readingThreshold) {
            aggregate.tempThresholdExceededCount += 1;
        }
        aggregate.averageReading =
                aggregate.temperatureSum / aggregate.numberReadings;

        return aggregate;
    }
}
```

Everything is simple with this code. It keeps a running count of the number of readings and the sum of the readings for calculating an average and the number of times the temperature goes above the established threshold. However, you quickly notice a limitation with your current approach (depicted in figure 9.7): the aggregation numbers continue to build on previous ones.

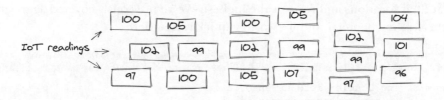

An aggregation will retain historical readings and continue to grow over time — at this point, 12 over 100 degrees.

Figure 9.7   The aggregations continue to build over time, retaining all historical information.

As you can see, it becomes hard to determine when the data shows an actual spike in temperature currently or if it is a previous one. While you could add some complexity to your aggregation code, there's a better way: add windowing to the aggregation (figure 9.8).

9.1.1   *Hopping windows*

With the addition of windowing to the aggregation, you are now segmenting your results into separate slots or windows of time. Now, let's look at the steps you'll take to add windowing (since I've covered aggregations before, I'm only going to describe the new actions).

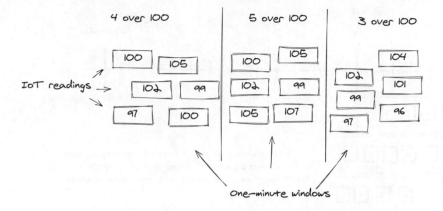

With windowing, you now break up the reading and can view the period with the highest temperatures.

Figure 9.8 By using windowing, you can separate the aggregations into discrete blocks of time.

Listing 9.3   Adding windowing to the aggregation for the Iot temperature readings

```
KStream<String,Double> iotHeatSensorStream =
  builder.stream("heat-sensor-input",
    Consumed.with(stringSerde, doubleSerde));
iotHeatSensorStream.groupByKey()                              Grouping
                                                             by key
    .windowedBy(TimeWindows.ofSizeWithNoGrace(
                  Duration.ofMinutes(1))                     Specifying the windowing—
    .advanceBy(Duration.ofSeconds(10)))                      this is what you've added
    .aggregate(
➡ () -> new IotSensorAggregation(tempThreshold),            The aggregation
      aggregator,
      Materialized.with(stringSerde, aggregationSerde))
    .toStream().to("sensor-agg-output",
      Produced.with(serdeString,
                  sensorAggregationSerde))                   Producing the aggregation
                                                             results to a topic
```

The steps you'll take are to add a `windowedBy` immediately after the `groupByKey` method. The `groupByKey` returns a `KGroupedStream`, and its API includes all the options to add windowing. After you add the `windowedBy` method, you'll need to provide its single parameter, the window instance for the aggregation. To do this, you'll use the `TimeWindows` class, which contains static factory methods for creating the window.

Now that you have some analysis, how do you want to see the data? Your team decides it would be best to see the average temperatures (plus the highest seen) for the last minute with updates every 10 seconds. In other words, every 10 seconds, you'll get temperature averages of the last minute. Figure 9.9 visualizes what this will look like.

The name for this type of windowing is called a hopping window. A hopping window has a fixed size, 1 minute in this case, but it advances or hops every 10 seconds.

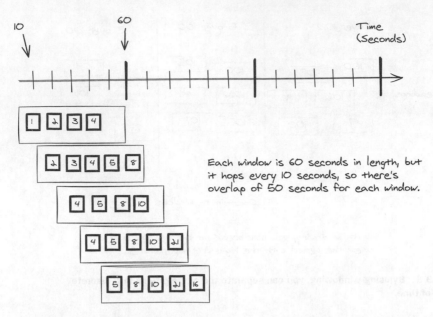

Figure 9.9   Average temperatures of the past minute, reported every 10 seconds

Since the advance is less than the size of the window, each one could have overlapping results, meaning some of the results in the previous window may be included in the next. In this case, the overlap is desirable for comparison purposes.

To achieve this hopping window, you'll use `TimeWindow.ofSizeWithNoGrace` `(Duration.ofMinutes(1))`. This specifies a window with a size of 1 minute. For now, I will defer what the `WithNoGrace` part means until we completely cover all the window types because the concept of grace applies equally to all of them. Next, to specify the advance time of 10 seconds, you'll chain an `advanceBy(Duration.ofSeconds(10))` method call immediately after defining the window.

But immediately after you update your code, you notice that you now have a compiler error at the `to("sensor-agg-output", Produced(..)` portion of the topology (or if you're not using an IDE like IntelliJ, you get an error compiling from the command line). This error is because when you do a windowed aggregation, Kafka Streams wraps the key in the `Windowed` class, which contains the key and the `Window` instance. Figure 9.10 helps you visualize what's going on.

Since the key type has changed, it doesn't match the expected type inference from the Serde used by the `Produced` configuration object. To remedy this compiler error, you can either unwrap the key from the `Windowed` instance or change the Serde type to work with the `Windowed` type and produce it to the output topic.

Before I show you the code for both solutions, the question arises: Which should you pick? The answer entirely depends on your preferences and needs. As I said before, the `Windowed` key contains not only the underlying key but the `Window`

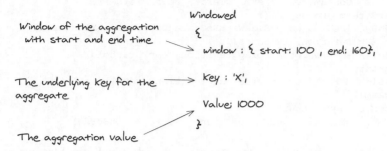

Figure 9.10 Windowed aggregations wrap the key in a `Windowed` class containing the key and the window of the aggregation.

instance for the aggregation as well and, as a result, includes the start and end time of the window, which is the very information needed to assess the effect. So, in the end, I recommend keeping the window time information contained in the `Windowed` key.

Let's first look at the first solution—unwrapping the underlying key.

Listing 9.4 Unwrapping the underlying key of a windowed aggregation

```
KStream<String,Double> iotHeatSensorStream =
      builder.stream("heat-sensor-input",
                Consumed.with(stringSerde, doubleSerde));
iotHeatSensorStream.groupByKey()
    .windowedBy(TimeWindows.ofSizeWithNoGrace(Duration.ofMinutes(1))
      .advanceBy(Duration.ofSeconds(10)))
    aggregate(() -> new IotSensorAggregation(tempThreshold),
      aggregator,
      Materialized.with(stringSerde, aggregationSerde))
      .toStream()
      .map((windowedKey, value) -> KeyValue.pair(windowedKey.key(),
      value))                                            ⟵──── Uses a map
      .to("sensor-agg-output",                                 to extract the
        Produced.with(serdeString, sensorAggregationSerde))    underlying key
```

To unwrap the underlying key, add a `map` operator, create a new `KeyValue`, and use the `Windowed.key()` method. To maintain the `Windowed` key, you'll take the steps in the following code listing.

Listing 9.5 Keeping the `Windowed` key and updating the Serde to produce to a topic

```
Serde<Windowed<String>> windowedSerdes =
        WindowedSerdes.timeWindowedSerdeFrom(String.class,   ⟵─── Parameter
                                          60_000L       ⟵───────     representing
                                          );                          the type of the
                                                                      underlying key
```

**Creates a
Serde for the
Windowed key**

The size of the window

```
KStream<String,Double> iotHeatSensorStream =
    builder.stream("heat-sensor-input",
              Consumed.with(stringSerde, doubleSerde));
iotHeatSensorStream.groupByKey()
   .windowedBy(TimeWindows.ofSizeWithNoGrace(Duration.ofMinutes(1))
     .advanceBy(Duration.ofSeconds(10)))
   aggregate(() -> new IotSensorAggregation(tempThreshold),
     aggregator,
     Materialized.with(stringSerde, aggregationSerde))
   .toStream()
   .to("sensor-agg-output",
       Produced.with(windowedSerdes, sensorAggregationSerde))
```

Providing windowSerdes to serialization to topic points to the last line.

Your first step is to use the `WindowedSerdes` class provided by Kafka Streams to create the Serde for the `Window` and the original aggregation key. When creating the Serde, you need to provide two parameters: the class of the aggregation key and the size of the window in milliseconds.

Then, you place that new Serde (`windowedSerdes`) in the `Produced.with` method as the first parameter since it's for the key. From this point forward, I may use either approach in the code examples, but my advice of maintaining the window in your results still stands.

Now that you've solved your problems and started running your new windowed aggregation application, you notice that overlapping results don't quite fit your analytical needs. You need to see which sensors are reporting higher temperatures per a unique period. In other words, you want non-overlapping results—a tumbling window.

9.1.2   *Tumbling windows*

The solution for you in this case is to make the advance time of the window the same as the size of the window; this way, you're guaranteed to have no overlapping results. Each window reports unique events—in other words, a tumbling window. Figure 9.11 illustrates how a tumbling window operates.

As you can see, since the window advances by the same amount of time as its size, each window contains distinct results; there is no overlap with the previous one. Windows with the same advance time as their size are considered tumbling windows. Tumbling windows are a particular case of hopping windows where the advance time is the same as the window size.

To achieve this in your code, you only need to remove the `advanceBy` call when creating the window. When you don't specify an advance time with a window, Kafka Streams uses a default approach of setting it to be the same as the window size. The following listing shows the updated code using tumbling windows.

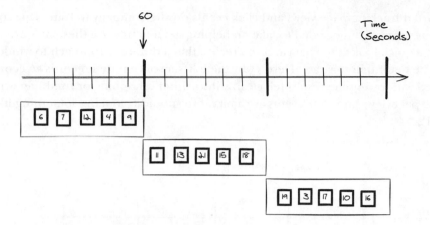

Each window is 60 seconds in length, and it
advances by 60 seconds, so there's
no overlap with a previous window.

Figure 9.11   Getting unique results per window by having the window advance by the size of the window—a tumbling window

Listing 9.6   Specifying a tumbling window by removing the `advanceBy` method

```
Serde<Windowed<String>> windowedSerdes =
        WindowedSerdes.timeWindowedSerdeFrom(String.class,
                                             60_000L
                                            );

KStream<String,Double> iotHeatSensorStream =
    builder.stream("heat-sensor-input",
                Consumed.with(stringSerde, doubleSerde));
iotHeatSensorStream.groupByKey()
   .windowedBy(TimeWindows.ofSizeWithNoGrace(Duration.ofMinutes(1))
    aggregate(() -> new IotSensorAggregation(tempThreshold),
     aggregator,
     Materialized.with(stringSerde, aggregationSerde))
     .toStream()
     .to("sensor-agg-output",
       Produced.with(windowedSerdes, sensorAggregationSerde))
```

> The window definition is now a tumbling one with the removal of the advanceBy clause

The code looks remarkably similar at this point, and it should; you only removed the `advanceBy` method, but everything else remained the same. Of course, you could keep the `advanceBy` and use the same value you did for the window size, but my advice is to leave it off in the case of a tumbling window, as it reduces any ambiguity.

Now that you're satisfied with the IoT sensor tracking approaches have developed, you move on to a new issue. The demand for the flux capacitor continues to grow. To help the business work more effectively with customers, the marketing department has started tracking page views and click events on the company website. Due to your previous success, you decided to take on helping get a handle on these new events.

But a quick look at the data and you realize that a different approach to windowing would be helpful in this case. Here's why the IoT sensors emitted records at a constant rate and windowing the results fell cleanly into either a hopping or tumbling window. But the pageviews and click events are a bit more sporadic, looking something like figure 9.12.

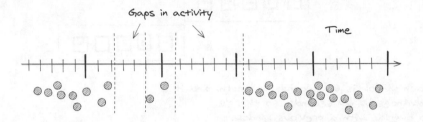

Figure 9.12   User behavior does not follow a particular pattern.

As you can see, user behavior is unpredictable and can span large or small amounts of time. It needs a different type of window that can continue to grow as long as the activity continues. Kafka Streams provides what you need—a session window.

9.1.3   *Session windows*

Session windows are different in that they will continue to grow in size as events arrive, but only up to a certain point. When there's been a gap in the activity of a defined amount of time, the session closes, and for any subsequent actions after the gap, a new session starts. Let's look at an illustration in figure 9.13 to help cement your understanding.

From this picture, you can see events arriving and extending the size of the window. However, when a gap in the activity exceeds a given amount of time, a new session starts for any newly arriving records. The inactivity gap is what makes session windows different. Instead of creating a fixed-size window, you specify how long to wait for a new activity before the session is considered closed. Otherwise, the window continues to grow.

Let's review what you need to implement a session windowing solution. You'll work with an aggregation similar to what you've done with tumbling or hopping windows.

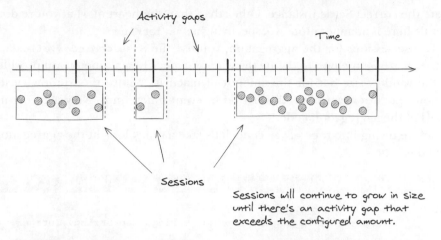

Figure 9.13 Session windows continue to grow in size up to an activity gap, and then a new session starts.

The difference is you'll need to specify using a `SessionWindow`. The `Aggregator` you provide will need a `Merger` instance and a functional interface that knows how to "merge" two sessions. It'll make more sense to look at the overall code first; then, I'll explain how you specify session windows and the concept of session merging. The following listing shows the code you'll use.

Listing 9.7 Using a `SessionWindow` for tracking pageviews by a customer

```
Serde<Windowed<String>> sessionWindowSerde =
            WindowedSerdes.sessionWindowedSerdeFrom(String.class);      ← Using a serde specific to session windows

KStream<String, String> pageViewStream = builder.stream("page-view",
                            Consumed.with(serdeString,serdeString ))
pageViewStream.groupByKey()
    .windowedBy(SessionWindow.ofInactivityGapWithNoGrace(
        Duration.ofMinutes(2))                                         ← Specifying a SessionWindow for the aggregation
    .aggregate(HashMap::new,                                           ←
        sessionAggregator,
        sessionMerger)                                                 Using a method handle for the intitializer instance
    .toStream()
    .to("page-view-session-aggregates",
        Produced.with(sessionWindowSerde, pageViewAggregationSerde))
```

Adding the aggregator with a SessionMerger

This aggregation code is similar to the others you've seen so far (the topic names are different, as are the Serdes, but those are implementation details). Note that for a `SessionWindow`, you need to use the `WindowedSerdes.sessionWindowedSerdeFrom` to create the correct Serde instance. Other than that, the heart of what you're doing differently here is at annotation two, specifying a `SessionWindow`.

To use sessions for the aggregation, you use the `SessionWindow.ofInactivityGapWithNoGrace` method to define the window type. The parameter you provide is not for the window size but for the amount of inactivity before the window closes. The window operator determines this period by event timestamps. Now, let's dig into the details of the aggregation code.

Before diving into the `Merger` code, let's take a quick look at the aggregation code since this is new.

Listing 9.8   Aggregation of pages viewed and count of times visited

```
public class PageViewAggregator
          implements Aggregator<String, String, Map<String, Integer>> {

    @Override
    public Map<String, Integer> apply(String userId,
                                      String url,
                                      Map<String, Integer> aggregate) {

        aggregate.compute(url, (key, count)
                             -> (count == null) ? 1 : count + 1);    <-
        return aggregate;
    }
}
```

Using the Map.compute method to keep a count of the number of times a user visits a page

The aggregation uses a `HashMap` and the `Map.compute` method to keep track of the pages viewed and a count of the number of times a user goes to each one in a given session.

Now, let's get into the `Merger` object. The `Merger` interface is a single abstract method that accepts three parameters, the aggregation key, and two aggregates. The two parameters are the current aggregate and the next aggregate to combine/aggregate into the overall aggregation for the session window. When Kafka Streams executes the `Merger.apply` method, it takes the two aggregates and combines or merges them into a new single aggregation. The following listing provides the code for our pageview session merger implementation.

Listing 9.9   Merging two session aggregates into a new single one

```
public class PageViewSessionMerger
          implements Merger<String, Map<String, Integer>> {

    @Override
    public Map<String, Integer> apply(String aggKey,
                                      Map<String, Integer> mapOne,
                                      Map<String, Integer> mapTwo) {

        mapTwo.forEach((key, value)->
            mapOne.compute(key, (k,v) -> (v == null) ? value : v + value
```

```
            ));
        return mapOne;
    }
}
```

The merger's action is simple: combining the pageview tracking `HashMap` contents into one that contains the combined information for both. The reason for merging session windows is that an out-of-order record could connect two older sessions into a single larger one. The process of merging sessions is interesting and is worth explaining. Figure 9.14 illustrates this session merging process.

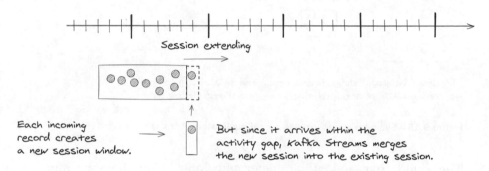

Figure 9.14   Merging session windows combines two sessions into one.

As you can see, each incoming record results in a new session (a `SessionWindow`) with a start and end time based on that record's timestamp. Kafka Streams then searches for all session windows for the given key with a timestamp range of current record time-stamp—`window.inactivityGap` to current record timestamp + `window.inactivity-Gap`. This search will likely yield one session window since previous records within the inactivity gap would have merged all previous windows.

It's important to note that since Kafka Streams fetches previous sessions by time, it guarantees that the merger is applied in order of arrival, meaning it computes merged aggregates in time order.

9.1.4   Sliding windows

You can now track user behavior with a `SessionWindow`, but you'd like to use one more type of analysis. You would also like to view the click events within 10 seconds of each other in a 1-minute window because it's essential to observe how users navigate to different pages on the site before purchasing. You can accomplish this with a `Sliding-Window` in Kafka Streams. It combines the characteristics of `TimeWindows` in that it's fixed in size and `SessionWindows` because the window start and stop times are determined by record timestamps. But you want this done continuously over the stream of records, as demonstrated in figure 9.15.

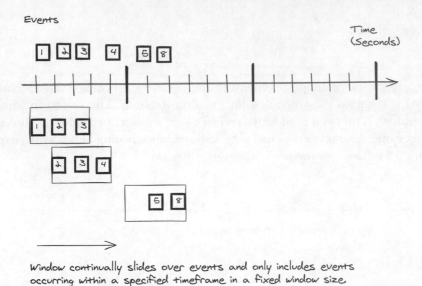

Figure 9.15   **Sliding windows provide a continuous view of changes in events.**

The window start and end are determined only by the timestamps on the records, but the size of the window remains fixed at 30 seconds. While you could simulate a sliding window by creating a hopping window of 30 seconds with a 1-millisecond advance, it would be very inefficient. I'll explain why after we review how you implement a sliding window.

You'll use the `SlidingWindows.ofTimeDifferenceWithNoGrace` method to implement sliding windows. It takes one parameter, the maximum time difference between records, expressed with a `Duration` object.

In sliding windows, the start and end times are inclusive, unlike hopping and tumbling windows, where the start is inclusive but the end time is exclusive. Another difference is that sliding windows "look back" for other invents to include in their windows. Let's look at another illustration in figure 9.16 to help you understand.

Let's step through this drawing: record 1 arrives, and it creates a new window with an ending of its timestamp and a start of timestamp minus the maximum record difference. But there are no events within 10 seconds, so it's by itself. Now, record 2 arrives, creating a new window, and this time, since record 1's timestamp is within the time difference, it's included in the window. So, with a sliding window, each incoming record results in a new window, and it will "look back" for other events with timestamps within the time difference to include. Otherwise, the code you'll write is the same as you've done for the other aggregations. Listing 9.10 shows the complete code example for a sliding window (some details are omitted for clarity).

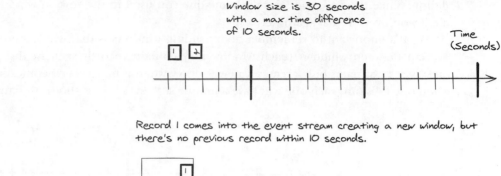

Window size is 30 seconds with a max time difference of 10 seconds.

Time (Seconds)

Record 1 comes into the event stream creating a new window, but there's no previous record within 10 seconds.

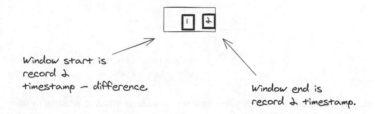

Window start is record 1 timestamp — difference.

Window end is record 1 timestamp.

Record 2 arrives, creating a new window, but it also includes record 1 since it occurred within the defined time difference.

Window start is record 2 timestamp — difference.

Window end is record 2 timestamp.

Figure 9.16   Sliding windows starting and ending times are inclusive and look back for events occurring within the defined time difference.

Listing 9.10   Page view aggregation with sliding windows

```
KStream pageViewStream = builder.stream("page-view",
                         Consumed.with(serdeString,pageViewSerde))
pageViewStream.groupByKey()
   .windowedBy(SlidingWindows.ofTimeDifferenceWithNoGrace(
      Duration.ofSeconds(30))                              ◁⎯ Specifies a SlidingWindow
    .aggregate(HashMap::new,                                     for the aggregation
pageViewAggregator)                                       ◁⎯
   .toStream()                                                 Adds the Aggregator instance
   .to("page-view-sliding-aggregates",
      Produced.with(windowedSerdes, pageViewAggregationSerde))
```

`SlidingWindows.ofTimeDifferenceWithNoGrace(Duration.ofSeconds(30))` is the central part of the code described in the previous paragraph, setting the window type to sliding. Since the `Aggregator` is the same one you used in the session window example, I won't review that again here.

What's important to remember about sliding windows is that while there may be overlap between windows (each window may contain records seen in the previous one), Kafka Streams only creates a new window when a new record comes into it or when one falls out, each one will have a unique set of results, as shown in figure 9.17.

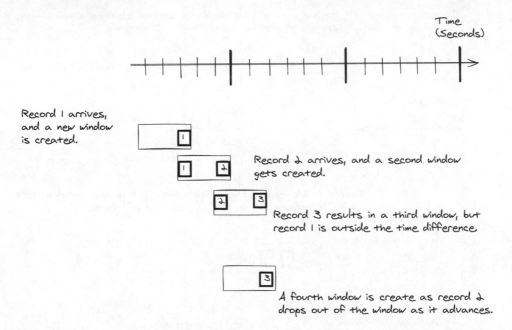

Figure 9.17   Sliding windows evaluate only when new records come into the window or when they fall out.

Following along in the illustration, the first record arrives, creating a new window. A second record comes, which again begins a new window and includes the first record. Then, a third record arrives, creating a third window. But since record 1 is outside the time difference, only record 2 is included in this latest window. Then, as time advances, the second record falls out, generating a fourth window.

Earlier in this section, I mentioned why a sliding window is highly preferred over a hopping window with minimal advanced times. Figure 9.18 helps with the explanation.

A sliding window only evaluates the aggregation when a new record enters or leaves the window; this is why each sliding window contains unique results. So, even though it advances by 1 millisecond, Kafka Streams only evaluates it when the contents of the window change.

But a hopping window will *always* evaluate the contents. Figure 9.18 shows that over time, the sliding window only performs an evaluation when a new record arrives

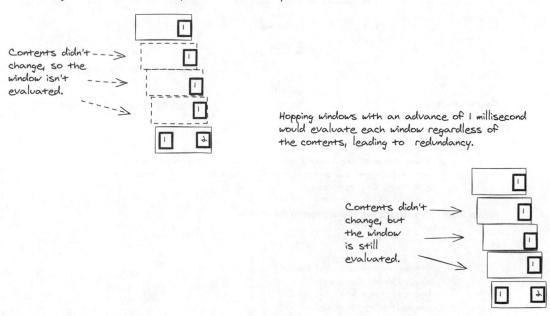

In sliding windows advance by 1 millisecond, but
it only evaluates the window when the contents
change — so there's overlap but no redundancy.

Contents didn't --->
change, so the
window isn't   --->
evaluated.

Hopping windows with an advance of 1 millisecond
would evaluate each window regardless of
the contents, leading to redundancy.

Contents didn't --->
change, but
the window   --->
is still
evaluated.   --->

Figure 9.18   Sliding windows only evaluate when the contents of the window change, but a hopping window always evaluates the contents.

or leaves the window. Meanwhile, the hopping window equivalent would have evaluated (aggregated) *N* times even though the content only changed once, leading to redundant calculations.

This wraps up our coverage of the window types available in Kafka Streams. Before we move on, I'd like to include a final summary discussion to cover additional details that will help you build windowed applications.

9.1.5   *Window time alignment*

First, let's discuss how the different windows align with time, meaning how Kafka Streams determines the starting and ending times. Any window created with the `TimeWindows` class aligns to the epoch. Let's look at two illustrations to help us understand what this means. First, figure 9.19 defines what "aligned to the epoch" means.

Windows aligned to the epoch means the first window starts with `[0, window-size)` until it reaches the current time. It's important to realize these windows are *logical*; there aren't that many window instances in Kafka Streams. That's what's meant by "windows align to the epoch." Time is represented in window-sized intervals based on the Unix epoch time (time elapsed from January 1, 1970). Now let's look at figure 9.20, which shows how this works in practice.

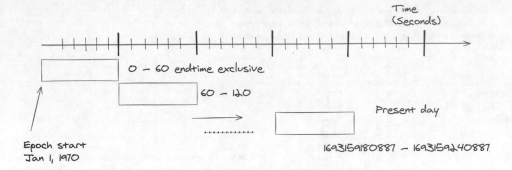

Time windows are aligned to the epoch meaning a 60-second window starts at time 0 and advances every 60 seconds. Start times are inclusive, but end times are exclusive.

Figure 9.19   `TimeWindows` **are aligned to the epoch.**

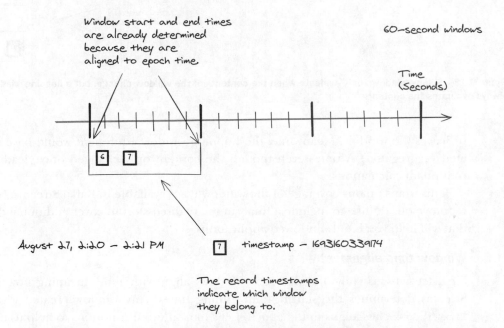

Figure 9.20   **With** `TimeWindows`**, the epoch time dictates the window.**

So, the event timestamps of records in tumbling or hopping windows don't determine the window's start or end but which one they belong to.

Now, let's take a look at sliding and session windows along with figure 9.21 to help us walk through the concept.

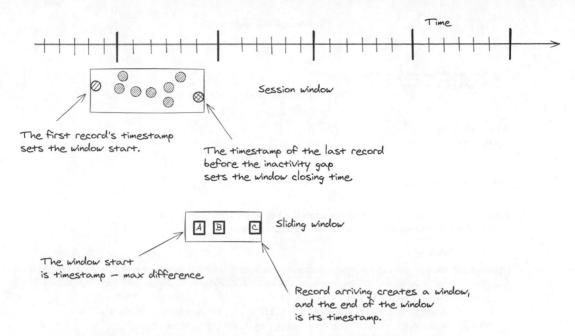

Figure 9.21   Event timestamps determine the start and end of sliding and session windows.

As you can see, event timestamps drive session and sliding windows. Even though `SlidingWindows` has a fixed size, the window start aligns with the event timestamps. The same is true of `SessionWindows`. The first record's timestamp sets the window start time. The end time of a sliding window is the specified max time difference plus the timestamp of the record creating the window. For a session window, the end time is the timestamp of the last record received preceding the inactivity gap.

9.1.6   Retrieving window results for analysis

Let's now take a moment to cover how to measure windowed aggregations. I mentioned this before in this chapter, but it's worth repeating. When you specify a windowed aggregation, Kafka Streams wraps the key with a `Windowed` instance that contains the key and a `Window` object, which includes the start and end time of the window. Let's review figure 9.22, which demonstrates a windowed key.

Since within the `Windowed` key, you have access to the start and end times of the window, in conjunction with the aggregation value, you have everything you need to analyze your aggregation within a given timeframe. Now, I'll show you some basic steps you can take to review the windowed results. Let's use the `TumblingWindow` example. First, we'll look over the aggregation code in listing 9.11, which only shows the aggregation itself (some details are omitted for clarity).

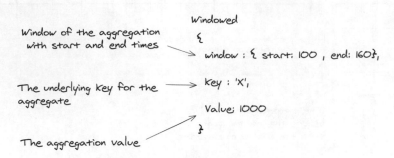

Figure 9.22 Windowed aggregations wrap the key in a `Windowed` class that contains the key and the `Window` of the aggregation.

Listing 9.11 `TumblingWindow` aggregation

```
iotHeatSensorStream.groupByKey()
    .windowedBy(TimeWindows.ofSizeWithNoGrace(Duration.ofMinutes(1))
    .aggregate(() -> new IotSensorAggregation(tempThreshold),
        aggregator,
        Materialized.with(stringSerde, aggregationSerde))
```

Remember, aggregation operations return a `KTable`. This makes sense as you'll want a newer aggregation result to replace the previous one for a given key (for a review, see chapter 8). The `KTable` API doesn't offer any way for you to view its contents, so the first step is to convert to a `KStream` using the `toStream()` method (some details are omitted for clarity).

Listing 9.12 Converting the aggregation `KTable` to a `KStream`

```
iotHeatSensorStream.groupByKey()
    .windowedBy(TimeWindows.ofSizeWithNoGrace(Duration.ofMinutes(1))
    .aggregate(() -> new IotSensorAggregation(tempThreshold),
        aggregator,
        Materialized.with(stringSerde, aggregationSerde))
    .toStream()        ◁──── Converts the aggregation
                              result KTable to a KStream
```

Now that you have a `KStream` object, you can use its API to access the results in several ways. Your next step is to add the `peek` method providing a lambda for the `ForEach-Action` interface that the method expects as a parameter.

Listing 9.13 Using `peek` to set up printing the start and end time of the window

```
iotHeatSensorStream.groupByKey()
        .windowedBy(TimeWindows.ofSizeWithNoGrace(Duration.ofMinutes(1))
        . aggregate(() -> new IotSensorAggregation(tempThreshold),
```

```
                    aggregator,
                    Materialized.with(stringSerde, aggregationSerde))
        .toStream()
        .peek((windowedKey, aggregation) -> {          ◁——┤ Declares the peek
                                                            method on the KStream
```

At this point, you've only declared the key-value pair provided as the parameters for the peek processor, but I'd like to point out the value in giving clear names for the key and value. Here, you've named the key windowedKey as that accurately describes the key object as an instance of the Windowed class; we covered this earlier in the chapter. The value is appropriately named aggregation because that's an accurate description. While the naming here is a minor point, it's helpful for others (or yourself after stepping away from the code!) to quickly understand what the key and value represent.

Next, get the Window object from the key and extract its starting and ending times. To do this, you will add code to pull the Window object first. Then you'll get the starting and ending times of the window, as shown in the following code listing (some details are omitted for clarity).

Listing 9.14 Printing windowed aggregation results for analytic purposes

```
iotHeatSensorStream.groupByKey()
        .windowedBy(TimeWindows.ofSizeWithNoGrace(Duration.ofMinutes(1)))
        . aggregate(() -> new IotSensorAggregation(tempThreshold),
                    aggregator,
                    Materialized.with(stringSerde, aggregationSerde))
        .toStream()
        .peek((windowedKey, aggregation) -> {
                Window window = windowedKey.window();       ◁——┤ Extracts the Window
                Instant start =                                 object from the key
                        window.startTime()
                                .truncatedTo(ChronoUnit.SECONDS);
                Instant end =
                        window.endTime()
                                .truncatedTo(ChronoUnit.SECONDS);
            })
        })
```

Window starts as Instant. ┤—————▷ (points to `window.startTime()`)

Window ends as Instant. ┤—————▷ (points to `window.endTime()`)

You've added the retrieval of the Window from the key, and you get the starting time with the Window.startTime() method, which returns a java.time.Instant object. We've also cleaned up the time by truncating the milliseconds from the time with the Instant.truncatedTo method. Finally, let's complete the code by logging the window start and end with the aggregation value.

Listing 9.15 Adding the log statement

```
iotHeatSensorStream.groupByKey()
        .windowedBy(TimeWindows.ofSizeWithNoGrace(Duration.ofMinutes(1)))
        . aggregate(() -> new IotSensorAggregation(tempThreshold),
                    aggregator,
                    Materialized.with(stringSerde, aggregationSerde))
        .toStream()
```

```
.peek((windowedKey, aggregation) -> {
    Window window = windowedKey.window();
    Instant start =
            window.startTime()
                    .truncatedTo(ChronoUnit.SECONDS);
    Instant end =
            window.endTime()
                    .truncatedTo(ChronoUnit.SECONDS);
    LOG.info("Window started {} ended {} with value {}",    ◁
            start,
            end,                                    Adds a log statement for
            aggregation);                          displaying the window open
        })                                        and close along with the value
    })
```

This final step adds the logging to display a view of the window start and end times, along with the aggregation value for that window. This simple example with KStream .peek prints the start and end of the window in a human-friendly format along with the aggregation. However, this is a good starting point for using your imagination to build your analytic feature.

One consequence of leaving an aggregation key in its original format is that you'll have to share the Serde with any application looking to consume the aggregation data. Additionally, having the window start and ending times in the key will alter the partition assigned to the record. This means records with the same key will end up on different partitions due to the different window times in the Windowed key. It's a good idea to partition the outgoing records by the underlying key. One approach is to implement a StreamPartitioner that will determine the partition for the aggregate result by the underlying key. The following code listing shows this process in action.

Listing 9.16   Determining correct partition by the underlying key

```
@Override
public Optional<Set<Integer>> partitions(String topic,
                                         Windowed<K> windowedKey,
                                         V value,
                                         int numPartitions) {
    if(windowedKey == null) {
        return Optional.empty();
    }
    byte[] keyBytes = keySerializer.serialize(topic,
                                             windowedKey.key());
    if (keyBytes == null) {
        return Optional.empty();
    }
    Integer partition =
    ➥ Utils.toPositive(Utils.murmur2(keyBytes)) % numPartitions;

    return Optional.of(Collections.singleton(partition));
}
```

I've left out some details in this listing but full example of this `StreamPartitioner` implementation can be found in the book source code at bbejeck.chapter_9.bbejeck .chapter_9.partitioner.WindowedStreamsPartitioner.java. To use the custom partitioner you'd add it to the `Produced` configuration object like in the following listing.

```
WindowedStreamsPartitioner<String, IotSensorAggregation>
➥ windowedStreamsPartitioner =
➥ new WindowedStreamsPartitioner<>(stringSerde.serializer());

....

  .to(outputTopic, Produced.with(
                      windowedSerdes, aggregationSerde)
➥  .withStreamPartitioner(windowedStreamsPartitioner));
```

Again I'm skipping some details in this listing. The full Kafka Streams example can be found in bbejeck.chapter_9.tumbling.IotStreamingAggregationStreamPartitioner TumblingWindows.java demonstrating using a `StreamPartitioner` to partition windowed aggregations by the original key.

However, the custom partitioner approach still leaves the window information in the key and the aggregation in the value. This is a "leaky abstraction" because consuming applications must know the key is a `Windowed` type. You can map the window time into the aggregation value, and then you'll have all the information in one object. To perform the mapping you'll first update the `IotSensorAggregation` object to have two new fields of type `long`, maybe named `windowStart` and `windowEnd`. Then you'll provide an implementation of a `KeyValueMapper`, which will pull the window time information from the key add, it to value, and return a `KeyValue` object with the original key and the updated aggregation value.

```
public class WindowTimeToAggregateMapper implements
➥ KeyValueMapper<Windowed<String>,IotSensorAggregation,
➥ KeyValue<String, IotSensorAggregation>> {
 @Override
 public KeyValue<String, IotSensorAggregation>
➥ apply(Windowed<String> windowed,
        IotSensorAggregation iotSensorAggregation) {

      long start = windowed.window().start();
      long end = windowed.window().end();
      iotSensorAggregation.setWindowStart(start);
      iotSensorAggregation.setWindowEnd(end);
      return KeyValue.pair(windowed.key(), iotSensorAggregation);
   }
}
```

Extracts the window start time

Extracts the window end time

Sets the start time on the aggregation

Sets the window end time on the aggregation

With the `KeyValueMapper` implemementation completed, you'll add it to the topology with a `KStream.map` operator, as in the following listing (some detail are omitted for clarity).

Listing 9.19   Adding the mapper

```
KeyValueMapper<Windowed<String>, IotSensorAggregation,
   KeyValue<String, IotSensorAggregation>> windowTimeMapper =
                       new WindowTimeToAggregateMapper();

iotHeatSensorStream.groupByKey()
        .windowedBy(TimeWindows.ofSizeWithNoGrace(Duration.ofMinutes(1))
        . aggregate(() -> new IotSensorAggregation(tempThreshold),
                    aggregator,
                    Materialized.with(stringSerde, aggregationSerde))
        .toStream()
        .map(windowTimeMapper)
        .to(outputTopic, Produced.with(
                     stringSerde, aggregationSerde));
```

Mapping the window times into the aggregation object and replacing the Windowed key with the original one.

Now your outgoing aggregation records will have the window time embedded in the aggregation and plain key.

Before we move on, let's take a look at table 9.1, which highlights what you've learned about the different window types.

Table 9.1

Name	Window Alignment	Fixed Size	Use Case
Hopping	epoch	Yes	Measures changes every *x* over last *y*
Tumbling	epoch	Yes	Measures events per time period
Sliding	Event timestamps	Yes	Captures changes over the continuous sliding window/rolling averages
Session	Event timestamps	No	Behavior events occurring within_ x_ of each other

Now, let's move on to areas related to all window types, handling out-of-order data and emitting only a single result at the end of the window.

9.2   *Handling out order data with grace—literally*

You learned in chapter 4 that the `KafkaProducer` will set the timestamp on a record when producing it to the broker. In a perfect world, the timestamps on the records in Kafka Streams should always increase. The reality of distributed systems is that anything can and will happen. Kafka producers communicate with brokers over a network connection, which makes them susceptible to network partitions disrupting the produce requests sent to the broker. Consider the situation in figure 9.23.

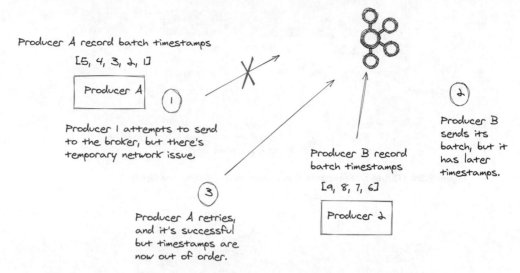

Producer A record batch timestamps

[5, 4, 3, 2, 1]

Producer A   ①

Producer 1 attempts to send
to the broker, but there's
temporary network issue

③

Producer A retries,
and it's successful
but timestamps are
now out of order.

Producer B record
batch timestamps

[9, 8, 7, 6]

Producer 2

②

Producer B
sends its
batch, but it
has later
timestamps.

Figure 9.23   Network partition causing out-of-order records between two producers

In this picture, producer A batches up a record to send to the broker, slightly ahead of producer B on another host, but right when producer A attempts to send the batch, there's a network partition for producer A, and it can't send its produce request. Since producer B is on another host, it has no connectivity problems, and its produce request proceeds.

The network disconnects for producer A and lasts about 10 seconds (which is no problem as the producer will keep trying to send up to the `delivery.timeout` expiration, which is 2 minutes by default). Still, eventually, the produce request goes through. But now the timestamps are out of order.

The producers set the event timestamp when it accepts a `ProducerRecord` from the `Producer.send` method, so any significant delay in completing the batch to the broker could result in timestamps being out of order. Because Kafka handles records in offset order, the potential for having later-arriving records with earlier timestamps is a reality.

In other cases, you may use a timestamp embedded in the record value payload. In that scenario, you can't guarantee that those timestamps are always in order since the producer is not in control of them. Of course, the network partition situation described in the previous paragraph also applies here.

Let's take a look at figure 9.24, which graphically demonstrates the concept of what out-of-order data is.

An out-of-order record is simply one where its timestamp is less (earlier) than the preceding one. Returning to a windowed aggregation, you can see how this out-of-order data can come into play. Figure 9.25 shows out-of-order data and how it relates to windows.

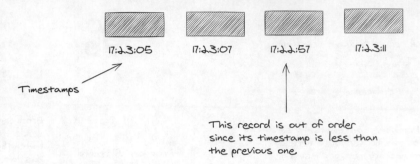

Figure 9.24   Out-of-order records didn't arrive in the correct sequence.

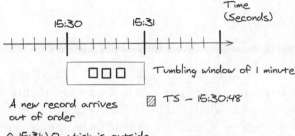

Figure 9.25   Out-of-order data could miss a window it would have made had its arrival been in order.

Being out of order means Kafka Streams excludes records that should have been in the window but aren't because the window closed.

Well, fear not because Kafka Streams has a mechanism to account for out-of-order data called Grace. Earlier in this chapter, you saw methods named `TimeWindows.ofSizeWithNoGrace` or `TimeWindows.ofSizeAndGrace`, and this is exactly how you apply (or not) Grace to windowing. Grace is the amount of time after a window closes that you're willing to allow an out-of-order record into the aggregation. Figure 9.26 demonstrates the concept.

From looking at the illustration, Grace allows records into an aggregation that Kafka Streams would have included were they on time and allows for a more accurate calculation. Once the grace period has expired, any out-of-order records are considered late and dropped. For the cases where you want to exclude any out-of-order data, you would use the `WithNoGrace` variants of any window constructor.

There are no actual guidelines on whether you should allow a grace period for your windows or how long to set it; you'll need to consider that on a case-by-case basis. But remember that Grace is a way to ensure you get the most accurate calculations by including records that arrive out of order.

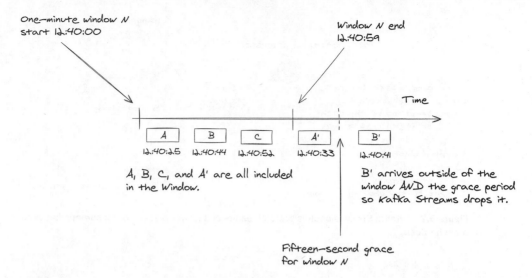

Figure 9.26   Grace is the amount of time you'll allow out-of-order records into a window after its configured close time.

You now have windowing of your aggregations, and you feel terrific about the insights and information your application provides. But you've discovered that it would be easier to analyze user behavior on the flux capacitor site and achieve more reliable results when you get the results for a single closed session, a final result.

If you remember from chapter 7 on KStream aggregations, Kafka Streams caches the results and emits updates when committing (every 30 seconds) or when it flushes the cache. This behavior applies to windowed aggregations as well. But given that windows have a defined start and end, you may want a single, final result when the window closes, and we'll cover that in the next section.

9.3   Final windowed results

As you've learned before (but it's worth reviewing again), Kafka Streams doesn't automatically forward each update for stateful operations like aggregations. Instead, Kafka Streams caches the results. Only when Kafka Streams commits or the cache is full will it forward the latest calculation from the stateful operation and persist the records to the state store. Let's review this process in figure 9.27.

Aggregation results will continue to build up in the cache, and at some point, the results are forwarded downstream to the other processors. At this point, Kafka Streams also persists the records to the state store and changelog topic.

This workflow is true for all stateful operations, including windowed ones. As a result, you'll observe intermediate results from a windowed operation, as shown in figure 9.28.

Even though your window has some time left before it's considered closed, it will emit an updated result when a new record arrives. But sometimes, receiving a final

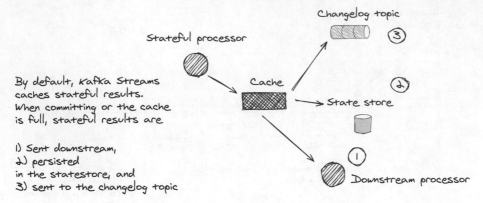

By default, Kafka Streams
caches stateful results.
When committing or the cache
is full, stateful results are

1) Sent downstream,
2) persisted
in the statestore, and
3) sent to the changelog topic

Figure 9.27   Kafka Streams caches stateful results and forwards them when committing or on a cache flush.

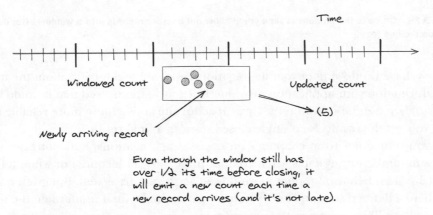

Figure 9.28   Windowed operations will emit intermediate results even when the window is still open.

result when the window closes for a windowed operation is desirable. Figure 9.29 is a visual of what it means for a final windowed result.

This illustration depicts a final result, which is just as it sounds: it emits no updates until the window is closed. I mentioned stream time in the figure, and we still need to cover it. For now, it's enough to know that's how Kafka Streams keeps time internally; we'll get to stream time in section 9.6.

NOTE   Final results are only available for windowed operations. With an event streaming application like Kafka Streams, the number of incoming records is infinite, so there's never a point where we can consider something final. However, since a windowed aggregation represents a discrete period, the available record when the window closes can be viewed as a final result for that time interval.

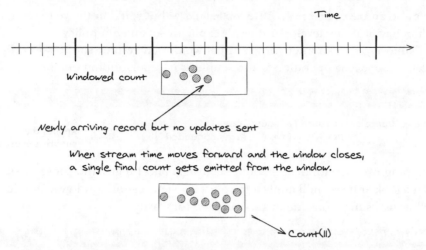

Figure 9.29   Only a single result is emitted when a window closes.

You have two choices in Kafka Streams to obtain a single result for windowed aggregations. One is to use the relatively recently introduced `EmitStrategy` or the `KTable` `.suppress` operation. The `EmitStrategy` approach is easier to understand, so I'll cover that first.

You've seen from our windowing examples when you want to add a window to an aggregation, you add a `windowedBy(..)` operation immediately after the `KStream` `.groupByKey()` method. It looks like the following code listing.

Listing 9.20   Adding windowing after the `groupByKey` call

```
iotHeatSensorStream.groupByKey()
            .windowedBy(TimeWindows.ofSizeWithNoGrace(
                                Duration.ofMinutes(1)))
```

This tiny bit of code is from the previous tumbling window example in listing 9.6. If you didn't want any intermediate results, you would set the `EmitStrategy` for that window to only emit on closing, as shown in the following listing (some details are omitted for clarity).

Listing 9.21   Setting the emit strategy to when the window closes

```
iotHeatSensorStream.groupByKey()
            .windowedBy(TimeWindows.ofSizeWithNoGrace(
                                Duration.ofMinutes(1)))
            .emitStrategy(EmitStrategy.onWindowClose())
            .aggregate(...)
```

Sets the emit strategy to window closing ⊢→

With that one line of code (`.emitStrategy(EmitStrategy.onWindowClose())`), you've set your tumbling window aggregation to only emit a result when the window closes.

There's also an `EmitStrategy.onWindowUpdate()`, but you don't have to set it; it is the default behavior if the developer doesn't explicitly set an emit policy.

Let's move on to another form of final windowed results: suppression. Enabling the `KTable.suppress` operation is also simple as it entails adding one line of code.

Listing 9.22   The suppression operator

```
.suppress(Suppressed.untilWindowCloses(
            StrictBufferConfig.unbounded())))    ⟵┤  Applies the
                                                    suppress operator
```

I'm skipping over several details about `suppress`, and I'll get into those next, but for now, let's look at how you'll apply it to your pageview session windowed application in the following listing (some details are omitted for clarity).

Listing 9.23   Adding suppression to the pageview session aggregation

```
KStream pageViewStream = builder.stream("page-view",
                            Consumed.with(serdeString,pageViewSerde))
pageViewStream.groupByKey()
                                                 Uses a session window
        .windowedBy(SessionWindow                  on the aggregation
            .ofInactivityGapWithNoGrace(
                            Duration.ofMinutes(2)))    ⟵┤
        .aggregate(HashMap::new,
                 sessionAggregator,
                 sessionMerger)
        .suppress(Suppressed.untilWindowCloses(unbounded()))
        .toStream()
        .to("page-view-session-aggregates",
             Produced.with(windowedSerdes, pageViewAggregationSerde))
```

Adds suppression to achieve a final result of the session window → (points to `.suppress` line)

You'll notice that the `suppress` operator accepts a parameter; the configuration class `Suppressed` and `Suppressed` itself receive a parameter named `BufferConfig`. Now you know why and how you can suppress intermediate windowed results. But I've thrown some new material at you, so let's get into the details of suppression's operation and configuration.

Suppression in Kafka Streams works by buffering the intermediate results, per key, until the window closes and there's a final result. This buffering occurs in-memory, so right away, you can see some tension between the two tradeoffs. You could eat up all the free memory if you fully retain records for all keys until a window closes. On the other hand, you could release some results before the window closes to free up memory at the cost of nonfinal results.

With these two scenarios in mind, there are two main options to consider about using suppression in Kafka Streams: strict or eager buffering. Let's discuss each option, starting with the strict option. Figure 9.30 helps you understand how it works.

With strict buffering, results are buffered by time, and the buffering is strictly enforced by never emitting a result early until the time bound is met. Now, let's go over what eager buffering is, along with a pictorial description in figure 9.31.

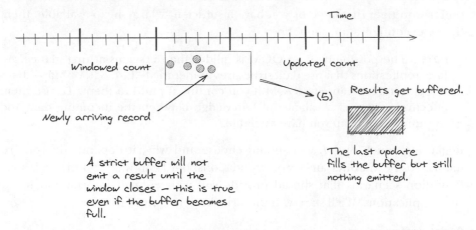

Figure 9.30   Strict buffering retains all results and never emits one before a window closes.

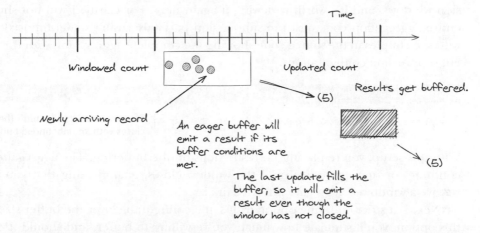

Figure 9.31   Eager buffering retains results up to the configured number of records or bytes and will emit results before the window closes.

With the eager buffering approach, results are buffered by size (number of bytes) or by number of records. When these conditions are met, the window operator will emit results downstream. This will reduce the number of downstream records but doesn't guarantee a final one.

The strict approach guarantees a final result and no duplicates. Eager buffering could also produce a final result, with the likelihood of emitting a few intermediate results. You can think of the tradeoff this way: with strict buffering, the buffer isn't bounded, so the possibility of getting an OutOfMemory (OOM) exists. Still, with eager buffering, you'll never hit an OOM exception, but you could end up with multiple results. While the possibility of incurring an OOM may sound extreme, if you feel the buffer won't get that large or you have a sufficiently large heap available, then using the strict configuration should be OK.

> **NOTE** The possibility of an OOM is milder than it seems at first glance. All Java applications that use data structures in-memory—List, Set, or Map—have the potential to cause an OOM if you continually add to them. To use them effectively requires a balance of knowledge between the incoming data and the amount of heap you have available.

Regarding which option you should choose and why, it depends on your requirements. So, I can't offer any recommendations, but I can walk you through some configuration scenarios that should provide you with enough information to apply to your applications. We'll start with the strict buffering options.

9.3.1 Strict buffering

Suppression windowed results with an unbounded buffer are something you've seen before. It's the suppression setting you used for our example in section 9.1.3 on session windows, and it's worth reviewing it again here. For clarity, I will not show the entire aggregation code, just the suppression part (also using static imports); there will be examples in the source code showing the complete aggregation with the different suppression configurations.

Listing 9.24 Suppressing all updates until the window closes with an unbounded buffer

```
.suppress(untilWindowCloses(unbounded()))     ⟵─┤ Suppresses all results until the window
                                                   closes with an unbounded buffer
```

With this setup, you're placing no restrictions on the buffering. The aggregation will continue to suppress results until the window closes, guaranteeing that you'll only receive a windowed aggregate when it's final.

   Next is a strict buffering option with some configurable over the buffer size. With this option, you'll stipulate how much you're willing to buffer, and should it exceed your established limits, your application will go through a controlled shutdown.

   You can configure the maximum number of records or bytes to store for the buffer constraints. The following listing shows the code for a max buffer size.

Listing 9.25 Setting up suppression for the final result, controlling the potential shutdown

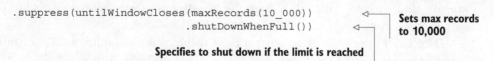

```
.suppress(untilWindowCloses(maxRecords(10_000))     ⟵       Sets max records
                          .shutDownWhenFull())       ⟵       to 10,000
```
Specifies to shut down if the limit is reached

Here, you're specifying the maximum number of records is 10,000, and should the buffering exceed that number, the application will shut down gracefully. A Buffer-Config.maxBytes option works similarly, except you specify total bytes instead of a record count.

But shutting down, even a controlled one, is a tough choice. We'll move to eager buffering next for those who favor having a possible early result (meaning there could be duplicates) versus shutting down.

9.3.2   *Eager buffering*

With eager buffering, the tradeoff is on the other side of the coin. You'll never experience an application shutdown, but your aggregation could emit a nonfinal result. This example has a couple of different concepts, so let's review figure 9.32 first to help understand how eager buffering works.

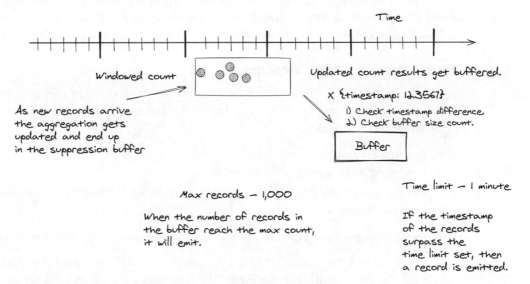

Figure 9.32   With eager buffering, you specify how long to wait for a new record to arrive before emitting a result.

What's different here is you specify how long to wait for a new record before emitting a result. If a new record does arrive, it will replace the existing one, but it does not reset the timer. You also specify a maximum size for the buffer, so should it fill up before the time has expired, the suppression operator forwards a record. Now, let's look at the code you'll use.

Listing 9.26   Using suppression with eager buffering

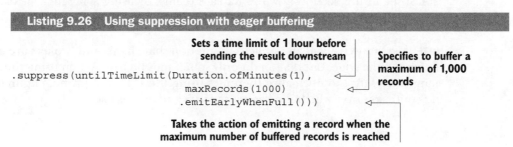

What's interesting about using eager buffering is that you can set the time limit to match the window size (plus any grace period), so depending on the volume of records, you may get a final result when the window closes. But since you're using an eager buffering strategy, should the number of records exceed the buffer size, the processor will forward a record regardless of whether the time limit is reached. Again, if you want the time limit to correspond to the window closing, you need to include the grace period, if any, in the time limit.

This wraps up our discussion on the suppression of aggregations in Kafka Streams. Even though the examples in the suppression section only demonstrated using the KStream and windowed aggregations, you could apply the same principle to nonwindowed KTable aggregations by using the time limit API of suppression.

Now, let's move on to the last section of this chapter, timestamps in Kafka Streams.

9.4   *Timestamps in Kafka Streams*

Earlier, in chapter 4, we discussed timestamps in Kafka records. In this section, we'll discuss the use of timestamps in Kafka Streams. Timestamps play a role in key areas of Kafka Streams functionality:

- Joining streams
- Updating a changelog (KTable API)
- Deciding when the Processor.punctuate() method is triggered (Processor API)
- Window behavior

With stream processing in general, you can group timestamps into three categories:

- *Event time*—A timestamp set when the event occurred, usually embedded in the object used to represent the event. We'll consider the timestamp set when the ProducerRecord is also created as the event time.
- *Ingestion time*—A timestamp set when the data enters the data processing pipeline. You can consider the timestamp established by the Kafka broker (assuming a configuration setting of LogAppendTime) to be ingestion time.
- *Processing time*—A timestamp set when the data or event record starts flowing through a processing pipeline.

Figure 9.33 illustrates these categories.

You'll see in this section how the Kafka Streams, by using a TimestampExtractor, allows you to choose which timestamp semantics you want to support.

> **NOTE**   So far, we've had an implicit assumption that clients and brokers are in the same time zone, but that might only sometimes be true. When using timestamps, it's safest to normalize the times using the UTC time zone, eliminating confusion over which time zones brokers and clients use.

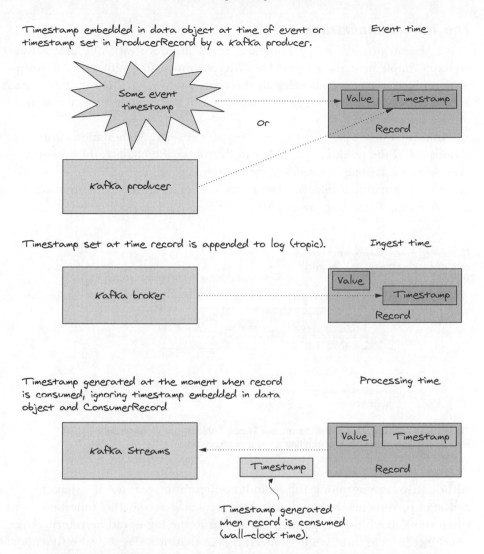

Figure 9.33   There are three categories of timestamps in Kafka Streams: event time, ingestion time, and processing time.

In most cases using event-time semantics, the timestamp placed in the metadata by the `ProducerRecord` is sufficient. But there may be cases when you have different needs. Consider these examples:

- You're sending messages to Kafka with events that have timestamps recorded in the message objects. There's some lag time when these event objects are made available to the Kafka producer, so you want to consider only the embedded timestamp.
- You want to consider the system time when your Kafka Streams application processes records instead of using the records' timestamps.

9.5   *The TimestampExtractor*

Kafka Stream provides a `TimestampExtractor` interface with one abstract and four concrete implementations to enable different processing semantics. If you need to work with timestamps embedded in the record values, create a custom `Timestamp-Extractor` implementation. Let's briefly look at the included extractors and implement a custom one.

Almost all of the provided `TimestampExtractor` implementations work with timestamps set by the producer or broker in the message metadata, thus providing either event-time processing semantics (timestamp set by the producer) or log append–time processing semantics (timestamp set by the broker). Figure 9.34 demonstrates pulling the timestamp from the `ConsumerRecord` object.

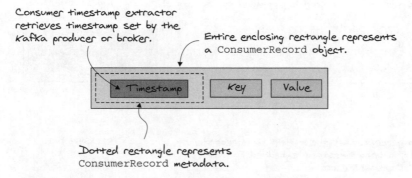

Figure 9.34   Timestamps in the `ConsumerRecord` object: the producer or broker sets this timestamp, depending on your configuration.

Although you're assuming the default configuration setting of `CreateTime` for the timestamp, remember that if you used `LogAppendTime`, the timestamp value for when the Kafka broker appended the record to the log would be returned. `Extract-RecordMetadataTimestamp` is an abstract class that provides the core functionality for extracting the metadata timestamp from the `ConsumerRecord`. Most of the concrete implementations extend this class. Implementors override the abstract method, `ExtractRecordMetadataTimestamp.onInvalidTimestamp`, to handle invalid timestamps (when the timestamp is less than 0).

Here's a list of classes that extend the `ExtractRecordMetadataTimestamp` class:

- `FailOnInvalidTimestamp`—Throws an exception for an invalid timestamp.
- `LogAndSkipOnInvalidTimestamp`—Returns the invalid timestamp and logs a warning message that Kafka Streams will discard the record due to the invalid timestamp.
- `UsePartitionTimeOnInvalidTimestamp`—In the case of an invalid timestamp, the new timestamp comes from the current stream-time.

We've covered the event-time timestamp extractors, but there's one more provided timestamp extractor to cover.

9.5.1   *WallclockTimestampExtractorSystem .currentTimeMillis() method*

WallclockTimestampExtractor provides process-time semantics and doesn't extract any timestamps. Instead, it returns the time in milliseconds by calling the System .currentTimeMillis() method. You'd use the WallclockTimestampExtractor when you need processing-time semantics.

That's it for the provided timestamp extractors. Next, we'll look at how you can create a custom version.

9.5.2   *Custom TimestampExtractor*

To work with timestamps (or calculate one) in the value object from the Consumer-Record, you'll need a custom extractor that implements the TimestampExtractor interface. For example, let's say you are working with IoT sensors, and part of the information is the exact time of the sensor reading. Your calculations need a precise timestamp, so you'll want to use the one embedded in the record sent to Kafka, not the one the producer set.

Figure 9.35 depicts using the timestamp embedded in the value object versus one set by Kafka (either producer or broker).

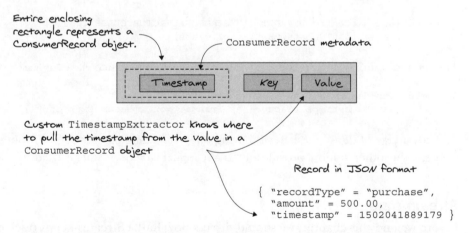

Figure 9.35   A custom TimestampExtractor provides a timestamp based on the value contained in the ConsumerRecord. This timestamp could be an existing value or one calculated from properties in the value object.

The following code listing is an example of a TimestampExtractor implementation (found in src/main/java/bbejeck/chapter_9/timestamp_extractor/Transaction-TimestampExtractor.java).

Listing 9.27   Custom `TimestampExtractor`

```
public class TransactionTimestampExtractor implements TimestampExtractor {

    @Override
    public long extract(ConsumerRecord<Object, Object>
                   consumerRecord,
                   long partitionTime) {
        PurchaseTransaction purchaseTransaction =
                   (PurchaseTransaction) consumerRecord.value();
        return purchaseTransaction.transactionTime();
    }
}
```

Retrieves the PurchaseTransaction object from the key-value pair sent to Kafka

Returns the timestamp recorded at the point of sale

In the join example, you used a custom `TimestampExtractor` to obtain the timestamps of the actual purchase time. This approach allows you to join the records even if delivery delays or out-of-order arrivals occur.

9.5.3   Specifying a TimestampExtractor

Now that we've discussed how timestamp extractors work, let's tell the application which one to use. You have two choices for specifying timestamp extractors.

The first option is to set a global timestamp extractor specified in the properties when setting up your Kafka Streams application. If no property is set, the default setting is `FailOnInvalidTimestamp.class`. For example, the following code would configure the `TransactionTimestampExtractor` via properties when setting up the application:

```
props.put(StreamsConfig.DEFAULT_TIMESTAMP_EXTRACTOR_CLASS_CONFIG,
        TransactionTimestampExtractor.class);
```

The second option is to provide a `TimestampExtractor` instance via a `Consumed` object:

```
Consumed.with(Serdes.String(), purchaseSerde)
        .withTimestampExtractor(new TransactionTimestampExtractor()))
```

The advantage of doing the latter is having one `TimestampExtractor` per input source, whereas the other option provides a `TimestampExtractor` instance used application-wide.

9.6   Stream time

Before we end this chapter, we should discuss how Kafka Streams keeps track of time while processing—that is, by using stream time. Stream time is not another category of timestamp; it is the current time in a Kafka Streams processor. As Kafka Streams selects the next record to process by timestamp, the values will increase as processing continues. Stream time is the largest timestamp seen by a processor and represents the current time for it. Since a Kafka Streams application breaks down into tasks and a task is responsible for records from a given partition, the value of stream time is not global in a Kafka Streams application; it's unique at the task (hence, partition) level.

Stream time only moves forward, never backward. Out-of-order records are always processed, except for windowed operations depending on the grace period, but its timestamp does not affect stream time. Figure 9.36 shows how stream time works in a Kafka Streams application.

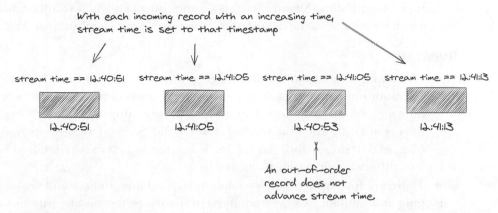

Figure 9.36   Stream time represents the highest timestamp seen so far and is the current time of the application.

As the illustration shows, the current time of the application moves forward as records go through the topology, and out-of-order records still go through the application but do not change stream time.

Stream time is vital for the correctness of windowed operations, as a window only advances and closes as stream time moves forward. If the source topics for your application are bursty or have a sporadic sustained volume of records, you might encounter a situation where you don't observe windowed results. Let's look at figure 9.37, a pictorial representation, to help understand what's going on.

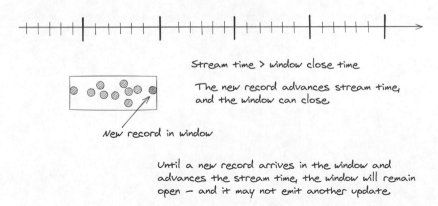

Figure 9.37   Stream time drives the behavior of windows, so low activity means you may not observe window updates.

This apparent lack of processing is because there haven't been enough incoming records to move stream time forward to force window calculations.

Timestamps' effect on operations in Kafka Streams is essential to remember when testing applications, as manually adjusting the value of timestamps can help you drive valuable tests to validate behavior. We'll talk more about using timestamps for testing in chapter 14 on testing. Stream time also comes into play when you have punctuations, which we'll cover in the next chapter when we discuss the Processor API.

Summary

- Windowing is a way to calculate aggregations for a given time. Like all other operations in Kafka Streams, new incoming records mean an update is released downstream. Still, windowed functions can use suppression to have a single final result only when the window closes. Kafka Streams window operators also offer `emitStrategy` for final results. Using the `emitStrategy` method is more straightforward than using suppression.
- There are four types of windows: hopping, tumbling, sliding, and session. Hopping and tumbling windows are fixed in size over time. Sliding windows are set in size by time, but record behavior drives record inclusion in a window. Session windows are entirely driven by record behavior, and the window can continue to grow as long as incoming records are within the inactivity gap.
- Timestamps drive the behavior in a Kafka Streams application, which is most apparent in windowed operations, as the timestamps of the records drive the opening and closing of these operations. Stream time is the highest timestamp viewed by a Kafka Streams application during processing.
- Kafka Streams provides different `TimestampExtractor` instances, so you can use different timestamps in your application—semantics event time, log-append time, or processing time.

The Processor API

This chapter covers

- Evaluating higher-level abstractions versus more control
- Working with sources, processors, and sinks to create a topology
- Digging deeper into the Processor API with a stock analysis processor
- Creating a co-grouping processor
- Integrating the Processor API and the Kafka Streams API

We've been working with the high-level Kafka Streams API until this point in the book. A DSL allows developers to create robust applications with minimal code. The ability to quickly assemble processing topologies is an essential feature of the Kafka Streams DSL. It will enable you to iterate rapidly to flesh out ideas for working on your data without getting bogged down in the intricate setup details some other frameworks may need.

But at some point, even when working with the best tools, you'll come up against one of those one-off situations—a problem that requires you to deviate

299

from the traditional path. Whatever the case, you need a way to dig down and write some code that isn't possible with a higher-level abstraction.

A classic example of trading off higher-level abstractions versus gaining more control is using object-relational mapping (ORM) frameworks. A good ORM framework maps your domain objects to database tables and creates the correct SQL queries for you at run time. When you have simple-to-moderate SQL operations (e.g., simple SELECT or JOIN statements), using the ORM framework saves you much time. But no matter how good the ORM framework is, there will inevitably be those few queries (i.e., very complex joins, SELECT statements with nested sub-select statements) that don't work the way you want. You need to write raw SQL to get the information from the database in the format you need. Here, you can see the tradeoff between a higher-level abstraction and more programmatic control. Often, you'll be able to mix the raw SQL with the higher-level mappings provided with the framework.

This chapter covers the situation where you want to do stream processing in a way that the Kafka Streams DSL doesn't make accessible. For example, you've seen from working with the KTable API that the framework controls the timing of forwarding records downstream. You may find yourself in a situation where you want explicit control over when Kafka Streams delivers a record to downstream processors. You might be tracking trades on Wall Street and only want to forward records when a stock crosses a particular price threshold. Or you want to scan the entries of a state store at regular given time intervals. To gain this type of control, you can use the Processor API. What the Processor API lacks in ease of development, it makes up for in power. You can write custom processors to do almost anything you want.

In this chapter, you'll learn how to use the Processor API to handle situations like the following that the Kafka Streams DSL does not address:

- Schedule actions, such as scanning the entries in a state store, to occur at regular intervals (either based on record timestamps or wall clock time).
- Gain complete control over when to forward records downstream.
- Forward records to specific child nodes.
- Create functionality that doesn't exist in the Kafka Streams API (you'll see an example when we build a data-driven aggregation processor in section 10.3).

First, let's look at how to use the Processor API by developing a topology.

10.1 Working with sources, processors, and sinks to create a topology

Let's say you own a successful brewery (Pops Hops) with several locations. You've recently expanded your business to accept online orders from distributors, including international sales to Europe. You want to route orders within the company based on whether the order is domestic or international, converting European sales from British pounds or euros to US dollars.

If you sketch out the flow of the business, it would look something like figure 10.1. In building this example, you'll see how the Processor API allows you to select specific child nodes when forwarding records. Let's start by creating a source node.

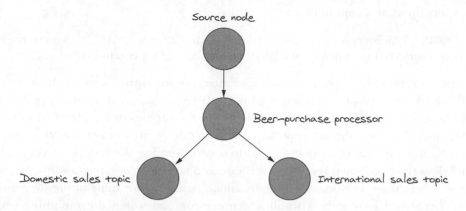

Figure 10.1   Beer sales distribution pipeline

10.1.1 Adding a source node

The first step in constructing a topology is establishing the source nodes. The following listing (found in src/main/java/bbejeck/chapter_10/PopsHopsApplication.java) sets the data source for the new topology.

Listing 10.1   Creating the beer application source node

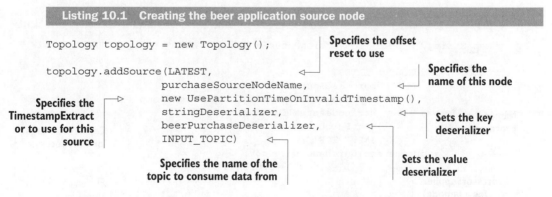

In the `Topology.addSource()` method, you see a different approach from the DSL. First, you name the source node. When you used the Kafka Streams DSL, you didn't need to pass in a name because the `KStream` instance generated a name for the node. But when you use the Processor API, you need to provide the names of the nodes in the topology. Kafka Streams uses the node name to connect a child node to a parent node. Additionally, in the DSL you provide optional source node parameters via a `Consumed` configuration object but here in the Processor API, you pass them in directly via different `Topology.addSource` method overloads.

Next, you specify the timestamp extractor to use with this source. Here, you're using the `UsePartitionTimeOnInvalidTimestamp` class, which, if the incoming timestamp is invalid (a negative number), will set it to be the highest value so far (stream time) for the partition of the incoming record. But for the most part, you should expect the timestamps to be valid.

> **NOTE**   When creating a source node with the Processor API, there are only two required parameters: the processor name and topic name(s) or pattern.

Next, you provide a key and value deserializer, representing another departure from the Kafka Streams DSL. When creating source or sink nodes, you supplied Serde instances in the DSL. The Serde contains a serializer and deserializer, and the Kafka Streams DSL uses the appropriate one, depending on whether you're going from object to byte array or from byte array to object. Because the Processor API is a lower-level abstraction, you directly provide a deserializer when creating a source node and a serializer when creating a sink node. Finally, you give the name of the source topic.

Let's next look at how you'll add processor nodes that do something with the incoming records.

10.1.2 *Adding a processor node*

Now, you'll add a processor to work with the records coming in from the source node in listing 10.2, found in src/main/java/bbejeck/chapter_10/PopsHopsApplication.java (some details are omitted for clarity). Let's first discuss how to wire up the processors, and then we'll cover the functionality of the added processor.

Listing 10.2   Adding a processor node

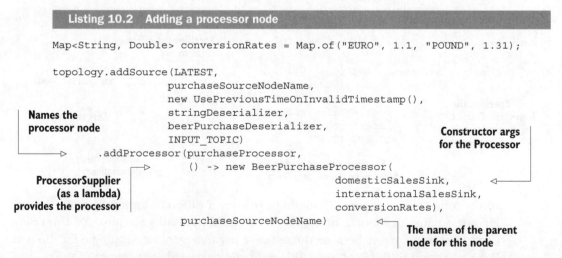

This code uses the fluent interface pattern for constructing the topology. The difference from the Kafka Streams API lies in the return type. With the Kafka Streams API, every call on a `KStream` operator returns a new `KStream` or `KTable` instance. In the Processor API, each call to `Topology` returns the same `Topology` instance.

In the previous code, you pass in a `ProcessorSupplier`. The `Topology.addProcessor` method takes an instance of a `ProcessorSupplier` interface for the second parameter, but because the `ProcessorSupplier` is a single-method interface, you can replace it with a lambda expression. The critical point is that the third parameter, `purchase-SourceNodeName`, of the `addProcessor()` method is the same as the second parameter of the `addSource()` method, as illustrated in figure 10.2.

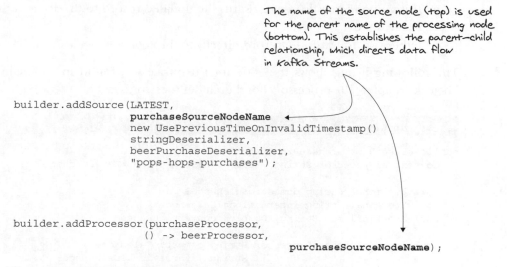

The name of the source node (top) is used for the parent name of the processing node (bottom). This establishes the parent–child relationship, which directs data flow in Kafka Streams.

```
builder.addSource(LATEST,
                  purchaseSourceNodeName
                  new UsePreviousTimeOnInvalidTimestamp()
                  stringDeserializer,
                  beerPurchaseDeserializer,
                  "pops-hops-purchases");



builder.addProcessor(purchaseProcessor,
                    () -> beerProcessor,
                                     purchaseSourceNodeName);
```

Figure 10.2   Wiring up parent and child nodes in the Processor API

Using the name of the source node establishes the parent–child relationship between nodes. The parent–child relationship, in turn, determines how records move from one processor to the next in a Kafka Streams application.

Now let's look at what you've built so far, shown in figure 10.3.

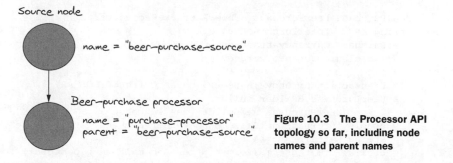

Source node

name = "beer–purchase–source"

Beer–purchase processor

name = "purchase–processor"
parent = "beer–purchase–source"

Figure 10.3   The Processor API topology so far, including node names and parent names

Let's take a second to discuss the `BeerPurchaseProcessor` functions. The processor has two responsibilities:

- Convert international sales amounts (in euros) to US dollars.
- Based on the origin of the sale (domestic or international), route the record to the appropriate sink node.

All of this takes place in the `process()` method. To quickly summarize, the `process()` method does the following:

1 Checks the currency type and, if it's not in dollars, converts it to dollars
2 If it's a nondomestic sale, forwards the updated record to the `international-sales` topic
3 Otherwise, forwards the record directly to the `domestic-sales` topic

The following listing shows the code for this processor, found in src/main/java/bbejeck/chapter_10/ processor/BeerPurchaseProcessor.java.

Listing 10.3   `BeerPurchaseProcessor`

```java
public class BeerPurchaseProcessor extends
    ContextualProcessor<String, BeerPurchase, String, BeerPurchase> {

    private final String domesticSalesNode;
    private final String internationalSalesNode;
    private final Map<String, Double> conversionRates;

    public BeerPurchaseProcessor(String domesticSalesNode,
                                 String internationalSalesNode,
                                 Map<String,Double> conversionRates) {

        this.domesticSalesNode = domesticSalesNode;
        this.internationalSalesNode = internationalSalesNode;
        this.conversionRates = conversionRates;
    }

    @Override
    public void process(
      Record<String, BeerPurchase> beerPurchaseRecord) {

        BeerPurchase beerPurchase = beerPurchaseRecord.value();
        String key  = beerPurchaseRecord.key();
        BeerPurchase.Currency transactionCurrency =
            beerPurchase.getCurrency();

        if (transactionCurrency != BeerPurchase.Currency.DOLLAR) {
            BeerPurchase.Builder builder =
                    BeerPurchase.newBuilder(beerPurchase);
            double internationalSaleAmount = beerPurchase.getTotalSale();
            String pattern = "###.##";
            DecimalFormat decimalFormat = new DecimalFormat(pattern);
            builder.setCurrency(BeerPurchase.Currency.DOLLAR);
```

Sets the names for different nodes to forward records to

The `process()` method, where the action takes place

```
                  builder.setTotalSale(Double.parseDouble(decimalFormat.format(
                       convertToDollars(transactionCurrency.name(),
  Converts  ┌──▷      internationalSaleAmount))));
international │   Record<String, BeerPurchase> convertedBeerPurchaseRecord =
sales to US dollars │       new Record<>(key,builder.build(),
             │                   beerPurchaseRecord.timestamp());
                  context().forward(convertedBeerPurchaseRecord,       ◁──────┐
                              internationalSalesNode);
           } else {                                                      Uses the
               context().forward(                                    ProcessorContext
                       beerPurchaseRecord,                          (returned from the
                       domesticSalesNode);      ◁──────┐             context() method)
           }                                                        and forwards records
       }                         Sends records for domestic         to the international
   }                            sales to the domestic sink node         sink node
```

This example extends `ContextualProcessor`, a class with overrides for `Processor` interface methods, except for the `process()` method. The `Processor.process()` method is where you perform actions on the records flowing through the topology.

> **NOTE** The `Processor` interface provides the `init()`, `process()`, and `close()` methods. The `Processor` is the primary driver of any application logic that works with records in your streaming application. You'll usually extend the `ContextualProcessor` class in the examples, so you'll only override the methods you want. The `ContextualProcessor` class initializes the `ProcessorContext` for you, so if you don't need to do any setup in your class, you don't need to override the `init()` method.

The last few lines of listing 10.3 demonstrate the main point of this example—the ability to forward records to specific child nodes. In these lines, the `context()` method retrieves a reference to the `ProcessorContext` object for this processor. All processors in a topology receive a reference to the `ProcessorContext` via the `init()` method, which is executed by the `StreamTask` when initializing the topology.

Now that you've seen how you can process records, the next step is to connect a sink node (topic) so you can write records back to Kafka.

10.1.3 Adding a sink node

By now, you probably have a good feel for the flow of using the Processor API. To add a source, you use `addSource`; to add a processor, you use `addProcessor`. As you might imagine, you'll use the `addSink()` method to wire a sink node (topic) to a processor node. Figure 10.4 shows the updated topology.

You can now update the topology you're building by adding sink nodes in the code, found in src/main/java/bbejeck/chapter_10/PopsHopsApplication.java.

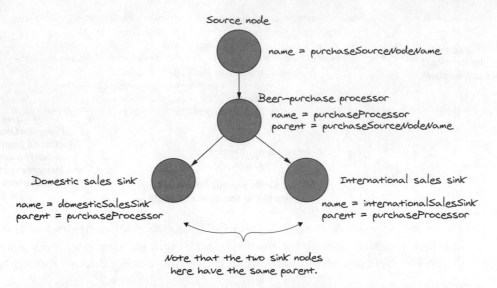

Source node
name = purchaseSourceNodeName

Beer-purchase processor
name = purchaseProcessor
parent = purchaseSourceNodeName

Domestic sales sink
name = domesticSalesSink
parent = purchaseProcessor

International sales sink
name = internationalSalesSink
parent = purchaseProcessor

Note that the two sink nodes here have the same parent.

Figure 10.4   Completing the topology by adding sink nodes

Listing 10.4   Adding sink nodes

```
String domesticSalesSink = "domestic-beer-sales";
String internationalSalesSink = "international-beer-sales";

topology.addSource(LATEST,
                   purchaseSourceNodeName,
                   new UsePreviousTimeOnInvalidTimestamp(),
                   stringDeserializer,
                   beerPurchaseDeserializer,
                   INPUT_TOPIC)
        .addProcessor(purchaseProcessor,
                  () -> new BeerPurchaseProcessor(
                                    domesticSalesSink,
                                    internationalSalesSink,
                                    conversionRates),
                  purchaseSourceNodeName)
        .addSink(internationalSalesSink,
                  "international-sales",
                  stringSerializer,
                  beerPurchaseSerializer,
                  purchaseProcessor)
        .addSink(domesticSalesSink,
                  "domestic-sales",
                  stringSerializer,
                  beerPurchaseSerializer,
                  purchaseProcessor);
```

Name of the sink

Serializer for the key

Parent node for this sink

The topic this sink represents

Serializer for the value

The topic this sink represents

Serializer for the value

Name of the sink

Serializer for the key

Parent node for this sink

In this listing, you add two sink nodes, one for domestic sales and another for international sales. Depending on the currency of the transaction, you'll write the records out to the appropriate topic.

The critical point to notice when adding two sink nodes here is that both have the same parent name. By supplying the same parent name to both sink nodes, you've wired them to your processor (as shown in figure 10.4).

In this first example, you've seen how you build Kafka Streams applications with the Processor API. While you walked through a specific example, the general principles of adding a source node(s), one or more processing nodes, and, finally, a sink node(s) apply to any application you build. Although the Processor API is a little more verbose than the Kafka Streams API, it's still easy to construct topologies. The following example will explore more of the flexibility the Processor API provides.

10.2 Digging deeper into the Processor API with a stock analysis processor

You'll now return to the world of finance and put on your day trading hat. As a day trader, you want to analyze how stock prices change to pick the best time to buy and sell. The goal is to take advantage of market fluctuations and profit quickly. We'll consider a few key indicators, hoping they'll indicate when you should make a move.

The list of requirements is as follows:

- Show the current value of the stock.
- Indicate whether the price per share is trending up or down.
- Include the total share volume so far and whether the volume is trending up or down.
- Only send records downstream for stocks displaying 2 percent trending (up or down).
- Collect at least 20 samples for a given stock before performing any calculations.

Let's walk through how you might handle this analysis manually. Figure 10.5 shows the decision tree you'll want to create to help make decisions.

You'll need to perform a handful of calculations for your analysis. Additionally, you'll use these calculation results to determine if and, if so, when you should forward records downstream.

This restriction on sending records means you can't rely on the standard mechanisms of commit time or cache flushes to handle the flow for you, which rules out using the Kafka Streams API. You'll also require a state store to keep track of changes over time. What you need here is the ability to write a custom processor. Let's look at the solution to the problem.

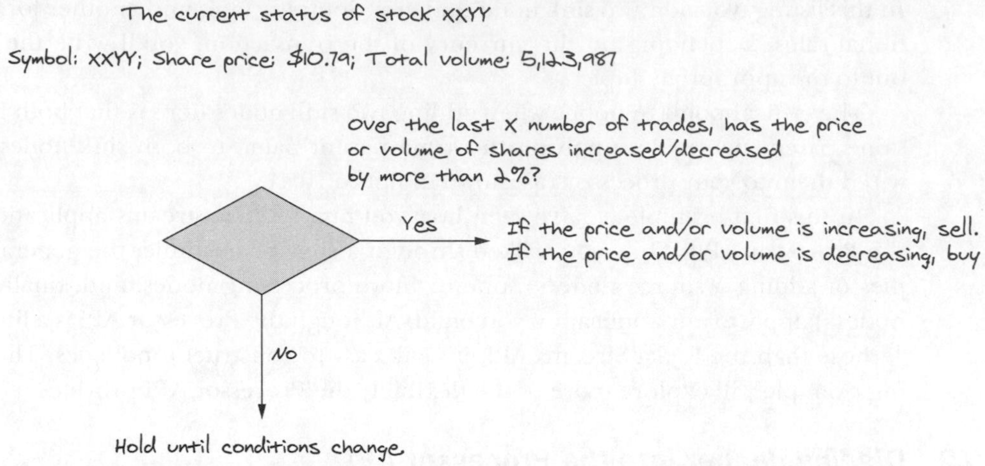

The current status of stock XXYY

Symbol: XXYY; Share price; $10.79; Total volume; 5,123,987

Over the last X number of trades, has the price
or volume of shares increased/decreased
by more than 2%?

Yes   If the price and/or volume is increasing, sell.
             If the price and/or volume is decreasing, buy

No

Hold until conditions change.

Figure 10.5   Stock trend updates

For demo purposes only

It goes without saying, but I'll state the obvious: these stock price evaluations are only for demonstration purposes. Please don't infer any actual market-forecasting ability from this example. This model bears no similarity to a real-life approach and is presented only to demonstrate a more complex processing situation. I'm certainly not a day trader!

10.2.1 *The stock-performance processor application*

The following listing shows the topology for the stock-performance application, found in src/main/java/bbejeck/chapter_9/StockPerformanceApplication.java.

Listing 10.5   Stock-performance application with a custom processor

```
Topology topology = new Topology();                            ┐ Sets the percentage
 String stocksStateStore = "stock-performance-store";          │ differential for forwarding
 double differentialThreshold = 0.02;              ◁───────────┘ stock information

KeyValueBytesStoreSupplier storeSupplier =                     ┐ Creates an in-memory
 Stores.inMemoryKeyValueStore(stocksStateStore);  ◁───────────┘ key-value state store
StoreBuilder<KeyValueStore<String, StockPerformance>> storeBuilder
 = Stores.keyValueStoreBuilder(
 storeSupplier, Serdes.String(), stockPerformanceSerde);   ◁──┐ Creates the
                                                              │ StoreBuilder
                                                              │ to place in the
  topology.addSource("stocks-source",                         ┘ topology
                 stringDeserializer,
                 stockTransactionDeserializer,
                 "stock-transactions")
```

Adds the processor to the topology using a ProcessorSupplier

```
.addProcessor("stocks-processor",
          new StockPerformanceProcessorSupplier(storeBuilder),
          "stocks-source")
.addSink("stocks-sink",
          "stock-performance",
          stringSerializer,
          stockPerformanceSerializer,
          "stocks-processor");
```

Adds a sink for writing out results

In this example, we've used a concrete implementation of the `ProcessorSupplier` instead of a lambda expression. This is because the `ProcessorSupplier` interface provides a `stores` method that will automatically wire up the processor with any `StoreBuilder` instances you provide. The following listing provides the `Stock-PerformanceProcessorSupplier` source code.

Listing 10.6 ProcessorSupplier implementation

```
public class StockPerformanceProcessorSupplier
        implements ProcessorSupplier<String, Transaction,
                                     String, StockPerformance> {
    StoreBuilder<?> storeBuilder;

    public StockPerformanceProcessorSupplier(StoreBuilder<?> storeBuilder) {
        this.storeBuilder = storeBuilder;
    }

    @Override
    public Processor<String, Transaction, String, StockPerformance> get() {
        return new StockPerformanceProcessor(storeBuilder.name());
    }

    @Override
    public Set<StoreBuilder<?>> stores() {
        return Collections.singleton(storeBuilder);
    }
}
```

Returns a new instance of the StockPerformanceProcessor

Returns the StoreBuilder instances to wire up to the processor returned by the supplier

With the `ProcessorSupplier.stores` method, you have a way to automatically wire up `StateStore` instances to processors, which makes building topology a bit simpler as you don't need to call `Topology.addStateStore` with the names of the processors having access to the store.

Since this stock performance topology has the same flow as the previous example, we'll focus on the new features in the processor. In the last example, you don't have any setup, so you rely on the `ContextualProcessor.init` method to initialize the `ProcessorContext` object. In this example, however, you need to use a state store, and you also want to schedule when you emit records instead of forwarding them each time you receive them.

First, let's look at the processor's `init()` method, found in src/main/java/ bbejeck/ chapter_10/processor/StockPerformanceProcessor.java.

Listing 10.7   `init()` method tasks

Initializes ProcessorContext via the AbstractProcessor superclass

Retrieving state store created when building topology

```
@Override
public void init(ProcessorContext<String, StockPerformance> context) {
  super.init(processorContext);
  keyValueStore = context().getStateStore(stateStoreName);
  StockPerformancePunctuator punctuator =
   new StockPerformancePunctuator(differentialThreshold,
                               context(),
                               keyValueStore);
  context().schedule(10000,
                  PunctuationType.STREAM_TIME,
                  punctuator);
  }
}
```

Initializing the Punctuator to handle the scheduled processing

Schedules Punctuator.punctuate() to be called every 10 seconds

First, you need to initialize the ContextualProcessor with the ProcessorContext, so you call the init() method on the superclass. Next, you grab a reference to the state store you created in the topology. All you need to do here is set the state store to a variable for use later in the processor. Listing 10.7 also introduces a Punctuator, an interface that's a callback to handle the scheduled execution of processor logic but is encapsulated in the Punctuator.punctuate method.

> **TIP** The ProcessorContext.schedule(Duration, PunctuationType, Punctuator) method returns a type of Cancellable, allowing you to cancel punctuation and manage more advanced scenarios, like those found in the "Punctuate Use Cases" discussion (http://mng.bz/YSKF). I don't have examples or a debate here, but I present some examples in src/main/java/bbejeck/chapter_10/cancellation.

In the last line of listing 10.7, you use the ProcessorContext (obtained via the context() method call) to schedule the Punctuator to execute every 10 seconds. The second parameter, PunctuationType.STREAM_TIME, specifies that you want to call Punctuator .punctuate every 10 seconds based on the timestamps of the data. Your other option is to select PunctuationType.WALL_CLOCK_TIME, which means the execution of Punctuator.punctuate is scheduled every 10 seconds but driven by the system time of the Kafka Streams environment. Let's take a moment to discuss the difference between these two PunctuationType settings.

10.2.2 *Punctuation semantics*

Let's start our conversation on punctuation semantics with STREAM_TIME, because it requires more explanation. Figure 10.6 illustrates the concept of stream time (note that I'm not showing some of the Kafka Streams internals).

In partitions A and B, the letter represents the record, and the number is the timestamp. For this example, we'll assume that punctuate is scheduled to run every 5 seconds.

Partition A

```
A:1
B:2
E:5
F:6
```

Partition B

```
C:3
D:4
G:10
```

Because partition A has the smallest timestamp, it's chosen first:
1) Process called with record A
2) Process called with record B

Now partition B has the smallest timestamp:
3) Process called with record C
4) Process called with record D

Switch back to partition A, which has the smallest timestamp again:
5) Process called with record E
6) punctuate called because time elapsed from timestamps is 5 seconds
7) Process called with record F

Finally, switch back to partition B:
8) Process called with record G
9) punctuate called again as 5 more seconds have elapsed, according to the timestamps

Figure 10.6 Punctuation scheduling using STREAM_TIME

Let's walk through some details to gain a deeper understanding of how the schedule is determined:

1 The `StreamTask` extracts the smallest timestamp from the `PartitionGroup`. The `PartitionGroup` is a set of partitions for a given `StreamThread`, containing all timestamp information for all partitions in the group.

2 During the processing of records, the `StreamThread` iterates over its `Stream-Task` object, and each task will end up calling `punctuate` for each of its processors that are eligible for punctuation. Recall that you must collect a minimum of 20 trades before you examine an individual stock's performance.

3 Suppose the timestamp from the last execution of `punctuate` (plus the scheduled time) is less than or equal to the extracted timestamp from the `Partition-Group`. In that case, Kafka Streams calls that processor's `punctuate()` method.

The critical point is that the application advances timestamps via the `Timestamp-Extractor`, so `punctuate()` calls are consistent only if data arrives at a constant rate. If your flow of data is sporadic, the `punctuate()` method won't get executed at the regularly scheduled intervals.

With `PunctuationType.WALL_CLOCK_TIME`, on the other hand, the execution of `Punctuator.punctuate` is more predictable, as it uses wall clock time. Note that system-time semantics is best-effort; wall clock time advances in the polling interval, and the granularity depends on how long it takes to complete a polling cycle. So, with the example in listing 10.7, you can expect the punctuation activity to execute closer to every 10 seconds, regardless of data activity.

Which approach you choose to use is entirely dependent on your needs. If you need some activity performed regularly, regardless of data flow, using system time is probably the best bet. On the other hand, if you only need calculations performed on incoming data and some lag time between executions is acceptable, try stream-time semantics.

> **NOTE** Before Kafka 0.11.0, punctuation involved the `ProcessorContext`
> `.schedule(long time)` method, called the `Processor.punctuate` method at
> the scheduled interval. This approach only worked on stream-time semantics;
> both methods are now deprecated. I mention deprecated methods in this
> book, but I only use the latest punctuation methods in the examples.

Now that we've covered scheduling and punctuation, let's move on to handling incoming records.

10.2.3 *The process() method*

The `process()` method is where you'll perform all your calculations to evaluate stock performance. There are several steps to take when you receive a record:

1 Check the state store to see whether you have a corresponding `StockPerformance` object for the record's stock ticker symbol.
2 If the store doesn't contain the `StockPerformance` object, create one. Then, the `StockPerformance` instance adds the current share price and share volume and updates your calculations.
3 Start performing calculations once you hit 20 transactions for any given stock.

Although financial analysis is beyond the scope of this book, we should look at the calculations. You will perform a simple moving average (SMA) for the share price and volume. In the financial trading world, SMAs are used to calculate the average for datasets of size N.

You'll set N to 20 for this example. Setting a maximum size means that as new trades come in, you collect the share price and number of shares traded for the first 20 transactions. Once you hit that threshold, you remove the oldest value and add the latest one. You get a rolling stock price and volume average over the last 20 trades using the SMA. It's important to note you won't have to recalculate the entire amount as new values come in.

Figure 10.7 provides a high-level walk-through of the `process()` method, illustrating what you'd do if you performed these steps manually. The `process()` method is where you'll perform all the calculations.

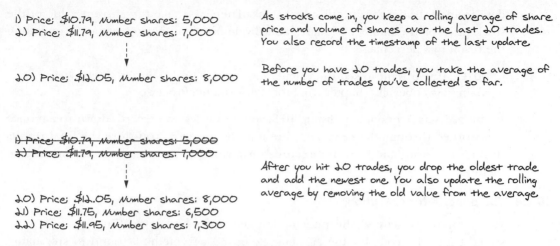

Figure 10.7   Stock analysis `process()` method walk-through

Let's look at the code that makes up the `process()` method, found in src/main/ java/ bbejeck/chapter_10/processor/StockPerformanceProcessor.java.

Listing 10.8   `process()` implementation

```
@Override
public void process(String symbol, StockTransaction transaction) {
    StockPerformance stockPerformance =
      keyValueStore.get(symbol);              ◄─┤  Retrieves previous performance
                                                   stats, possibly null

    if (stockPerformance == null) {
      stockPerformance = new StockPerformance();   ◄─   Creates a new
    }                                                   StockPerformance object if
                                                        one isn't in the state store

    stockPerformance.updatePriceStats(
        transaction.getSharePrice());         ◄─   Updates the price
    stockPerformance.updateVolumeStats(            statistics for this stock
        transaction.getShares());            ◄─
    stockPerformance.setLastUpdateSent(Instant.now());   Updates the volume
                                                         statistics for this stock
    keyValueStore.put(symbol, stockPerformance);  ◄─
}                                                     Places the updated
                                                      StockPerformance
   Sets the timestamp                                 object into the
   of the last update                                 state store
```

In the `process()` method, you take the latest share price and the number of shares involved in the transaction and add them to the `StockPerformance` object.

There are two essential calculations: determining the moving average and calculating the stock price/volume differential from the current one. You want to calculate an average once you've collected data from 20 transactions, so you defer doing anything until the processor receives 20 trades. You calculate your first average when you have data from 20 trades for an individual stock. Then, you take the current value of the stock price or the number of shares and divide it by the moving average, converting the result to a percentage.

> **NOTE** If you want to see the calculations, you'll find them in streams/src/ main/java/bbejeck/chapter_10/StockPerformanceProcessor.

In the `Processor` example in listing 10.3, you forwarded the records downstream once you worked through the `process()` method. In this case, you store the final results in the state store and leave the records forwarding to the `Punctuator-.punctuate` method.

10.2.4 *The punctuator execution*

We've already discussed the punctuation semantics and scheduling, so let's jump straight into the code for the `Punctuator.punctuate` method, found in src/main/ java/bbejeck/chapter_10/processor/punctuator/StockPerformancePunctuator.java.

Listing 10.9   Punctuation code

```
@Override
public void punctuate(long timestamp) {
    try (KeyValueIterator<String, StockPerformance> performanceIterator
                        = keyValueStore.all()) {          ◁──┐ Retrieves the
                                                             │ iterator to go
        while (performanceIterator.hasNext()) {              │ over all key
            KeyValue<String, StockPerformance> keyValue =    │ values in the
                        performanceIterator.next();          │ state store
            String key = keyValue.key;
            StockPerformance stockPerformance = keyValue.value;

            if (stockPerformance != null) {
                if (stockPerformance.getPriceDifferential() >=
                                differentialThreshold ||
                    stockPerformance.getShareDifferential() >=
                                differentialThreshold) {
                    context.forward(new Record<>(key,
                                        stockPerformance,
                                        timestamp));
                }
            }
        }
    }
}
```

Checks the threshold for the current stock →

If you've met or exceeded the threshold, forwards the record →

The procedure in the `Punctuator.punctuate` method is simple. You iterate over the key-value pairs in the state store, and if the value has crossed over the predefined threshold, you forward the record downstream.

In contrast, before, you relied on a combination of committing and cache flushing to forward records; now, you define the terms for when records get forwarded. Additionally, even though you expect this code to execute every 10 seconds, that doesn't guarantee you'll emit records. They must meet the differential threshold. Also, note that Kafka Streams doesn't run the `Processor.process` and `Punctuator.punctuate` methods concurrently.

> **NOTE** Although we're demonstrating access to a state store, it's an excellent time to review Kafka Streams' architecture and review a few key points. Each `StreamTask` has its own copy of a local state store, and `StreamThread` objects don't share tasks or data. As records make their way through the topology, each node is visited in a depth-first manner, meaning there's never concurrent access to state stores from any given processor.

This example has given you an excellent introduction to writing a custom processor. You can also take writing custom processors a bit further by creating an entirely new way of aggregating data that doesn't currently exist in the API. With this in mind, we'll move on to adding data-driven aggregation.

10.3 Data-driven aggregation

We discussed aggregation in chapter 7, when we covered stateful operations in Kafka Streams. But imagine you have different requirements for creating an aggregation. Specifically, instead of having windows based on time, you want to have a "window" based on specific aspects of the incoming events. You'll want only to include events that meet a given criteria in an aggregation and forward the results once an incoming record no longer meets that criteria.

For example, you are responsible for a production line at a prominent manufacturer, Waldo Widgets. To control costs and improve efficiency, you installed several sensors that continually send information about essential indicators in the manufacturing process. All the sensors transmit their data to Kafka. You've determined over time that the temperature sensors are one of the best leading indicators for trouble in the manufacturing process. Prolonged temperature spikes are almost always followed by a costly production line shutdown, sometimes for hours, until a site reliability engineer (SRE) can service the machine or, in the worst case, replace it.

So, you've determined that you'll need to create a Kafka Streams application to help monitor the temperature sensor readings. From your experience, you've developed some requirements for the exact information you need. What you've been able to put together over the years is that when a machine is about to have trouble, it will start with smaller spikes in temperature reading, leading to progressively more prolonged periods of increased temperature over time.

What you need is an aggregation of only the increased temperatures. Getting a constant flow of all temperature readings is counterproductive; it's simply too much information to be worthwhile otherwise. You spend some time coming up with the precise requirements you want from your new application:

- If the temperature is below a given threshold, ignore it.
- Once it rises above the threshold, start an aggregation window.

- Continue the aggregation as long as the readings exceed the threshold.
- When the readings reach a given number, emit the aggregation.
- Once a reading drops below the threshold, cease the aggregation and immediately emit the aggregation result.
- Since the sensors have a known history of spotty network connections, if more than 10 seconds elapse without either a closing reading or an additional high reading, go ahead and emit the aggregation.

This approach will enable the dashboard application to display the information your team needs to take action. You start looking at the Kafka Streams DSL API and focus on the windowed aggregations. From reading, you decide you'd like to use an aggregation with `SessionWindow` as it's driven by behavior instead of time. However, it would be best to have fine-grained control of what to include in the aggregation and when to emit the results. So, you turn your attention to the Processor API and write your own data-driven windowing behavior.

You've built Kafka Streams Processor API applications before, so you dive right in and start on the `Processor` implementation you'll need to complete the task. The following listing provides the start of the `ProcessorSupplier` you've come up with.

Listing 10.10 `ProcessorSupplier`

The StoreBuilder required for the state store

A Predicate for determining the start of an aggregation

```
public class DataDrivenAggregate implements
      ProcessorSupplier<String, Sensor, String, SensorAggregation> {

    private final StoreBuilder<?> storeBuilder;
    private final Predicate<Sensor> shouldAggregate;
    private final Predicate<Sensor> stopAggregation;

    public DataDrivenAggregate(final StoreBuilder<?> storeBuilder,
                      final Predicate<Sensor> shouldAggregate,
                      final Predicate<Sensor> stopAggregation) {
      this.storeBuilder = storeBuilder;
      this.shouldAggregate = shouldAggregate;
      this.stopAggregation = stopAggregation;
    }
```

The Predicate used to indicate the aggregation should stop

So when creating the `ProcessorSupplier`, you provide two `Predicate` instances to determine when to start an aggregation or include a record in the current aggregation. The second one chooses when to reject a record from the aggregation, which shuts the aggregation off. Let's look at the `Processor` implementation that the supplier returns.

Listing 10.11 Processor Implementation

```
private class DataDrivenAggregateProcessor extends
      ContextualProcessor<String, Sensor, String, SensorAggregation> {
```

```
KeyValueStore<String, SensorAggregation> store;
long lastObservedStreamTime = Long.MIN_VALUE;
```

Retrieves the state store

```
@Override
public void init(ProcessorContext<String, SensorAggregation> context) {
    super.init(context);
    store = context().getStateStore(storeBuilder.name());
    context().schedule(Duration.ofSeconds(10),
                PunctuationType.WALL_CLOCK_TIME,
                this::cleanOutDanglingAggregations);
}
```

Schedules a punctuation to make sure to emit aggregations that have been stalled

Initializing the `Processor` is straightforward and looks very similar to the other processor implementations you've built before. However, the interesting logic occurs in the `process` method.

Listing 10.12 `DataDrivenAggregateProcessor` process method implementation

```
@Override
public void process(Record<String, Sensor> sensorRecord) {
    lastObservedStreamTime =
        Math.max(lastObservedStreamTime,
                sensorRecord.timestamp());
    SensorAggregation sensorAgg = store.get(sensorRecord.key());
    SensorAggregation.Builder builder;
    boolean shouldForward = false;

    if (shouldAggregate.test(sensorRecord.value())) {
        if (sensorAgg == null) {
            builder = SensorAggregation.newBuilder();
            builder.setStartTime(sensorRecord.timestamp());
            builder.setSensorId(sensorRecord.value().getId());
        } else {
            builder = sensorAgg.toBuilder();
        }
        builder.setEndTime(sensorRecord.timestamp());
        builder.addReadings(sensorRecord.value().getReading());
        builder.setAverageTemp(builder.getReadingsList()
                                    .stream()
                                    .mapToDouble(num -> num)
                                    .average()
                                    .getAsDouble());
        sensorAgg = builder.build();
        shouldForward =
            sensorAgg.getReadingsList().size() % emitCounter == 0;
        store.put(sensorRecord.key(), sensorAgg);

    } else if (stopAggregation.test(sensorRecord.value())
                    && sensorAgg != null) {
        store.delete(sensorRecord.key());
        shouldForward = true;
    }
```

Sets stream time for the processor

The first predicate; determines whether the record goes into the aggregate

Checks whether the number readings require an emit of the aggregation

The second predicate; determines whether to shut down the aggregation

```
      if (shouldForward) {
          context().forward(new Record<>(sensorRecord.key(),
                                          sensorAgg,
                                          lastObservedStreamTime));
      }
}
```

Guard condition; checks whether the processor should forward the aggregate

Since the logic for the aggregation is straightforward, I'm only going to discuss the main points of this method. We update the `lastObservedStreamTime` variable. Stream time only moves forward, so we don't unquestioningly accept the timestamp from the incoming record; instead, we'll reuse the existing stream time value should an out-of-order record arrive. You'll see the importance of the `lastObservedStreamTime` variable in the next section when we cover the method punctuation executes.

The second annotation in listing 10.12 gets to the crux of this processor: Does the record belong in an aggregation? If yes, you'll create a new aggregation object or add the record to an existing one. Additionally, a few lines of code further down, you'll see where the `shouldForward` variable is set to `true` or `false`, indicating whether there are enough readings to require forwarding the aggregation before it closes. Since we can only count on sometimes having a small number of readings, we'll want to emit intermediate aggregation results. The final line in this section stores the aggregation in the state store.

If the first `if` block doesn't evaluate to `true`, we'll end up at the second predicate. If the record matches the `stopAggregation` condition, the temperature reading dropped below the threshold. If there is an existing aggregate, it's deleted from the state store, and the `shouldForward` gets set to `true`. The processor ignores the record if it doesn't meet the threshold and there's no current aggregation. Finally, if the `shouldForward` variable evaluates to `true`, the processor forwards the aggregation to any downstream processors.

At this point, we've covered the main requirements of this processor, creating a "windowed" (setting a start and ending timestamp) aggregation determined by some aspects of the record data—in this case, the temperature sensor reading. We have one last requirement to take care of. We'll want to clear the aggregation and forward the results if we don't receive any updates for a given period. The scheduled punctuation handles this final requirement. Let's take a look at the punctuation code now.

Listing 10.13   Punctuation code

Gets an iterator over the contents of the aggregation store

```
void cleanOutDanglingAggregations(final long timestamp) {
  List<KeyValue<String, SensorAggregation>> toRemove = new ArrayList<>();
  try (KeyValueIterator<String, SensorAggregation> storeIterator
                                          = store.all()) {
      while (storeIterator.hasNext()) {
        KeyValue<String, SensorAggregation> entry = storeIterator.next();
```

```
        if (entry.value.getEndTime() <
                        (lastObservedStreamTime - 10_000)) {    ◁─┐
            toRemove.add(entry);                                    │
        }                                                           │
    }                                                               │
}                                                                   │
toRemove.forEach(entry -> {                                         │
    store.delete(entry.key);                                        │
    context().forward(new Record<>(entry.key,                       │
                        entry.value,                                │
                        lastObservedStreamTime));                   │
    });                                                             │
}                                                                   │
```

> **Checks whether the end timestamp of the aggregation is within 10 seconds of stream time**

For each entry that hasn't received an update in time, removes it from the store and forwards it

The scheduled punctuation will be executed every 10 seconds (wall clock time) and will examine all the records in the state store. If a timestamp is more than 10 seconds behind the current stream time (meaning it hasn't had an update in that amount of time), it's added to a list. After iterating over the records in the store, the resulting list iterates over its contents, removing the record from the store and forwarding the aggregation.

This concludes our coverage of using the Processor API to provide functionality unavailable to you in the DSL. While it's a made-up example, the main point is that when you want finer-grained control over emitting records from an aggregation and custom calculations, the Processor API is the key to achieving that objective.

10.4 Integrating the Processor API and the Kafka Streams API

So far, our coverage of the Kafka Streams and the Processor APIs has been separate, but you can still combine approaches. Why would you want to mix the two methods?

Let's say you've used the Kstream and Processor APIs for a while. You've come to prefer the Kstream approach, but you want to include some previously defined processors in a Kstream application because they provide the lower-level control you need. Or it could be that you need some specific behavior not offered by the DSL but only in a portion of the topology. You can complete the rest with the DSL.

The Kafka Streams API offers a method, KStream.process(), that allows you to plug in functionality built using the Processor API. Introduced in version 3.0.0, it represents a new approach for combining the Processor API in the Kstream DSL from previous versions. The new process is a significant improvement, as it brings the much-desired functionality of forwarding records directly to downstream processors, where previously only transformValues allowed for direct forwarding.

> **NOTE** There is a deprecated KStream.process method, but its return type is void, and it takes an org.apache.kafka.streams.processor.Processor-Supplier; also, all the various transformXXX methods are deprecated as well.

To get an understanding of how to combine the DSL and the Processor API, let's look at an example, found in src/main/java/bbejeck/chapter_10/StockPerformance-DslAndProcessorApplication.java. Since I've already described the functionality of the stock performance application previously in this chapter, I won't repeat it here (some details are omitted for clarity). We'll see how to combine the Kstreams DSL and the Processor API.

Listing 10.14   Example of mixing the DSL and the Processor API

```
StreamsBuilder builder = new StreamsBuilder();
StoreBuilder<KeyValueStore<String, StockPerformance>> storeBuilder =
            Stores.keyValueStoreBuilder(storeSupplier,
                                          Serdes.String(),
                                          stockPerformanceSerde);

builder.stream(INPUT_TOPIC,
            Consumed.with(stringSerde, stockTransactionSerde))
        .process(
  new StockPerformanceProcessorSupplier(
                              storeBuilder))
        .peek(printKV("StockPerformance"))
        .to(OUTPUT_TOPIC,
            Produced.with(stringSerde, stockPerformanceSerde));
```

Creates a stream with the DSL → `builder.stream`

Adds in the stock performance processor ← `storeBuilder`

So here, you've created the KStream instance as you would typically with the DSL. Still, you've injected your custom StockPerformanceProcessor (provided by the Stock-PerformanceProcessorSupplier) in the middle of the topology. Essentially, you've simplified your topology to use the DSL for everything other than the custom processor. In many cases, using the new KStream.process method will be the best approach for adding custom logic when confined to one processor.

At this point in the book, we've covered how you can build applications with Kafka Streams. Our next step is to look at how to configure these applications optimally, monitor them for maximum performance, and spot potential problems.

Summary

- The Processor API gives you more flexibility at the cost of more code. The tradeoff for being able to provide any logic you want is that you have to wire the entire topology together manually explicitly.
- Although the Processor API is more verbose than the Kafka Streams API, it's still easy to use, and the Processor API is what the Kafka Streams API uses under the covers.
- When deciding which API to use, consider using the Kafka Streams DSL and integrating the process() method to provide your custom processor. If you only have one, the mix-in approach is the way to go. If you have several custom processors or specific routing requirements, going wholly with the Processor API may be the better approach.

ksqlDB

This chapter covers

- Understanding ksqlDB
- More about streaming queries
- Building streaming applications with SQL statements
- Creating materialized views over streams
- Using ksqlDB advanced features

At this point in the book, we've learned about several components of the Kafka event streaming platform—Kafka Connect for integrating external systems and Kafka Streams for building an event streaming application. These two components together form the bedrock of building event streaming applications. In this chapter, you will learn about ksqlDB, which allows you to use those components by writing SQL to construct an event streaming application. ksqlDB is a "streaming database purpose-built for streaming applications" (https://ksqldb.io/). It will enable you to build powerful streaming applications with a few SQL statements.

So why would you use ksqlDB? For starters, it vastly simplifies the application development process. With ksqlDB, you're not working with code or configuration files. You write your SQL queries and execute them, which launches a continually running application where you can get instant notification of events.

Imagine you're working with business analysts at the Fintech company Big Short Equity. They've seen the applications you've built using Kafka Streams. They want the ability to construct near-real-time financial analysis applications, but their primary skill is not writing Java code. While you can support the analysts' needs and build the required Kafka Streams apps, it would be far more efficient if the analysts could make them independently. The analysts are experts in SQL, as most of their work involves writing queries on relational databases. You have the idea of introducing ksqlDB to them; it provides scalable, distributed stream processing, including aggregations, joins, and windowing. But unlike running a SQL query against a typical relational database, where the query will return results and stop, the results of a ksqlDB query are continuous.

You'll want to use ksqlDB because you can quickly build a powerful event-streaming application in the same time it takes you to write a SQL query! So, what you're going to learn in this chapter is how to apply what you've learned so far about building event-streaming applications by using the familiar syntax of an SQL query. ksqlDB uses Kafka Streams under the covers, so all the concepts from the previous chapters also apply here. Additionally, the ksqlDB server provides direct integration with Kafka Connect so that you can build an entire end-to-end solution without any code. We'll start with the basic concepts of a stream and a table and how ksqlDB handles different data formats, including JSON, Avro, and Protobuf. Finally, we'll explore more advanced aggregations, joins, and windowed operations options.

11.1 *Understanding ksqlDB*

Earlier in the book, we discussed the concepts of an event-stream and an update-stream. Remember, an event stream is an infinite sequence of independent events. Records in an event stream with the same key aren't related; each stands alone as an event. But an update stream is a little different. An update stream is also infinite, but events with the same key as a previous event are considered an update to that event. ksqlDB has the same concept: you can query from a `Stream` or a `Table`. Additionally, you can perform a point-in-time query against a materialized view or a stream.

So, let's get started with your first ksqlDB query. To make the comparison to a Kafka Streams application more accessible, we'll repurpose the Yelling app, introduced in chapter 6.

> **NOTE** You can deploy ksqlDB in one of three ways: in standalone mode, as a cluster on-premises, or in Confluent Cloud. In this book, you'll work with ksqlDB in standalone mode via Docker. Later in the chapter, we'll cover the ksqlDB architecture in more detail.

First, we need to create a STREAM from a Kafka topic.

Listing 11.1 Creating a STREAM in ksqlDB

```
ksql> CREATE STREAM input_stream (phrase VARCHAR) WITH
 (kafka_topic='src-topic', partitions=1, value_format='KAFKA');
```

So, the first step is to create a ksqlDB STREAM. In this case, you've made a STREAM named `input_stream` based on the topic `src-topic`.

> **NOTE**   For all the ksqlDB examples, I'm going to assume you are using one of the `docker compose` files (arm64_ksqldb-docker-compose.yml or x86_ksqldb-docker-compose.yml, depending on the architecture of your laptop) included in the source code. Once you've run the `docker compose -f <file name> up` command, you'll want to open a new terminal window and run `docker exec -it ksqldb-cli ksql http://ksqldb-server:8088`. This command will start a ksqlDB command-line interface (CLI) session. Consult the README for chapter 11 in the source code for complete instructions. Also, these commands should work if you choose to use ksqlDB on Confluent Cloud.

After you execute the CREATE STREAM... command, you should see something like the following listing on the ksqlDB CLI screen.

Listing 11.2   Result of the CREATE STREAM... statement

```
Message
----------------
 Stream created
----------------
```

> **TIP**   You can confirm the STREAM objects you've created by running the command `show streams;` from the CLI.

Now that you have a STREAM, the next step is to populate the underlying topic so you can start yelling! You could use a `KafkaProducer` to produce records to the topic, but let's stick with SQL commands for now. Run a few inserts like the following listing.

Listing 11.3   Insert statements for loading data into a stream

```
INSERT INTO input_stream (phrase) VALUES (
  'Chuck Norris finished World of Warcraft');
INSERT INTO input_stream (phrase) VALUES (
  'Chuck Norris first program was kill -9');
INSERT INTO input_stream (phrase) VALUES (
  'generate bricks-and-clicks content');
INSERT INTO input_stream (phrase) VALUES (
  'brand best-of-breed intermediaries');
.....
```

Using INSERT INTO.. statements is a great way to get going quickly with ksqlDB. But in practice, after you've completed some quick prototyping, you'll want to get data into the topic more efficiently, like a `KafkaProducer` or the results of another ksqlDB query.

The next step is to create a continuous or push query. Before we do that, let's go over some background information. In ksqlDB, since the queries are based on data from a Kafka topic, they run continuously, as an event stream never stops. So, once you start a query, it will continue evaluating the incoming data until you explicitly stop it. These are push queries because the results are "pushed" to the client that issued the query. We'll cover the clients you have to give a ksqlDB query to later in the chapter. You'll also learn about another query type called a pull query that yields point-in-time results from a materialized store. If these terms seem unfamiliar to you, don't worry; we'll clarify them later in the chapter when we cover them.

With our background information complete, let's get back to writing the query that will be your streaming application.

Listing 11.4   A continuous query that becomes your streaming application

```
CREATE STREAM yelling AS                         Creates a new STREAM
  SELECT UCASE(phrase) AS SHOUT
  FROM input_stream                              Selects columns and
  EMIT CHANGES;                                  apply the UCASE function

        The EMIT CHANGES clause
```

Congratulations, with a simple query statement, you have a streaming application! Notice that you used a ksqlDB built-in function to perform the uppercase of the phrase. There are several built-in functions available. We'll cover functions as we go along in the chapter.

Before going on, let's compare this application to your first Kafka Streams application, the Yelling app. By comparing the two, you'll understand what ksqlDB is and what it can do when developing a streaming application. Let's start by looking at the Kafka Streams version.

Listing 11.5   Kafka Streams version of the Yelling app

```
Serde<String> stringSerde = Serdes.String();
StreamsBuilder builder = new StreamsBuilder();

builder.stream("src-topic",
               Consumed.with(stringSerde, stringSerde))
       .peek(sysout)
       .mapValues(value -> value.toUpperCase())
       .peek(sysout)
       .to("out-topic",
               Produced.with(stringSerde, stringSerde));
```

As far as Kafka Streams applications go, this one is straightforward. One of the benefits of using the Kafka Streams DSL is that it's primarily declarative versus imperative. This means you're specifying what you want to do (declarative) instead of how to do it. What's the significance of this difference? The declarative approach is more straightforward to express and understand.

For example, let's say you've invited a friend over for dinner, and they ask if you need something, and you ask them to pick up a bottle of red wine. That would be a declarative statement. You left the decision of where and when to pick it up to your friend. Now, contrast that with having to give directions to a beer and wine store, where to find it in the store, what to do if your favorite brand isn't available, etc.—that level of instruction is imperative. I could go on, but you get the point I'm trying to make here.

Using Kafka Streams versus raw or plain `KafkaConsumer` and `KafkaProducer` instances greatly simplifies your application development effort, but ksqlDB takes that to another level. The following listing shows what you used to create the same stream application with ksqlDB.

Listing 11.6 ksqlDB version of the Yelling app

```
CREATE STREAM yelling AS SELECT UCASE(phrase) AS SHOUT FROM input_stream
    EMIT CHANGES;
```

Just one line of text, a SQL statement, is all it took to create the Yelling application in ksqlDB.

While the Kafka Streams version is still elementary to build, you must provide `Serde` instances, create the `StreamBuilder` instance, compile and run the code, etc. This effort of developing a streaming application gets more significant as you start to build more complex applications. Another benefit, and arguably the most important, to using ksqlDB is that since it uses SQL, it opens the door to nondevelopers to create streaming applications.

Note that under the covers, a ksqlDB SQL statement compiles down to a Kafka Streams application. So, while you're using SQL, it's beneficial to understand Kafka Streams when using ksqlDB.

Does this mean that ksqlDB will solve all your problems and you no longer need Kafka Streams? Certainly not. No single tool can do everything, and Kafka Streams will always have a place in building powerful event streaming applications. ksqlDB is another powerful means at your disposal.

Next, let's dive into a more complex and realistic example and simultaneously learn about using the `TABLE` abstraction, which tracks the latest record for a given key.

11.2 More about streaming queries

Earlier in the book, you learned about the concepts of `KStream` and `KTable` in Kafka Streams (see chapters 6 and 7, respectively). The `KStream` is an event stream, where records (key-value pairs) with the same key are independent events, but with the `KTable`, an update stream, events with the same key are updates to a previous event with the same key. In ksqlDB, the `STREAM` and `TABLE` concepts follow the same rules. Let's dig deeper with a more realistic example than the Yelling application. We'll dig into more details of working with ksqlDB along the way.

Let's say you've started a fitness website that encourages members to get in as many "steps" per day as possible. But members can do other activities that count toward steps, not just running or walking. After some influences endorsed your application, your traffic has grown significantly, and you'd like to encourage the growth by offering the ability to display the "step" leaders in near-real time and have it updated regularly.

You have a mobile application that updates user results to a backend server running in the cloud. The server takes the information from the mobile applications and produces user activity updates to a Kafka topic named `user_activity`. This topic is the starting point enabling you to publish updates to the website, and your first step is to create a STREAM based on it.

Listing 11.7   Creating the stream

```
CREATE STREAM user_activity (first_name VARCHAR,          ◁─────  Columns of
                             last_name VARCHAR,                   the STREAM
                             activity VARCHAR,
                             event_time VARCHAR,
                             steps INT

     ) WITH (kafka_topic='user_activity',                        The field to use for the
             partitions=4,                                       activity timestamp
             value_format='JSON',
             timestamp = 'event_time',             ◁─────┐      The format
             timestamp_format = 'yyyy-MM-dd HH:mm:ss'  ◁─┤      for parsing the
     );                                                          timestamp field
```

You have a SQL statement here creating the `user_activity` stream. The underlying value in the topic is in JSON, and you've specified the JSON's structure as the stream's column names. The WITH statement contains properties ksqlDB uses for handling the stream, and this time, we've added a couple of new items, properties named `time-stamp` and `timestamp_format`.

You're using these properties to instruct ksqlDB to use a field embedded in the record itself as the timestamp for the record. If you recall from chapter 4, a `Kafka-Producer` embeds a timestamp into a record when it produces to Kafka. The producer added a timestamp, which is the event time for the record. Usually, the producer timestamp is close enough to serve as the official time of an event. But in some cases, you may not want to use the timestamp set by the producer but one set on the record's value.

Why would you want to use a value-embedded timestamp instead? While there could be many reasons, the chief reason would be when the timestamp on the record represents a more accurate reflection of when the event occurred. In our case, the timestamp on the record is when the user made the entry on your fitness application. The timestamp from the producer reflects when it received the record from your application server. While these two should be very close in practice when awarding points or prizes, tracking when the user entered on their mobile device is more important.

So, to enable using the embedded timestamp, you provide the `timestamp` property in the WITH clause, and it specifies the column to use for the event time versus the one Kafka provides. Since you have the `event_time` field as a string, you need to tell ksqlDB its format, which you've done with the `timestamp_format` property.

In practice, it's prevalent to use a `long` primitive for the timestamp in event objects since it represents the number of milliseconds in the Unix epoch time (time in milliseconds since January 1, 1970). In those cases where you have a `long` for the timestamp, you would use a type of `BIGINT` for the timestamp column, and in the `WITH` clause, you'd only need to specify the column name since ksqlDB can work with the field directly. So, in that case, you'd update the query to create the stream like the following listing.

Listing 11.8   Creating the stream with event time as a `long`

```
CREATE STREAM user_activity (first_name VARCHAR,
                             last_name VARCHAR,
                             activity VARCHAR,
                             event_time BIGINT,          ← Timestamp field
                             steps INT                      represented as
                                                            a BIGINT
) WITH (kafka_topic='user_activity',
        partitions=4,                          Tells ksqlDB to use the
        value_format='JSON',                   event_time column for
        timestamp = 'event_time'    ←          the timestamp
);
```

You're using a type of `BIGINT` as that is the numeric type for an 8-byte number in ksqlDB. The backing type in Java is `long`. But that raises the question: What if you want to use the producer-provided timestamp? ksqlDB has a system column for each record called `ROWTIME`, containing the timestamp from the underlying Kafka record. The term *system column* means it's provided automatically by ksqlDB. You don't need to do anything with it. We'll see an example of using `ROWTIME` in section 11.3. Before we move on with our example of the fitness steps application, we should cover some additional information about the `WITH` clause. For review, the following listing provides the `WITH` clause from the `user_activity` stream.

Listing 11.9   `WITH` clause from the `user_activity` stream

```
Specifies the Kafka topic        The number of partitions
for the stream to use            of the underlying topic
                                                            The format of the value
                                                            in the key-value pair
 └─▷ WITH (kafka_topic = 'user_activity',
           partitions = 4,                ←
           value_format = 'JSON',              ←           The timestamp column
           timestamp = 'event_time',               ←
           timestamp_format = 'yyyy-MM-dd HH:mm:ss'    ←   Tells ksqlDB the format
          );                                                of the timestamp
```

NOTE   We've already discussed the timestamp-related properties earlier in this section, so I won't review them again.

You'll notice that we specify the topic name. This topic will be the data source for the stream of user activity records. Remember, ksqlDB continuously evaluates the query over the incoming records of a topic and not a relational database table. The `partitions` entry specifies the number of partitions of the topic. Take note that the `kafka_topic` property is always required.

Since ksqlDB uses Kafka Streams under the covers, why are you telling it the number of partitions of the input topic? After all, you only provide the topic's name, and Kafka Streams takes care of everything needed to consume from the topic properly. If the underlying topic doesn't exist when you create the stream, ksqlDB will attempt to make it.

But if the topic does exist, ksqlDB will not overwrite it. An important point to note is that when you provide the `partitions` property for the `STREAM` and the topic exists, the number of partitions you specify must match the actual number of partitions. Otherwise, you'll get an error. You can safely omit the `partitions` property if the topic exists.

When quickly developing or prototyping in a local environment, having ksqlDB create the topics can be a time saver. But in a production setting, it's best always to create the topics you need ahead of time to ensure clarity and communication about the structure of your applications.

The last part of setting stream properties we'll cover here is the `value_format` setting. As you might have guessed, this tells ksqlDB the value format in the key-value pair. There is a setting for the key as well—`key_format`. The acceptable entries for either the `key_format` and `value_format` properties include:

- `JSON`
- `JSON_SR`
- `AVRO`
- `PROTOBUF`
- `PROTOBUF_NOSR`
- `NONE`
- `KAFKA`
- `DELIMITED`

While I won't go into a whole discussion here on the different types (I'll refer you to the ksqlDB documentation instead: http://mng.bz/ZEZP), it will be worthwhile to point out a few critical points. The `JSON_SR`, `AVRO`, and `PROTOBUF` formats use Schema Registry. To use Schema Registry with ksqlDB, you must provide the HTTP endpoint (as you did with either the Kafka clients or Kafka Streams) via the `ksql.schema.registry.url` property.

We'll cover configuration options for ksqlDB in a later section. If you're using Schema Registry, ksqlDB will automatically register a schema as needed. But this raises an important point: if you have an existing schema for a given value or key, you can provide the schema ID for the key or value in the `WITH` statement when setting the properties for your stream. For example, let's say you used Avro for the values in the fitness step application. In that case, you would use the stream in the following listing.

Listing 11.10 Specifying the schema ID when constructing the stream

```
CREATE STREAM user_activity                          Specifies the
WITH ( kafka_topic = 'user_activity',                value format
       value_format = AVRO,
       value_schema_id = 1                           Provides the ID for
     )                                                the existing schema
```

You specify that the value uses Avro for the serialization. Still, you also provide the schema ID so it can retrieve the precise version of the schema from Schema Registry. This ability to use the exact schema version is essential when ksqlDB either consumes from or writes to a topic that other applications work with, as all clients must use the same schema; otherwise, your application will likely experience errors due to data formatting errors. If you don't provide the schema ID, ksqlDB will retrieve the latest version of the schema. ksqlDB will assume the subject name for the schema is <topic-name-key> or <topic-name-value>. We covered Schema Registry in chapter 3, so you can refer back to it if you need to refresh your understanding of Schema Registry concepts.

You'll also notice that you didn't define any columns for the stream. That's because when using serialization format supported by Schema Registry, ksqlDB will infer the column names and field types from the schema. In our current example, we specified the column names and types because we're using plain JSON for the data format, and there's no inferencing available. We'll cover using ksqlDB with Schema Registry in more detail in section 11.5, but for now, let's get back to building the fitness step application. For review, the following listing shows the initial query.

Listing 11.11 Initial query for building the fitness step application

```
CREATE STREAM user_activity (first_name VARCHAR,
                             last_name VARCHAR,
                             activity VARCHAR,
                             event_time VARCHAR,
                             steps INT

    ) WITH (kafka_topic='user_activity',
            partitions=4,
            value_format='JSON',
            timestamp = 'event_time',
            timestamp_format = 'yyyy-MM-dd HH:mm:ss'
    );
```

With this stream in place, you're primed to build the often-requested new feature, a leaderboard. You add up the steps for a given activity and present the results in sorted order from highest to lowest. To create the first leaderboard stat, you must perform an aggregation, which is, in this case, just a simple sum of the `steps` column.

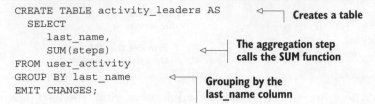

Listing 11.12   Adding a sum aggregation for calculating leaders in a category

```
CREATE TABLE activity_leaders AS        ◁──┐  Creates a table
  SELECT
     last_name,
     SUM(steps)                         ◁──┐  The aggregation step
FROM user_activity                         │  calls the SUM function
GROUP BY last_name                      ◁──┐  Grouping by the
EMIT CHANGES;                              │  last_name column
```

When you use aggregation in ksqlDB, the query's result is a TABLE, so we'll need to explicitly create one with our query. Also note that just like when you query a relational database, you must include the columns in the SELECT statement in the GROUP BY clause. So you've created a table or materialized view of an aggregation, a SUM. It will continue to provide results since it's selecting from the underlying stream user_activity, so whenever new results end up on the user_activity stream, your table here updates with new results. Additionally, ksqlDB creates a changelog topic to ensure the aggregation records are backed up should the ksqlDB server running the table query experience any issues.

You should also note that under the covers, creating this table creates a new topic in Kafka named activity_leaders by default since that's the table's name. If you want to use a different name for the underlying topic supporting the table, you can provide the topic name in a WITH statement, and we'll see just how to do that next.

Initially, this query works for you, and your customers in your demo preview respond well to the leader dashboard. But in the solicited feedback, there are several comments for you to update the dashboard with more details. First, the results combine the steps score across all activities. They are only displayed with the user's last name, making it difficult to disambiguate the differences between your app members with the same last name.

Based on user feedback, you decide to put the release on hold and pursue the required changes. Since you're in a development environment for the preview, you'd like to clean up the table query. Because you plan on a complete revamp of the activity_leaders table, removing the backing topic would also be a good idea. So, your next step to clean things up is to run the command in the following listing.

Listing 11.13   Command to remove the table and the underlying topic

```
DROP TABLE activity_leaders DELETE TOPIC;
```

This command will delete the table from ksqlDB and the backing topic. Note it only marks the topic for deletion, the actual removal of the topic is asynchronous and occurs eventually when the broker cleans up resources.

NOTE   The DELETE TOPIC clause at the end is optional. If you want to keep the records in the topic, consider leaving the delete clause off the command. Also, you can add IF EXISTS to your command so it won't error out should a table you're trying to delete not exist. So the command would look like DROP TABLE IF EXISTS activity_leaders DELETE TOPIC.

Now that you've cleaned up the existing table, you go about redefining your table query and decide that adding the first name and the activity type should satisfy the user comments. You'll get to distinguish between users and see the leaders for each activity type. Since you're proficient in SQL, it doesn't take you long to come up with the query.

Listing 11.14 Updating the sum aggregation for leaders per activity and full name

```
CREATE TABLE activity_leaders AS
  SELECT
      first_name,
      last_name,
      activity,
      SUM(steps)
FROM user_activity
GROUP BY first_name, last_name, activity
EMIT CHANGES;
```

Your updated query is essentially the same, but now you're selecting additional columns, first_name and activity. Since you've added two columns in the select portion of the SQL, you'll need to add those columns in the GROUP BY clause. This is required because, with an aggregation, we'll need to group the records by the selected fields to form unique results. The grouping acts as a "key" for the aggregation. You execute the new table statement in the ksqlDB CLI, and you see an unexpected error.

Listing 11.15 Error creating a GROUP BY with multiple columns

```
Key format does not support schema.
format: KAFKA
schema: Persistence{columns=[`FIRST_NAME` STRING KEY, `LAST_NAME`
  STRING KEY, `ACTIVITY` STRING KEY], features=[]}
reason: The 'KAFKA' format only supports a single field.  Got:
  [`FIRST_NAME` STRING KEY, `LAST_NAME` STRING KEY, `ACTIVITY` STRING KEY]
```

This detailed error message explains things well, but let's add more information. When ksqlDB attempts to create the underlying topic for the table, the primary key type will default to KAFKA, meaning it must be a scalar type supported by Kafka—a string or integer, for example. But here, you've provided three columns to make up a composite key. You need to do this because you're grouping by these three columns to ensure the results are unique per row, so three fields won't work with a schema-less key.

Fortunately, there is a simple solution to getting your updated query to work. You'll need to specify the key format for the new table to one that supports a schema; in this case, we'll use JSON. To do this, we'll make a slight change to the query and add a WITH statement.

Listing 11.16   Updating the CREATE TABLE to specify a key format that supports a schema

```
CREATE TABLE activity_leaders WITH (KEY_FORMAT = 'JSON') AS
  SELECT
      first_name,
      last_name,
      activity,
      SUM(steps)
FROM user_activity
GROUP BY first_name, last_name, activity
EMIT CHANGES;
```

By adding WITH (KEY_FORMAT = 'JSON'), your new table will now use JSON and create a composite key containing the first_name, last_name, and activity columns. But you're not quite done yet, as this change will make the query successful in performing the aggregation, but the group-by columns are in the key and won't show up in the final results.

So you'll need to instruct ksqlDB that you want those columns in the value portion of the results (remember, Kafka works in key-value pairs). Again, you'll quickly achieve this using the AS_VALUE function, which also instructs ksqlDB to copy a row's key into its value. You'll need to keep the original columns in the select statement and add an AS_VALUE function for each key you want copied into the value.

Listing 11.17   Specifying key format and copying keys to value

```
CREATE TABLE activity_leaders WITH (KEY_FORMAT = 'JSON') AS
  SELECT
      first_name as first_name_key,          Gives an alias
      last_name as last_name_key,            to the keys
      activity as activity_key,
      AS_VALUE(first_name) as first_name,    Copies a part of the key
      AS_VALUE(last_name) as last_name,      to the value and setting
      AS_VALUE(activity) as activity,        a column name alias
      SUM(steps) as total_steps
FROM user_activity
GROUP BY first_name, last_name, activity
EMIT CHANGES;
```

You now have a working aggregate query with multiple GROUP BY items, essentially a composite key for each row. Note that this is not something you need to do with all aggregations in ksqlDB. But having multiple columns you are grouping by is familiar enough that it's worth us explaining how to handle the situation.

What you've seen so far is just scratching the surface of what you can do with ksqlDB. More complex queries are possible, and several built-in functions exist to learn about. But before we continue exploring the capabilities of ksqlDB, let's take a quick pause to discuss some of the possible conceptual types of queries.

11.3   Persistent vs. push vs. pull queries

So far, we've built a streaming application that shows updates as the `user_activity` stream continues receiving user input. Once a client executes the query, the results are continually *pushed* to that client unless the query is specifically terminated. But there could be a situation where you want to issue a single query to retrieve a specific result instead of a constant stream of updates. Additionally, you may need a continuous query that doesn't serve a particular client but can be used by any client issuing a query, something more permanent. We'll discuss how to implement both of these approaches next.

ksqlDB has three categories of query types. One of the types is a push (continuous query), where the stream or table constantly executes against the incoming records. It returns the results to the original client issuing the query. The `activity_leaders` table you created in the previous section exemplifies a persistent query.

Push queries are an excellent choice for an asynchronous workflow; you issue a command or request but don't expect an answer immediately; it executes in the background, and the answer will come later. A concrete example of an asynchronous workflow is sending an email; you write your text and then send it, knowing you'll get a response at some point, but you don't expect an immediate answer.

For more of a synchronous, request–response workflow, ksqlDB offers a *pull* query. An example of a synchronous type of workflow is when you have contractors at your house, and they need your permission to remove a section of the wall. All work stops until they receive an answer. You can issue a pull query against a stream or a table, and there are some restrictions, which we'll get into in a moment.

So the question is, which one should you use? To answer that question, let's compare all three query types (persistent, push, and pull) against each other (figure 11.1).

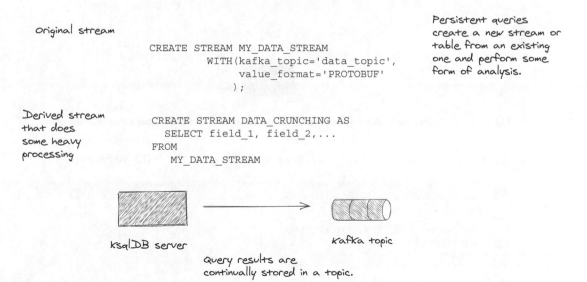

Figure 11.1   Persistent queries run on the server and persist results to a topic.

A persistent query runs on the server; it stores the results of the query in a Kafka topic, so the results persist for the duration of the topic's configured retention time. Also, once you have a persistent query running, you can easily share the outcome because any Kafka consumer client can read the records from the topic. You can think of a continuous query as the workhorse or backbone, and it carries the full load of performing the analysis of your streaming application. You can use the full range of ksqlDB SQL syntax with a persistent query. Persistent queries take the form of CREATE TABLE|STREAM AS SELECT...

A push query, on the other hand, does not persist its results to a topic. A push query returns its results to the client issuing the query. But the results are continually pushed to the client. You can think of a push query as a subscription for changes to the persistent query (figure 11.2).

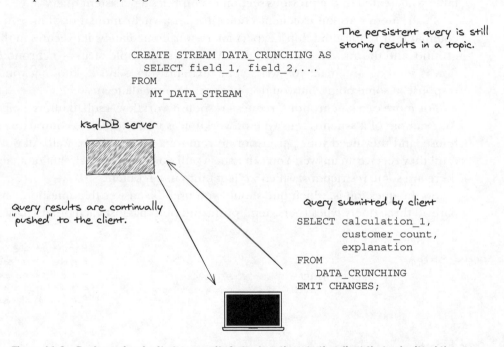

The persistent query is still storing results in a topic.

```
CREATE STREAM DATA_CRUNCHING AS
    SELECT field_1, field_2,...
FROM
    MY_DATA_STREAM
```

ksqlDB server

Query results are continually "pushed" to the client.

Query submitted by client

```
SELECT calculation_1,
       customer_count,
       explanation
FROM
    DATA_CRUNCHING
EMIT CHANGES;
```

Figure 11.2   Push queries don't store results but return them to the client that submitted the query.

A push query against `activity_leaders` could look like the following listing.

Listing 11.18   Push query example on an Activity Leaders table

```
SELECT
    last_name, activity, total_steps
FROM activity_leaders
EMIT CHANGES;
```

With this query, you will receive updates for each user as they update their activity. But these changes aren't stored anywhere. The results of this query return to the original client executing the query, whether it's from the ksqlDB CLI, the REST API, or the available ksqlDB Java client. We'll cover the client options available for ksqlDB later in this chapter. This query returns all updates to the table, but you could refine the results further with a WHERE clause.

Listing 11.19   Push query with WHERE clause

```
SELECT,
  last_name, activity, total_steps
FROM activity_leaders
WHERE total_steps > 1000
EMIT CHANGES;
```

Now, you'll only receive updates where the total number of steps is more than 1,000. The conditions in the WHERE clause can refer to any column the stream or table defines, including the pseudo columns ROWTIME, ROWPARTITION, and ROWOFFSET defined by ksqlDB. These columns are injected or attached to each row for an incoming record into a Kafka topic backing a stream or table. Let's take a moment to define each of these:

- ROWTIME—This is the timestamp associated with the Kafka record, which either the producer or the broker sets, depending on your configuration.
- ROWPARTITION—The records' partition it belonged to from the backing topic.
- ROWOFFSET—Each record in a Kafka topic has an offset, representing its logical position.

When would you use any of these pseudo-columns in a query? Sometimes, you may find it helpful to filter results by external factors about a record. For example, you may only want to view events within a given timeframe. Even if the record doesn't define a timestamp as one of its columns, you can still filter results by the values in the ROWTIME for each one.

You could specify more conditions to refine the results. I won't enumerate them all here, but I'm sure you understand what you do. At this point, you should be able to see the relationship between persistent and push queries. The persistent query carries the entire load and allows you to issue a push query to receive a subset of the information you're interested in. Since the push query doesn't persist its results, it makes sense to use push queries as an alerting source until reaching a given state.

Then, once getting to that state—say, where you observe a given user reaching 10,000 steps—you can terminate your query. You can place a hard cap on the number of results using the LIMIT clause. For example, you are testing a query from the ksqlDB CLI and only want to see whether it works appropriately and then exit. You can change the push query in listing 11.19 to the following listing.

Listing 11.20   Push query with a LIMIT

```
SELECT
  last_name, activity, total_steps
FROM activity_leaders
```

```
WHERE total_steps > 1000
EMIT CHANGES
LIMIT 10;
```

Now, the query will terminate once it emits 10 result records. I should note at this point that push queries can use the full range of SQL commands ksqlDB provides. There are two main differences between persistent and push queries:

1 Push queries don't persist results; they are emitted to the console or back to the client executing the query.
2 Push queries aren't shared. A persistent query evaluates against the incoming records *once* and stores the results in a Kafka topic. But with push queries, if separate clients issue the same queries, ksqlDB evaluates each one independently, even if they provide the same output (see figure 11.3).

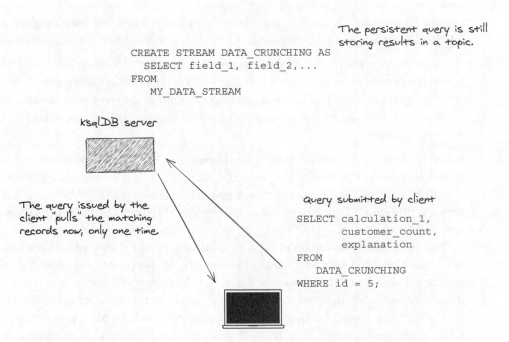

Figure 11.3   Pull queries return point-in-time results once; to get updates, you need to resubmit the query.

Now, let's move on to the final query type, the pull query. While persistent and push queries constantly evaluate the incoming records, a pull query evaluates its statement once and terminates afterward. However, the pull query does not persist in its results in a topic like the push query. ksqlDB returns the outcome of the query to the client. You can think of a pull query as reaching out and pulling down a result at a particular time. A pull query is best when you need an immediate answer, like in a request–response workflow.

Pull queries only support a subset of the ksqlDB SQL statements. You can use a pull query with any stream or any table created with `CREATE TABLE as SELECT`, but currently, not tables created directly against a backing topic. Additionally, pull queries don't support the use of `JOIN`, `GROUP BY`, `PARTITION BY`, and `WINDOW` clauses.

Out of the box, there are limitations to what you can do in a `WHERE` clause. The restrictions in the `WHERE` clause are that you must use a key column, and the comparison needs to be against a literal value. For example, the following listing shows a pull query against the `activity_leaders` table you created earlier.

Listing 11.21 Example of a pull query `WHERE` clause with a key lookup

```
SELECT last_name, activity, total_steps
FROM activity_leaders
WHERE key_1 = 'Smith'
```

Now you can execute a query against the `activity_leaders` table and get results where the last name equals "Smith" and the query terminates. To get any additional updates for this query, you'd have to execute it again.

NOTE There is a configuration you can set that allows for more liberal use of the `WHERE` clause. If you set `ksql.query.pull.table.scan.enabled` to `true` in either a CLI session or a ksqlDB server, it allows for several enhancements, like using non-key columns or comparing them to other columns. See the ksqlDB documentation for more information on table scans: http://mng .bz/lVld.

We've covered a lot of ground on the different query types available, so let's wrap things up with a table to compare each situation where persistent, push, and pull queries are most effective. (table 11.1).

Table 11.1 Where to use persistent, push, and pull queries

Type	Syntax	Best use	Running mode	Results stored in topic
Persistent	`CREATE [STREAM TABLE] AS SELECT.. EMIT CHANGES`	Asynchronous response, heavy work on server	Continual updates	Yes
Push	`SELECT [items] FROM [STREAM TABLE]… EMIT CHANGES`	Asynchronous response, more refined queries	Continual updates	No
Pull	`SELECT [items] FROM [STREAM Materialized TABLE]… WHERE`	Evaluate query and terminate, point in time query	No updates must re-issue for additional results	No

To summarize our table here, persistent queries take the form of `CREATE STREAM | TABLE AS SELECT.. EMIT CHANGES`, and the query persists its results in a Kafka topic. Since different clients can share the results in a topic, persistent queries are best suited for doing heavy or more complex queries. Changes are continually emitted as new records arrive in the stream or table.

A push query starts with SELECT [items] FROM.. EMIT CHANGES, but the results do not persist in a topic; they are returned continually to the client. A push query can use the full range of SQL available in ksqlDB. Since the results do not persist, ksqlDB always evaluates the same query from different clients. The push query is optimal for subscribing to changes in a stream or table as in an event-driven architecture, but usually, the query is much simpler.

Finally, the pull query retrieves a distinct result and terminates; there are no updates as new records arrive. A pull query takes the form of SELECT [items] FROM.. The results do not persist and are returned to the issuing client. The best use of a pull query is obtaining a single result in a request-response format. A pull query has limitations on the SQL statements it can use, most notably in the WHERE clause.

The general pattern then is to have persistent queries running on the ksqlDB server and use a combination of push and pull queries to extract a subset of information from them in your applications. We've concluded our coverage on query types, but before we go on, let's formalize how you can create a stream or table with the different query types.

11.4   *Creating Streams and Tables*

So far, we've established that you can create streams and tables in ksqlDB, but we've done it without categorizing the different ways you can do so. But we'll take care of that in this section. This will also be an excellent time to discuss integrating with Schema Registry because when you define a stream or table where the backing topic has a schema, defining the columns in the create statement is optional. ksqlDB will infer the names and types based on the schema. We'll discuss creating streams and tables without schemas first and then move on to integration with Schema Registry later in this section.

The first category of a stream or table you can create could be considered a "base" stream or table. You generate these base streams or tables directly against a backing Kafka topic. We've seen creating a base stream earlier with the user_activity stream, but it's repeated in the following listing.

Listing 11.22   Creating the user_activity stream

```
CREATE STREAM user_activity (first_name VARCHAR,
                             last_name VARCHAR,
                             activity VARCHAR,
                             event_time VARCHAR,
                             steps INT

     ) WITH (kafka_topic='user_activity',
             partitions=4,
             value_format='JSON',
```

```
                    timestamp = 'event_time',
                    timestamp_format = 'yyyy-MM-dd HH:mm:ss'
        );
```

This statement creates a stream with the backing topic of `user_activity`. If you recall, a Kafka topic stores key-value pair records, but as defined here, this stream contains only values. The keys for each record are `null`. Defining a key on a stream is optional in ksqlDB; in this case, there are no keys for the `user_activity` topic. But what if a topic does have keys? How would you change the `CREATE STREAM` statement here? Let's say the `user_activity` topic has populated keys—an integer representing the user ID; you would update the create statement, as in the following listing.

Listing 11.23   Creating the user_activity stream

```
CREATE STREAM user_activity ( user_id INT KEY,          ◁──┐ Declaring the key
                              first_name VARCHAR,           │ for the stream
                              last_name VARCHAR,
                              activity VARCHAR,
                              event_time VARCHAR,
                              steps INT

    ) WITH (kafka_topic='user_activity',
            partitions4,
            key_format='KAFKA'           ◁──┐ Specifying the data
            value_format='JSON',            │ format of the key
            timestamp='event_time',
            timestamp_format='yyyy-MM-dd HH:mm:ss'
    );
```

So, for adding a key, the only change you needed to make was adding a column declaring its type and adding the `KEY` reserved word telling ksqlDB this is the key of the key-value pair. You also need to tell ksqlDB how the key is formatted, which you've done by specifying the format of `KAFKA`, indicating it's one of the basic types supported by Kafka—for example, `String`, `Long`, and `Integer`. Just like we've seen with Kafka clients and Kafka Streams, having a key will drive the partitioning for the incoming records, and without a key, records are evenly distributed across partitions.

    You can also create a table directly with a backing topic. Let's take our `user_activity` stream and make it a table named `user_activity_table`.

Listing 11.24   Creating the `user_activity_table` table

```
CREATE TABLE user_activity_table (user_id INT PRIMARY KEY,     ◁──┐ Declares the
                                  first_name VARCHAR,             │ primary key
                                  last_name VARCHAR,              │ for the table
                                  activity VARCHAR,
                                  event_time VARCHAR,
                                  steps INT

    ) WITH (kafka_topic='user_activity',
            partitions4,
```

```
        key_format='KAFKA'
        value_format='JSON',                          Defines the format
        timestamp='event_time',                       of the key
        timestamp_format='yyyy-MM-dd HH:mm:ss'
    );
```

Creating a table is similar to creating a stream but with one significant difference. While the KEY column on the stream is optional, it's necessary for a table. You specified the key format, which follows the same rules as the key format specification of the previous stream. Key format values can also be AVRO, PROTOBUF, and JSON_SR (for JSON Schema), but for our purposes, as we've done throughout the book, the examples you'll use will only use keys of type KAFKA.

There are also several differences between a stream and a table with the semantics of keys and values. In a stream, we've established a valid record can have a null key, but a table will drop any incoming record with a null key. Another difference we've covered before is that records with the same key don't affect each other in a stream. They remain independent of each other.

However, in a table, just like in a relational database table, you can only have one primary key, so an incoming record with the same key as a previous record will be an update replacing it. There's also a difference in semantics with null values between a stream and a table. A null value in a stream holds no special meaning, but a null value in a table is considered a *tombstone*. The significance of a tombstone record marks that row for deletion from the table and its backing topic. Let's summarize these differences in table 11.2 for quick reference.

Table 11.2 Stream vs. table

Stream	Table
Key optional	Key required; otherwise, the record is dropped.
Keys don't have to be unique; records with the same key aren't related.	Keys must be unique; records with the same key are updates.
null values have no effect or meaning.	null values are tombstones, marking the row for deletion from the table.
Type KEY	Type PRIMARY KEY

As a side note, it's essential to understand that a row with a null value does not get immediately deleted. Instead, it's marked for deletion. Under the covers, a table uses a backing topic that is a *compacted* topic, and only when the log cleaner runs will the record get removed. The log cleaner runs at regular intervals as configured.

When you create either a stream or table with a CREATE statement directly against a Kafka topic, you don't directly query them. These streams and tables are how you bring a Kafka topic into ksqlDB. It's the subsequent streams and tables you create by selecting from these base streams and tables that provide results either returned to a client in the case of push or pull queries (remember, push queries run indefinitely until terminated, and pull queries evaluate once and end) or persisting their results to a topic.

To summarize, you have three basic types of streams and tables:

1 The base streams or tables that expose a topic to ksqlDB for further queries
2 Persistent queries that publish results to a Kafka topic
3 Push and pull queries that you can run against either the base stream or table or persistent queries

Now that we've wrapped up our coverage on query types, let's move on to the formats of keys and values and Schema Registry integration.

11.5 Schema Registry integration

Schema Registry integrates relatively seamlessly with ksqlDB, and it offers a sizable advantage when defining a stream or a table where there is a schema. Since the schema contains the field names and the types, you can omit the column definitions when you create a stream or table. For example, let's take another look at the `user_activity_table` definition and assume the value used Protobuf.

Listing 11.25 `user_activity_table` with Schema Registry: Protobuf schema

```
CREATE TABLE user_activity_table (user_id INT PRIMARY KEY          Defines the
                                                                   Primary Key
    ) WITH (kafka_topic='user_activity_proto',
            partitions4,
            key_format='KAFKA'                          Specifies the key format
            value_format='PROTOBUF',
            timestamp='event_time',                     Specifies the
            timestamp_format='yyyy-MM-dd HH:mm:ss'       value format
    );
```

We still need to provide the primary key definition since it's a basic KAFKA type, but for the columns, we omit the definitions, and ksqlDB infers the names and types from the schema. A situation like this, where you use a KAFKA type and a type supported by Schema Registry, is known as partial schema reference. If the key were also of type Protobuf, you could simplify the table definition further.

Listing 11.26 `user_activity_table` and Protobuf schema for key and value

```
CREATE TABLE user_activity_table
    WITH (kafka_topic='user_activity_proto',
            partitions4,                                 Specifies the
            key_format='PROTOBUF'                        key format
            value_format='PROTOBUF',
            timestamp='event_time',                      Specifies the
            timestamp_format='yyyy-MM-dd HH:mm:ss'       value format
    );
```

I'm only showing the updated table definition for the `user_activity_table` because the `user_activity` stream would have identical changes for creating the stream with schema-enabled values or keys. There is an exception to this rule of omitting column definitions with schemas, which happens when you only want to use a subset of the columns. For example, let's revisit the `user_activity_table`, but now let's say you only want to use three columns: `last_name`, `activity`, and `steps`. Now, the table definition would look like the following listing.

Listing 11.27   `user_activity_table` and a subset of columns

```
CREATE TABLE user_activity_table (user_id INT PRIMARY KEY,          ◁────    Defines the
                                  last_name VARCHAR,      ◁─────            key column
                                  activity VARCHAR,
                                  steps INT                                 Specifies the
                                                                           value columns
     ) WITH (kafka_topic='user_activity_proto',
             partitions4,
             key_format='KAFKA',
             value_format='PROTOBUF'
     );
```

So here, even though the value is in Protobuf format, we need to declare the names and types of columns since we're only selecting a subset of them. Although the inferencing done by ksqlDB reduces the amount of writing you need to do for a stream or table definition, it reduces the clarity as now you'll have to either issue a DESCRIBE statement or view the physical schema to understand the column names and types.

I have two final points to consider in our Schema Registry section: when ksqlDB infers data types and writes a schema and the conversion of data types.

When you create a persistent query—for recall, that means using the syntax of CREATE STREAM AS SELECT…—it will inherit the key and value format of the base stream or table (figure 11.4). Remember, persistent queries store their results in a topic with the name of the stream or table.

This means if the value is in `Protobuf` format, ksqlDB will register a new schema with Schema Registry using a subject of the stream or table followed by `value`. For example, let's revisit the `user_activity` stream, but this time, we'll say the value is `Protobuf`.

Listing 11.28   The `user_activity` stream in Protobuf

```
CREATE STREAM user_activity (first_name VARCHAR,
                             last_name VARCHAR,
                             activity VARCHAR,
                             event_time VARCHAR,
                             steps INT

     ) WITH (kafka_topic='user_activity',
             partitions=4,
             value_format='PROTOBUF',
             timestamp = 'event_time',
             timestamp_format = 'yyyy-MM-dd HH:mm:ss'
     );
```

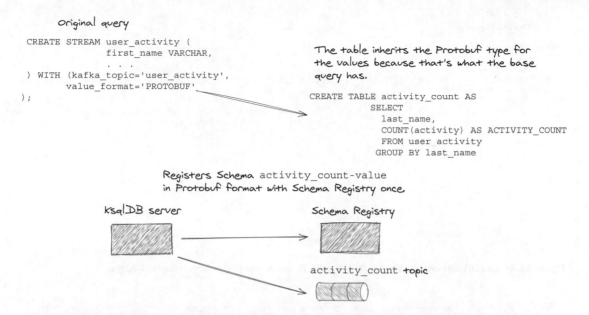

Figure 11.4 **When creating a new query from a persistent one, it will inherit the data format by default.**

Even though we don't have to have the column definitions, we'll have them there, as it should help clarify the example. Let's say you want a persistent query to count the number of activities by the user's last name. You'd end up with a persistent query looking like the following listing.

Listing 11.29 **Persistent query inheriting value format**

```
CREATE TABLE activity_count AS
  SELECT
    last_name,
    COUNT(activity) AS ACTIVITY_COUNT
  FROM user_activity
  GROUP BY last_name
EMIT CHANGES;
```

From creating this table, ksqlDB registers a schema named `activity_count-value`; its format is `Protobuf` since the source stream is in that format. But let's now say that the materialized topic from the query needs to be in `JSON` format, as some downstream clients can't support another data format, as shown in figure 11.5.

That's not a problem for ksqlDB, as you can seamlessly change the data format from the underlying persistent query by overriding the value data format, as in the following listing.

Original query

```
CREATE STREAM user_activity (
          first_name VARCHAR,
          . . .
) WITH (kafka_topic='user_activity',
       value_format='PROTOBUF'
);
```

New query

```
CREATE TABLE activity_count WITH (value_format = 'JSON') AS
         SELECT
            last_name,
            COUNT(activity) AS ACTIVITY_COUNT
         FROM user_activity
         GROUP BY last_name
```

ksqlDB server

activity_count topic

Protobuf -> JSON

Figure 11.5   ksqlDB can change the datatype on the fly when creating a new persistent query.

Listing 11.30   Overriding the value format of a source stream

```
CREATE TABLE activity_count WITH (value_format = 'JSON') AS
  SELECT
    last_name,
    COUNT(activity) AS ACTIVITY_COUNT
  FROM user_activity
  GROUP BY last_name
 EMIT CHANGES;
```

The resulting topic, `activity_count`, will contain records with values in JSON. You can use this ability to transform the record format in ksqlDB with source streams/tables and persistent queries. Since push and pull queries output their results directly to the client in a deserialized form, there's no option or need to convert the resulting format.

As we conclude this chapter, I'd like to give a final example of converting the serialization format. Let's say you have a stream of IoT data in Avro format, and you need to have all records in the stream converted to Protobuf to support more downstream clients. The following lists shows the stream you have in AVRO.

Listing 11.31   Stream defined in AVRO

```
CREATE STREAM IoT_TEMP_AVRO (device_id INT KEY, temp DOUBLE)
  WITH (kafka_topic = "iot_temp", value_format 'AVRO');
```

You want to create an identical stream but in Protobuf format. To do this, you can create a new stream by selecting everything from the `IoT_TEMP_AVRO` stream, as in the following listing.

Listing 11.32 Creating a new stream in a different serialization format

```
CREATE STREAM IoT_TEMP_PROTOBUF WITH (value_format 'PROTOBUF') AS
    SELECT * FROM IoT_TEMP_AVRO;
```

So, you have created an identical stream in a different serialization format with a single line of SQL! Now that we've covered the core of ksqlDB, let's move on to more advanced features, including joins and aggregations.

11.6 *ksqlDB advanced features*

So far in this chapter, you've learned how to use ksqlDB to create streams and tables, but usually, you'll need to use more advanced features to solve complex problems. Consider our scenario from a previous chapter where we have two streams of different purchases, one for coffee bought at the internal store cafe and the other representing all other purchases made in the store. To refresh your memory, we wanted to join purchases made within 30 minutes of each other to create a promotion for the customer. Let's get started by first creating a stream for each category.

Listing 11.33 Creating ksqlDB streams for each purchase category

```
CREATE STREAM coffee_purchase_stream (custId VARCHAR KEY,
                                      drink VARCHAR,
                                      drinkSize VARCHAR,
                                      price DOUBLE,
                                      purchaseDate BIGINT)
    WITH (kafka_topic = 'coffee-purchase',
          partitions = 1,
          value_format = 'PROTOBUF',
          timestamp = 'purchaseDate'
    );


CREATE STREAM store_purchase_stream(custId VARCHAR KEY,
                                    credit_card VARCHAR,
                                    purchaseDate BIGINT,
                                    storeId VARCHAR,
                                    total DOUBLE)
    WITH (kafka_topic = 'store-purchase',
          partitions = 1,
          value_format = 'PROTOBUF',
          timestamp = 'purchaseDate'
    );
```

Now that you have your two streams, the next step is to set up the join itself. Before we do that, let's take a moment to discuss the requirements. Like in Kafka Streams, to perform a join, both streams need to be *co-partitioned*, meaning that the underlying topics must have the same number of partitions and are keyed the same (keys are the same field and type). In our case, both topics have four partitions, and the key for both streams is the customer ID, so they are all set for joining. Now let's take a look at the SQL for the join.

Listing 11.34   Creating a stream-stream join for potential customer rewards

```
CREATE STREAM customer-rewards-stream AS           Selects the customer ID
  SELECT c.custId AS customerId,
       s.total as amount,                          Selects the total amount
       CASE
          WHEN s.total < 25.00 THEN 15             Sets a CASE statement for
          WHEN s.total < 50.00 THEN 50             determining reward points
          ELSE 75
        END AS reward_points
  FROM coffee-purchase-stream c                    Sets the join window
                                                   of 30 minutes with a
    INNER JOIN store-purchase-stream s             2-minute grace period
    WITHIN 30 MINUTES GRACE PERIOD 2 MINUTES
    ON c.custId = s.custId                         The join condition customer
                                                   IDs are the same.
```

So you've selected one field from each stream, the customer ID, and the total amount of the store purchase. To determine the amount of reward points, you use a CASE statement, assigning different point levels depending on the total amount the customer spent in the store. CASE statements are an elegant way of evaluating other conditions based on the value of a field that's part of the query.

> **NOTE**   For stream–stream joins in ksqlDB, other join types—LEFT OUTER, RIGHT OUTER, and FULL OUTER—follow the same semantics you learned about with KStream joins.

But you're not limited to joining streams in ksqlDB. You can also perform stream–table joins. When using a stream–table join in ksqlDB, only new records on the stream side will trigger a result (just like Kafka Streams). So, typically, you'll use a stream–table join when you want to enrich the stream side by performing a lookup in the table.

For example, consider the results of the stream–stream join you just implemented. One of the columns you projected into the join was the customer ID, but you'd like more complete information about the customer. To add additional information, join the customer-rewards-stream with a fact table of members, which contains full details of all the shoppers who participate in the rewards program.

Let's say you have a sink connector that is exporting all the records from a members table into a Kafka topic named rewards-members, so the first thing you'll need to do is create a table in ksqlDB.

Listing 11.35   Creating a lookup table in ksqlDB from an existing topic

```
CREATE TABLE rewards_members (member_id VARCHAR PRIMARY KEY,
                             first_name VARCHAR,
                             last_name VARCHAR,
                             address VARCHAR,
                             year_joined INT)
```

```
    WITH (kafka_topic = 'rewards-members',
          partitions = 1,
          value_format = 'PROTOBUF'
    );
```

Now that you've created your table, you can set up a join with the customer-rewards-stream. But in this case, not all customers are members of the rewards program, so you're going to set up a LEFT OUTER join, which will enable you to filter records later on that don't contain customer information.

Listing 11.36 LEFT OUTER JOIN **to add customer information**

```
CREATE STREAM enriched-rewards-stream
    WITH (kafka_topic='customer-rewards-stream',
          value_format='PROTOBUF') AS
    SELECT crs.custID as customer_id,
           rm.first_name + ' ' + rm.last_name as name,
           rm.year_joined as member_since
           crs.amount as total_purchase,
           crs.reward_points as points
    FROM customer-rewards-stream crs
    LEFT OUTER JOIN rewards-members rm
          on crs.customerId = rm.member_id
```

The CREATE STREAM statement

Specifies the output topic name since we want the name to be different from the stream name

Selects fields from both the stream and table

The LEFT OUTER JOIN statement where the IDs from the stream and table match

You've created an enriched stream by joining the rewards stream with a customer information table. You've specified this as a LEFT OUTER JOIN, meaning you'll still get a join result if the customer-rewards-stream does not find a corresponding record in the rewards-members table. Still, any of the fields representing the table will be null.

This has been an example of a stream–table join, and ksqlDB also supports table–table joins. What's unique about table–table joins in ksqlDB is that it supports foreign and primary key joins. You'll need a foreign key join when the primary key of one table matches a nonprimary key column on another table. For example, let's take the activity-count table you created earlier in the chapter and join it against the rewards-members table from the stream–table join example. I'll repeat the definition of both tables here.

Listing 11.37 Two tables for performing a foreign key join

```
CREATE TABLE activity_count WITH (value_format = 'JSON') AS
  SELECT
    last_name,
    COUNT(activity) AS ACTIVITY_COUNT
  FROM user_activity
  GROUP BY last_name
 EMIT CHANGES;



CREATE TABLE rewards_members (member_id VARCHAR PRIMARY KEY,
                             first_name VARCHAR,
                             last_name VARCHAR,
```

```
                    address VARCHAR,
                    year_joined INT)
     WITH (kafka_topic = 'rewards_members',
           partitions = 3,
           value_format = 'PROTOBUF'
);
```

Let's say that the same retail store purchased your fitness app for which we built the stream-stream and stream-table joins, and they'd like to award members of the rewards club points for store use based on their participation in the fitness app. Since you already have the `reward-members` table, you should be able to join it against the `activity-counts` table.

Now the `activity-counts` table has a primary key of `last_name`, and the `rewards-members` primary key is the member's ID, but it does have a `last_name` column so that we can join on the `rewards-members.last_name` as a foreign key. Since we're joining on a column that's part of the value, we won't have the restrictions of co-partitioning. This is because we are joining against a value, and there's no way to deterministically know which partition it belongs to since we put records on a partition by the key.

> **TIP** If you need to change a key for a stream or table in ksqlDB, use `SELECT` `FROM` and a `partitionBy=<column>` in the `with` clause to get the correct key.

So, let's create the join between these two tables.

Listing 11.38   Foreign key join between `activity_count` and `rewards-members`

```
CREATE TABLE rewards-members-fitness-count AS
SELECT * FROM
activity_count ac JOIN rewards-members rm
  ON ac.last_name = rewards-members.last_name
EMIT CHANGES;
```

Without much effort, you've joined the `activity_count` table with members' information using a nonprimary key on another table.

Before we conclude this chapter on ksqlDB, we should discuss one of its more powerful features. We've discussed the different data formats ksqlDB supports: Avro, Protobuf, and JSON. So far, all the objects we've worked with have been flat, meaning there is one top-level object, and all the fields we want to access are attributes of that object. But what would you do when the data has nested structures? Consider the JSON schema in the following listing.

Listing 11.39   JSON with nested structures

```
"event_id": 1234,
  "school_event": {
      "type": "registration",
      "date": "2023-02-18",
```

```
        "student": {
               "first_name": "Rocky",
               "last_name": "Squirrel",
               "id": 1234567,
               "email": "rsquirl@gmail.com"

        },
        "class": {
               "name": "Geology-100",
               "room": "23RF",
               "professor": {
                       "first_name" : "Bullwinkle",
                       "last_name"  : "Moose"
                       "other_classes" : ["Geology-200", "Rocks-400", "Earth
   Minerals-304"]
                       }


        }
   }
```

This is a deeply nested JSON structure. As you can see from the schema here, that information is available, but how do we model and access it? Fortunately, ksqlDB makes accessing nested data easy.

To access nested data, ksqlDB uses a data type of a STRUCT, which maps string (VARCHAR) keys to arbitrary values. When defining a stream, you will use a STRUCT to describe the schema of the nested data. For example, using the JSON Schema, we just looked at, you'd define a stream for it as shown in the following listing.

Listing 11.40   Using the STRUCT datatype to represent nested data

```
CREATE STREAM school_event_stream (
   event_id INT,
   event STRUCT<type VARCHAR,                    ← Outermost
               date VARCHAR,                       structure
               student STRUCT<first_name VARCHAR,   ← The first nested
                             last_name VARCHAR,       structure
                             id BIGINT,
                             email VARCHAR
                             >,
               class STRUCT<name VARCHAR,
                           room VARCHAR,
                           professor STRUCT<first_name VARCHAR,
                                           last_name VARCHAR,
                                           other_classes ARRAY<VARCHAR>
                                           >
                           >

               >
   )
WITH (kafka_topic='school_events',
     partitions=1,
     value_format='JSON'
   )
```

As seen from this code listing, you define a nested object each time in the same manner you would for any column. First, you provide the name followed by the type, which is STRUCT< followed by the names and types of the fields on the object. When you reach the last field for the object, you will close it with a > character. You repeat this process each time you encounter an object in the schema.

To query the nested fields, you provide the name of the outermost key and use a -> to dereference from the object, again repeating as necessary. To see this in action, let's say you want to write a query that would make course suggestions for each given student based on the other courses taught by a professor of a class the student is currently attending. You'd write a query such as in the following listing.

Listing 11.41   Query on nested data for course recommendations

```
SELECT
      event->student->id as student_id,
      event->student->email as student_email,
      event->class->professor->other_classes as suggested
FROM
   school_event_stream

EMIT CHANGES
```

To access the nested data, you use the name→ pattern until you get to the fields you'd like to retrieve in your query. You can follow the same design to access individual elements of an array or map. For example, let's say you only want to offer one suggestion. So, access the first entry of other_classes and update the query to the following listing.

Listing 11.42   Query nested data and access a specific array entry

```
SELECT
      event->student->id as student_id,
      event->student->email as student_email,
      event->class->professor->other_classes[1] as suggested
FROM
   school_event_stream

EMIT CHANGES
```

Now, your query will only offer one suggested course to take.

> **TIP**   To access individual elements from a nested Map, you'd use name->map_name['key'], and if the map entry was another STRUCT, you could drill down using the same dereferencing syntax.

The book's source code contains SQL files, the required docker compose files, and instructions for running examples shown in this chapter, found in streams/src/main/java/bbejeck/chapter_11.

Summary

- ksqlDB is an event streaming database where you can build event streaming applications using the familiar syntax of SQL. The queries you write will continually evaluate events coming into a Kafka topic and may persist the results to a topic or return them directly to a client application.

- You can create a STREAM or a TABLE in ksqlDB, and they have the same semantics as the corresponding streams and tables in Kafka Streams. A STREAM is an unbounded stream of independent events, and the TABLE is an update stream where event key-value pairs with the same key are an update to a previous one with the same key.

- There are different query types in ksqlDB—source queries where you create a STREAM or TABLE with a backing Kafka topic and persistent queries that select some or all of the columns from a source query and persist their results to a Kafka topic. Persistent query results can be shared with multiple clients. A push query selects a subset of columns from a persistent query, but the results are streamed directly to the client. Push queries will run indefinitely until the client terminates the connections. Pull query results aren't shared; ksqlDB will execute identical queries from different clients. A pull query executes once and terminates. A pull query has some limitations on the SQL statements it supports.

- ksqlDB seamlessly supports Schema Registry serialization formats. Streams and tables inherit their backing topics' key and value format or source streams and tables. You can easily change the format of a stream or table by using a WITH clause and providing a different key and/or value format. ksqlDB automatically registers a schema when creating a stream or table based on a persistent query. You don't have to declare the column names and types in the definition when defining a stream or table and using Avro, Protobuf, or JSON Schema format for your events.

- ksqlDB offers rich library aggregation functions like COUNT, SUM, and AVE out of the box.

- You can create new streams using stream-stream joins or enrich a stream with a stream–table join. Stream–stream and stream–table joins require that both sides are co-partitioned (same key type and number of partitions). You can also perform table–table joins with primary keys, but you can also perform foreign key joins for cases where you want to join on a field in the value of one of the tables.

- You can query arbitrarily nested data using the STRUCT datatype to model a stream or table schema and then use the -> operator to dereference objects and drill down to the desired field.

Spring Kafka

This chapter covers

- Learning about Spring and when to use it with Kafka
- Understanding dependency injection
- Using Spring Kafka for building Kafka applications
- Building Kafka Streams applications with Spring

In this chapter, you will learn about using another open source library, Spring, to build Kafka and Kafka Streams applications. But before we get into that, let's give some quick background on what Spring is and why you'll use it. Spring originates as an IoC (inversion of control) container developed initially by Rod Johnson.

The inversion of control principle means that the main program or application does not control where its dependencies are coming from. If you are familiar with Spring, you can skip this introduction and go directly to the next section, where we use Spring Kafka to build a Kafka producer and consumer application.

12.1 Introducing Spring

A non-IoC application, say a payment processing system, would directly instantiate all of the collaborating components in the application. Essentially, it's in control of

352

the other parts used. This control includes being aware of the concrete types of the components vs. the interface. However, with IoC, the main application only has references to the interface types of the collaborators, and instead of directly instantiating them, the container injects them into the application. This process makes for more flexible and testable applications, as you can change the implementation as needed. Testing is more straightforward since you inject interfaces and can supply mock instances into the application.

Dependency injection is one way of achieving IoC. An external mechanism injects dependencies via constructors, setter methods, or directly at the field level. That's what Spring provides: a mechanism for wiring up applications where each component only uses interfaces. Spring has grown into a sizeable thriving project, providing much more than an IoC container.

As for why you'd want to use Spring, let's look at a concrete example of a payment system mentioned before. Without a doubt, your payment application will require a network connection to receive and send payment information. Additionally, you'll want to use something other than raw network sockets for this network communication. Instead, you'll wish to wrap it up in a software component so that your payment processor doesn't need to know the details of connecting to the network and communicating with it. So, a basic skeleton of your payment processor class could look like the following listing.

Listing 12.1 PaymentProcessor class

```
public class PaymentProcessor implements Processor {        PaymentProcessor
                                                            class
  private NetworkClient networkClient;

                                                            The PaymentProcessor
                                                            class has a collaborator
                                                            (dependency) on the
  public PaymentProcessor(NetworkClient networkClient) {    NetworkClient.
    this.networkClient = networkClient;
  }                                                         Satisfies the dependency
                                                            by passing the
                                                            NetworkClient instance
  public void handlePayment() {                             via the constructor
    payment = networkClient.receive();
    // do some work
    networkClient.send(processedPayment);
  }
                        Uses the NetworkClient
                        to get work done
}
```

From looking at the PaymentProcessor class, you see that it depends on the NetworkClient to complete its job. But notice that the code knows nothing of the NetworkClient other than the exposed methods on the interface. This lack of knowledge is very beneficial, as you don't want the PaymentProcessor to have any knowledge beyond the contract specified by the interface. Why is this important? As time goes on with your project, you'll make changes, and it might include changing the implementation of the NetworkClient, but from how you've written the code here, that won't matter. The `PaymentProcessor` only needs something that implements the expected interface. Otherwise, you'd have to find all uses of the specific implementation and update the usage.

By following this approach, you reap an additional benefit when testing. Ideally, you only want to validate the logic of the `PaymentProcessor` in a test, and you don't need a real network connection to do that. So you can inject a "mock" `NetworkClient` that implements the interface but doesn't connect to the internet. It will simply provide the canned information you've provided. In chapter 14, we'll go into more detail about mocks and different testing approaches.

So now, you can see the benefit of using a dependency injection approach to composing software applications. Still, the question remains: How do I inject the different required classes? Enter the Spring container. Spring provides different annotations that you use to "annotate" the various relationships between types have with each other. Let's take a look at an example.

Listing 12.2   Annotations on classes specifying the relationship between them

```
@Component
public class PaymentProcessor implements Processor {        ◁──── Annotates the
                                                                  PaymentProcessor as a
                                                                  component participating
    private NetworkClient networkClient;                          in the Spring container


    @Autowired                                              ◁──── Specifies to inject
    public PaymentProcessor(NetworkClient networkClient) {        the NetworkClient
        this.networkClient = networkClient;                       in the constructor of
    }                                                             PaymentProcessor
    ...
}

...                                                               Annotates the
                                                                  SecureNetworkClient
                                                                  as a component in
                                                                  the container
@Component
public class SecureNetworkClient implements NetworkClient {  ◁

    public SecureNetworkClient(...) {
        ...
    }


    ....
}
```

By providing these annotations when you start the application, Spring will scan all the classes and wire up the dependencies based on their annotations. So, Spring takes care of putting all the pieces together for you.

This concludes our brief introduction to the Spring framework. The following section will dive into using Spring to create Kafka-enabled applications.

12.2 Using Spring to build Kafka-enabled applications

When building Kafka applications with Spring, you have two choices: you can use standard Spring, which requires more configuration, or you can use Spring Boot. Spring Boot is an extension of the Spring framework. It provides a more opinionated approach by following a convention over the configuration approach. Spring Boot handles many of the details required to use Kafka with Spring. While using either approach is fine, I will only cover using Spring Boot as it makes creating an application more effortless and provides excellent default options like a built-in web server.

Let's say you work for a startup specializing in online loan applications for mortgages, car, and business loans. The company, called Dime, offers substantially lower interest rates and plans to become profitable by having a large volume of loans to compensate for the reduced interest income. The goal is to provide quick turnaround on loan applications by automating the loan application process as much as possible (figure 12.1).

Your application forwards loan applications completed on the company website to a Kafka topic, and a sophisticated underwriting application will process the loans. The underwriting application sends the results to three potential topics, one each for accepted applications, rejected applications, and a quality assurance department that will audit loans selected at random to ensure the rigor of the loan process. Let's look at how you configure a Spring Boot application using Kafka.

Listing 12.3 Class declaration for configuration class with Spring Boot

```
@Configuration
public class LoanApplicationProcessingApplication {

    ...
}
```

The @Configuration annotation marking this as a configuration class for the Spring container

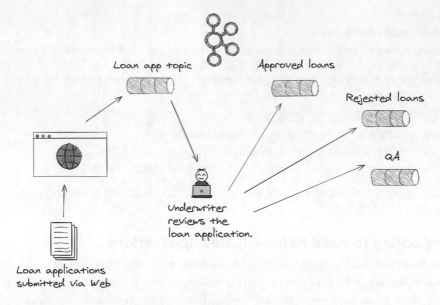

Figure 12.1   Online loan application process

Looking at the class-level annotation, you're specifying this class as the configuration for your application, and Spring Boot looks in the `src/main/resources` directory for a file named `application.properties`. The `application.properties` file contains all values for any fields with the `@Value` annotation. Note that if you give the file a different name or place it in a different location, you'll need to tell Spring where to find it.

Listing 12.4   Basic configuration, which is substantially less with Spring Boot

```
@Configuration
public class LoanApplicationProcessingApplication {
    @Value("${application.group}")
    private String groupId;

    @Value("${loan.app.input.topic}")          Injected
    private String loanAppInputTopic;          configuration

    // Other configurations left out for clarity

    @Bean
    public NewTopic loanAppInputTopic() {          NewTopic bean for creating
        return new NewTopic(loanAppInputTopic,     the input source topic
                          partitions,
                          replicationFactor);
    }

    // Other NewTopic beans left out for clarity
```

Other than some additional fields for injected property values and `NewTopic` beans for creating the required topics, this is all there is to the configuration class for the Spring Boot application! As you can see, Spring Boot removes most of the infrastructure configuration required for a Spring Kafka application. This simplification is only sometimes the case, as we'll see later in the chapter when we change the application requirements. But for the issues where you need the Kafka infrastructure classes as they come straight out of the box, using Spring Boot is a faster path for development.

Before we look into the specific Spring Kafka components, let's look at how you'd start the Spring Boot application.

Listing 12.5 Main class for starting a Spring Boot application

```
@SpringBootApplication(scanBasePackages =            ◁─┐  Specifies a Spring
➥ "bbejeck.spring.application")                         Boot application
public class LoanApplicationProcessingApplication {

public static void main(String[] args) {

SpringApplicationBuilder applicationBuilder =            Creates a
  new SpringApplicationBuilder(                          SpringApplicationBuilder
    LoanApplicationProcessingApplication.class)      ◁─  instance
      .web(WebApplicationType.NONE);         ◁─┐
  applicationBuilder.run(args);       ◁─┐       Sets no WebApplicationType
                                       │       for the application
}                                      │
}                              Starts the application
```

To create a Spring Boot application, you add a `@SpringBootApplication` annotation at the top of the class with the `main` method for starting it. We also provide the base packages containing the different components we want the Spring context to pick up and include.

You see the `SpringApplicationBuilder` class, which we use for building the application; this also creates an `ApplicationContext`, which is the primary interface for configuring the different components specified for the Spring Container. `.web(WebApplicationType.NONE);` establishes that you don't want a web application started.

By default, Spring Boot starts a web server (Apache Tomcat is the default), but we won't need it for our purposes here, so we set the type to `NONE`. Later in the chapter, you'll build another Spring Boot application where you'll need the embedded web server. You start the application in listing 12.5 by executing the `SpringApplicationBuilder.run` method.

> **TIP** You can also have a Spring Boot skeleton application generated for you by going to https://start.spring.io/.

So far, we've covered how to configure a Spring-Kafka application and how you would start it. What's next for us to go over is the different components of the application and how it all ties together. We won't focus too much on the application's business logic, as that's unimportant here; you'll add that to the applications you develop.

12.2.1 *Spring Kafka application components*

There are two main components for your loan processing application: one is to receive the application information and apply the approval algorithm to each incoming record, which contains all the data to process it. Once this first processor uses the algorithm for a given application, meaning it's either approved or not, the result gets produced to a Kafka topic—one for approvals, another for rejections, and a third topic for quality assurance. A certain number of processed loan applications are chosen randomly for review (by a human!) to ensure the algorithm performs as expected.

With that background in mind, let's take a look at the `NewLoanApplication-Processor` starting with the class declaration and constructor first.

Listing 12.6   `NewLoanApplicationProcessor` declaration and constructor

```
@Component
public class NewLoanApplicationProcessor {          ◁   Component annotation marking
                                                        this class as a member of the
                                                        Spring container
  @Value("${accepted.loans.topic}")            ◁
  private String acceptedLoansTopic;
                                                   Injects the names of different
                                                   output topics via properties
  @Value("${rejected.loans.topic}")
  private String rejectedLoansTopic;

  @Value("${qa.application.topic}")
  private String qaLoansTopic;
                                                        Instance
  private final KafkaTemplate<String, LoanApplication>  variable for a
  ⇥ kafkaTemplate;                                  ◁   KafkaTemplate

  @Autowired                                    ◁   Autowired for setting any
  public NewLoanApplicationProcessor(               dependencies via the constructor
          KafkaTemplate<String, LoanApplication> kafkaTemplate) {
     this.kafkaTemplate = kafkaTemplate;
  }
}
```

At the top, where you declare the class name, there is a `@Component` annotation. When starting the application, the Spring container scans for any classes with this annotation and includes them in the application context. Having a `@Component` at the top allows for dependencies to be injected into it or for Spring to use it as a dependency where other classes reference it.

Next, you see the injection of properties for the variables containing the names of the different output topics, and this should look familiar to you as you saw the same thing in the configuration class. Finally, take a look at the last lines of listing 12.6. `private final KafkaTemplate<String, LoanApplication> kafkaTemplate;` is the variable declared for a `KafkaTemplate` instance, and we've decorated the constructor with an `@Autowired` annotation, which instructs the Spring container we want to provide any of the parameters found there. In our case, we'll get the `KafkaTemplate` created at container startup.

> **NOTE** You can also have autowired dependencies at the field level.

Now let's take a look at where the "rubber meets the road," so to speak, in the method where you handle the loan application processing (some details are omitted for clarity).

Listing 12.7 Loan processing handling in the method

Sets this method to receive Kafka records with the @KafkaListener annotation

Sets the consumer group ID

Determines the output topic by loan approval status

Uses the KafkaTemplate to produce the processed loan results

Selects some loans at random for quality assurance

```
@KafkaListener(topics = "${loan.app.input.topic}",
                groupId = "${application.group}")
public void doProcessLoanApplication(LoanApplication loanApplication) {

    boolean loanApproved = debtRatio <= 0.33 &&
        loanApplication.getCreditRating() > 650;
    String topicToSend = loanApproved ? acceptedLoansTopic :
        rejectedLoansTopic;

    LoanApplication processedLoan = LoanApplication.Builder.newBuilder(
        loanApplication).withApproved(loanApproved).build();
    kafkaTemplate.send(topicToSend, processedLoan.getCustomerId(),
        processedLoan);
    if (random.nextInt(100) > 75) {
        kafkaTemplate.send(qaLoansTopic, processedLoan.getCustomerId(),
            processedLoan);
    }
}
```

By placing the `@KafkaListener` annotation at the top of this method, the Spring container uses the `ConcurrentKafkaListenerContainerFactor` to create a `KafkaListener` instance that wraps a `KafkaConsumer` that will consume records from the topic(s) specified in the `@KafkaListener` declaration. We'll explore the relationship between listeners, consumers, and topics soon, but let's continue for the moment by specifying the listener itself.

`@KafkaListener(topics = "${loan.app.input.topic}"` specifies the topic name for the listener, which, under the covers, a `KafkaConsumer` subscribes to. You're setting the consumer group ID for the underlying consumer with `groupId = "${application.group}")`. For both attributes, you'll notice you're not supplying hard-coded values, but instead, you're utilizing property replacements, which again gives you more flexible code. Should the topic(s) or group ID need to change, you only need to update the properties file versus changing and recompiling your code.

You also see here the idea of a message-driven POJO (Plain Old Java Object); there's nothing specific to Kafka about this class. But by applying one line of code, you've taken a simple Java class and converted it into one that can handle receiving messages from a Kafka broker.

After you've processed the loan application (I've intentionally left the code out here as it's not important to learning how to use Spring), you configure the destination topic for the loan based on its approval status. Then, you use `KafkaTemplate` to return the processed loan application record to a Kafka topic.

Here, you'll notice some of the conveniences `KafkaTemplate` provides. You're only providing a topic name, key, and value. If you recall from chapter 4 on Kafka clients, when sending a record with the `KafkaProducer`, you first need to create a `Producer-Record` instance to send to Kafka. The `KafkaTemplate#send` method returns a `ListenableFuture<SendResult<K, V>>` object, and to get the result of the send, you'd need to wait for the future to complete by executing the `ListenableFuture#get` method, but that would block the application until the `get` method returned a result. You can provide the `ListenableFuture` a callback that will process the result of the send asynchronously.

In our example, you're not capturing the returned `ListenableFuture`, but we will revisit how you use the `KafkaTemplate` and update the loan processing application. But before we do that, we have one more process to consider: the post-loan processing.

Listing 12.8   Post-loan processing class

```
@Component
public class CompletedLoanApplicationProcessor {          ← Uses the @Component annotation

@KafkaListener(topics = "${accepted.loans.topic}",          ←
              groupId = "${accepted.group}")
public void handleAcceptedLoans(LoanApplication acceptedLoan) {
    ....
}

@KafkaListener(topics = "${rejected.loans.topic}",          ← Establishes a listener container for each post-processing topic
              groupId = "${rejected.group}")
public void handleRejectedLoans(LoanApplication rejectedLoan) {
    ....
}

@KafkaListener(topics = "${qa.application.topic}",          ←
              groupId = "${qa.group}")
public void handleQALoans(LoanApplication qaLoan) {
    ....
}

}
```

We won't spend much time here as you've already learned about the `@Component` and `@KafkaListener` annotations, but there is some additional information we'll get into about the `@KafkaListener`. I said before that the `@KafkaListener` annotation means the Spring container creates `KafkaListenerContainer` for each annotation it encounters when starting the application. Even if you provide another method with `@Kafka-Listener` (and a different group ID), it will create a separate listener container.

Since you have four methods decorated with `@KafkaListener`, there will be four `KafkaListenerContainer` instances running for your application. Each listener container will wrap one `KafkaConsumer` subscribed to the topic(s) specified in its declaration. The consumer created by the container factory will consume from all topic partitions. Before we move on to the next section, there's one final subject to cover with the `@KafkaListener`.

You've placed the annotation at the method level in all the examples you've seen so far. You can also set it at the class level, which involves some additional work. Let's look at a simple example.

Listing 12.9 Class level `KafkaListener` requires method annotations

Sets the KafkaListener at the class level

```
@Component
@KafkaListener(topics = "${loan.app.input.topic}", groupId =
    "${application.group}")
public class NewLoanApplicationProcessorListenerClassLevel {


    @KafkaHandler
    public void doProcessLoanApplication(LoanApplication loanApplication) {
        // Handle the loan application
    }

    @KafkaHandler(isDefault = true)
    public void handleUnknownObject(Object unknown) {
        // Handle the unknown object
    }
```

KafkaHandler for the loan applications

Default handler for unknown types or types other than the expected LoanApplication

As you can see here, setting the class as the `KafkaListener` is as simple as placing the `@KafkaListener` at the class declaration along with the `@Component` annotation. You must ensure that you have at least one method annotated with `@KafkaHandler`. We added `@KafkaHandler` to the `doProcessLoanApplication` and another `handleUnknown-Object` marked as the default method with the `isDefault` attribute. When using the entire class as a `KafkaListener`, Spring determines the way to use it based on the parameter type of its signature.

This selection process means that all `@Handler` methods must have a single parameter, and the different ones can't have any ambiguity between their parameter types. In this class level `@KafkaListener` example, we've added a handler for a Kafka record that contains a value with a type other than the `LoanApplication`. However, it's not representative of the best use of `KafkaListener` at the class level; I include this here for completeness. I'm going to assert that the canonical use of a listener at the class level would be when you are consuming multiple types from a single topic. Consider the following example.

Listing 12.10 `KafkaListener` at the class level for consuming multiple types

```
@Component
@KafkaListener(topics = "${multi.topic.input}", groupId =        ◁── Sets the
    "${multi.input.group}")                                          KafkaListener
public class NewLoanApplicationProcessorListenerClassLevel {        at the class level

  @KafkaHandler                                                 ◁── KafkaHandler
  public void doSomethingWithLong(String someString) {             for a String
      // Do something with a String type
  }

  @KafkaHandler                                                 ◁── KafkaHandler
  public void doSomething(Long longNumber) {                       for a Long
      // Do something with the Long object
  }

  @KafkaHandler                                                 ◁── Method for
  public void doSomething(Double doubleNumber) {                   processing Doubles
      // Do something with the double
  }

  @KafkaHandler(isDefault = true)                               ◁── Default handler for
  public void handleUnknownObject(Object unknown) {                types other than String,
      // Handle the unknown object                                 Long, or Double
  }
```

Each method has a distinct type for Spring to select for handling the records. So the question is, when would you choose to place the @KafkaListener at the class level versus the method level? While there isn't a hard and fast answer, I take the opinionated approach that one should always favor using a listen at the method level. Topics should represent a single event type and have an appropriate name that makes it easy to reason why that topic exists. Of course, there are always exceptional situations that you'll need to account for, and if one of them is multiple types in a single topic, then using a listener at the class level is one way to handle it.

So, that wraps up our coverage of building your first Spring–Kafka application. Still, there's more to cover, including more advanced functionality that should broadly apply to you when creating an event stream application. A single consumer for all partitions may be sufficient for single-partition topics or topics with low traffic levels. But what if we want a faster way to consume (i.e., not on a single thread)? Additionally, what about getting the key of the Kafka record and other metadata (timestamp, partition) or sending records of different types? That's what we're going to cover next.

12.2.2 *Enhanced application requirements*

Your online loan operation is running, and things are going well. But some changes will make the application better. The changes you'll need to make are as follows:

1 Add more partitions to the input topic.
2 With more partitions, parallelize the application so you have a consumer per partition.

3 Capture the key and timestamp of the incoming record.

4 Track the offset and timestamp of records produced.

While that seems like quite a list, fortunately, the changes needed to accommodate them are easily achievable. We'll only focus on the changes you must make for the Spring-Kafka application. I will assume changes like partition number and domain objects you already know how to do (the updates will be in the source code for you to examine).

Let's take on increasing the partition count and how you will increase the concurrency of your application to improve throughput. Chapter 4 covered the Kafka clients, and we discussed the unit of parallelization of a Kafka topic: the partition. Generally speaking, to increase the throughput of a Kafka application, you add more partitions (or overpartition at the beginning with growth in mind) to assign a single consumer for each partition. By using a dedicated consumer for each partition, you can maximize the application's throughput (I'm generalizing here; of course, there are always exceptions and different considerations that can occur).

You've done some analysis and determined that the input topic for the loan application should have three partitions for optimal throughput. This number considers the current level of applications and the increase you expect soon. You're also going to increase the partitions of the post-processing topics but to a lesser degree by two partitions each. The thought process behind the minor partition increase is that each loan application will only exist in one of two states—approved or rejected—so with an approval rating hovering at 50%, each path in the post-processing will only need to accommodate half of the expected max loan application traffic.

Earlier in the chapter, you saw that by using Spring Boot and the `@EnableKafka` annotation, the Spring container automatically created a `ConcurrentKafkaListener-ContainerFactory` for you. To refresh your memory, the listener container factory creates a `KafkaListenerContainer` for each method decorated with `@KafkaListener`. So if you increase the number of partitions in your input topic, as things stand now, you'll have a single consumer for all partitions, but you want to have a consumer for each one. The good news is that while Spring Boot provides much functionality out of the box, you're always free to alter the configurations, which we'll do now.

The first step is to create a Spring Bean for a `ConcurrentKafkaListenerContainer-Factory`, but we're going to set a specific property, `concurrencyLevel`.

Listing 12.11 Increasing the concurrency level for a container listener factory

> **Creates a new**
> **ConcurrentKafkaListenerContainerFactory**

```
@Bean
public ConcurrentKafkaListenerContainerFactory<String, LoanApplication>
            kafkaListenerContainerFactory() {
    ConcurrentKafkaListenerContainerFactory<String, LoanApplication>
        kafkaListenerContainerFactory =
            new ConcurrentKafkaListenerContainerFactory<>();
    kafkaListenerContainerFactory.setConsumerFactory(consumerFactory());
```

```
        kafkaListenerContainerFactory.setConcurrency(partitions);
        return kafkaListenerContainerFactory;
}
```

**Explicitly sets the level of
concurrency to the partition count**

This configuration here is similar to the one you created in the non–Spring Boot application earlier, with one distinct difference. We've used the `setConcurrency` method, setting it to the number of configured partitions. The effect of putting the concurrency level to the same number of partitions means you get a `KafkaConsumer` per partition, which is what you'll need for maximum throughput.

Another point to consider is that we've used the same name for the method as the default expected by the container. Why is that? By using `kafkaListenerContainer-Factory` for the method name (remember, without providing a `name` attribute, the method name becomes the bean name), the Spring container will pick up your custom container factory instead of creating the default factory with the same name. Using this "shadow" bean naming process, your processing class will automatically use the updated container factory with no code changes, which keeps your code flexible as it picks up and uses either container.

That's not to say you should always use the same name when overriding a Spring Boot default. You can give it any name you want when using a custom factory. You'll need to explicitly tell the `KafkaListener` which container factory to use by adding an attribute, `containerFactory` to the `@KafkaListener` annotation.

Listing 12.12   Adding the custom container factor for the listener

```
// In the configuration class
@Bean("custom-container")
public ConcurrentKafkaListenerContainerFactory<K, V>
                                customContainerFactory()
```

**Tells the KafkaListener which
container factory to use**

```
// In the @Component class
@KafkaListener(groupId = "${application.group}",
            containerFactory = "custom-container",
```

**Explicit reference
to use the custom
container**

This code is the alternative to providing a custom container. In the configuration class, you create the customized container, and you supply a name to the `@Bean` annotation, and in the `@Component` class, you use the same name for the `containerFactory` attribute.

> **NOTE** To increase the concurrency of the `ConcurrentKafkaListener-ContainerFactory`, you can also provide the configuration `spring.kafka.listener.concurrency` and set it equal to the desired number. It could be something like `spring.kafka.listener.concurrency=${num.partitions}`, assuming the `num.partitions` comes earlier in the `application.properties` file. I've chosen to create a custom container factory to serve as an example of providing an override to the defaults provided by Spring Boot. Another

approach would be directly providing a `concurrency` attribute with the `@Kafka-Listener` annotation.

Now that you've achieved the desired consumer-per-partition ratio, there's an additional consideration you'll have to take into account: you now have a multithreaded processing application.

The Spring container creates a separate thread equal to the number you set for the concurrency level. Creating new threads is expected and is essential for increased throughput. But this means each thread will call the listener method *concurrently*. The concurrent calls are acceptable if the method is *thread safe* (i.e., no shared mutable state exists).

> **TIP** It's easy to find the threads associated with a particular Kafka listener by adding an `id` attribute to the `@KafkaListener` annotation. Spring will use the ID you provide as part of the thread name, making it easy to identify which threads consume from different topics.

In the example code, the application shares a single `KafkaTemplate` instance across all threads, but since the underlying `KafkaProducer` is thread safe, the template is as well. It's good to keep the Kafka listener methods stateless; otherwise, you must add synchronization to your code to ensure you get deterministic results.

> **NOTE** I'm not going to cover Java threading or synchronization here, but a quick Google search on the topic will yield plenty of resources for you to explore the subject.

Before moving on, we have two more application improvements you need to implement from our previous list. The remaining requirements are as follows:

1. Capture and log the key and timestamp of the incoming loan application.
2. When producing the processed loan result, log the offset and timestamp of the produced record. Also, if there's an error producing the record, you won't have an offset and timestamp, so you'd like to log out the error as well.

To retrieve the key, timestamp, and other metadata associated with the incoming Kafka record, you'll apply `@Header` annotations to additional parameters to your listener method. So, the changes you'll make to your loan processing class will look like the following code listing.

Listing 12.13 Getting the message key and timestamp from the incoming Kafka record

```
public void doProcessLoanApplication(LoanApplication loanApplication,
        @Header(Kafka Headers.RECEIVED_TIMESTAMP)
    long timestamp,                                         ← Extracts the
        @Header(Kafka Headers.RECEIVED_MESSAGE_KEY)           original timestamp
    String key)      ←                                       for the record
                        Gets the key for
                        the Kafka record
```

By adding the `@Header` annotation with the desired specific header, you can retrieve the original timestamp and message key. Other header values, such as the topic, partition, and offset, are available.

> **TIP** When extracting header information for consumed records, use `Kafka Headers.RECEIVED_X`. The `KafkaHeaders` class provides constants for both producer and consumer records, and the ones for the consumer records start with `RECEIVED`.

Now that you've retrieved the message key and the timestamp of the incoming record, your next task is to log the offset and timestamp of the *produced* record after loan processing. Earlier in the chapter, we demonstrated that the loan application processing class, `NewLoanApplicationProcessor`, after processing a loan, produces a record back to Kafka with the approval status of the loan determining the topic. To refresh your memory, the following listing shows the line of code responsible for producing the record.

Listing 12.14   Producing a processed loan

```
kafkaTemplate.send(topicToSend, processedLoan.getCustomerId(),
    processedLoan);
```

We also mentioned that the `KafkaTemplate.send` method returns a `Listenable-Future<SendResult<K, V>>`, and you could extract the timestamp and offset directly at that point. But that is the drawback of needing to call the `ListenableFuture.get` method, which will block your application's main thread until the `get` method returns, meaning waiting until the produce request is complete. Instead, you'd not take that approach as that would affect performance. So what you can do in this case is supply the returned `ListenableFuture`, a callback that will execute asynchronously when the produce request is complete, whether it's a success or failure.

Listing 12.15   Callback for information after the produce request completes

Creates an instance of the
ListenableFutureCallback

```
private final ListenableFutureCallback<SendResult<String, LoanApplication>>
        produceCallback = new ListenableFutureCallback<>() {
  @Override
  public void onFailure(Throwable ex) {                          ◁── The onFailure method
      LOGGER.error("Problem producing a record", ex);                  for logging errors
  }

  @Override
  public void onSuccess(SendResult<String, LoanApplication> result) {  ◁──
      RecordMetadata metadata = result.getRecordMetadata();
      LOGGER.info("Produced a record to topic {} at offset{} at time {}",
              metadata.topic(), metadata.offset(), metadata.timestamp());
  }
};
```

Logs the required metadata from a
successful produce request

Now that you have your callback created, it's simply a matter of adding it to the resulting `CompletableFuture` after executing a `KafkaTemplate.send` action.

Listing 12.16 Adding a callback for notification of completed produce requests

```
ListenableFuture<SendResult<String, LoanApplication>>
    produceResult =
    kafkaTemplate.send(topicToSend,
                       processedLoan.getCustomerId(),
                       processedLoan);

produceResult.addCallback(produceCallback);
```

Captures the
ListenableFuture in a
variable after the send call

Adds the callback for execution
when the future completes

Now, you'll get notification of completed produce requests even in the case of a failure. This process is similar to what you saw in the chapter on Kafka clients. Still, when working directly with a `KafkaProducer`, you add the callback now as a parameter to the `KafkaProducer.send` method.

We've wrapped up our coverage of building a Kafka application using Spring Kafka, and now we'll move on to using Spring Kafka with Kafka Streams applications.

12.3 *Spring Kafka Streams*

Just as we saw how Spring Kafka with Spring Boot could simplify building Kafka-based applications, there is also support for building Kafka Streams applications. It's different since Kafka Streams already abstracts away a lot of the details of working with Kafka. As with using any tool or framework, there are tradeoffs to consider. In this case, I'm referring to simplified application development versus more control over the application-building process.

When using Spring Boot with Kafka Streams, there are some advantages to dealing with less infrastructure code. Still, there's some loss of control over how to build the application versus if you were to use Spring for its dependency injection capability only. We'll look at both approaches so you can decide which direction is best for you. We'll also cover one of the great features of using Spring Boot with Kafka Streams; Spring Boot applications will, by default, launch a web server when you start them.

Having a web server automatically with our Kafka Streams application is a real advantage when using interactive queries (IQs). IQs enable you to query the results of stateful operations in Kafka Streams directly. So, within your Spring–Kafka Streams application, you can include web-based classes to handle incoming requests for serving up IQ queries. This has the potential to simplify your application architecture. We'll explain how that works later.

Let's start with a simple Kafka Streams application, first using all the Spring Boot utilities and then just using Spring to wire up the application. You've decided to redo your Kafka loan application to use Kafka Streams. It will still use the same logic for loan approval, but instead of using listeners and the `KafkaTemplate`, it's all handled in Kafka Streams. This will allow you to perform aggregations on loan approvals and rejections directly in one place. Your first step toward using Spring Kafka with Kafka Streams is that you'll add an annotation to the configuration class.

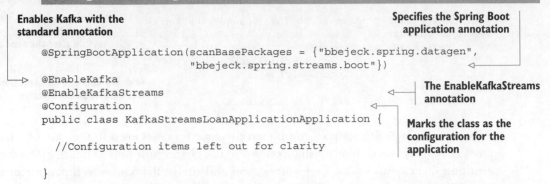

Listing 12.17 Adding annotation to the configuration class

Enables Kafka with the
standard annotation

Specifies the Spring Boot
application annotation

```
@SpringBootApplication(scanBasePackages = {"bbejeck.spring.datagen",
                       "bbejeck.spring.streams.boot"})
@EnableKafka
@EnableKafkaStreams
@Configuration
public class KafkaStreamsLoanApplicationApplication {

  //Configuration items left out for clarity

}
```

The EnableKafkaStreams
annotation

Marks the class as the
configuration for the
application

You've seen some of this code before, but the @EnableKafkaStreams is new and required to activate Spring Boot's support for running Kafka Streams applications.

TIP You'll notice in listing 12.17 that we've specified the packages to scan for components to include in the container. You won't need to do this if you have everything required for the application in the same package as the annotated configuration class.

Applying the @EnableKafkaStreams annotation, Spring will create a wrapper around the KafkaStreams instance and control the lifecycle (starting and stopping) of the streams application. Before we build the Kafka Streams application itself, you'll need to provide one more bit of configuration.

Listing 12.18 Configurations required for Kafka Streams with @EnableKafkaStreams

```
@Bean(name =
    KafkaStreamsDefaultConfiguration.
  DEFAULT_STREAMS_CONFIG_BEAN_NAME)
KafkaStreamsConfiguration kafkaStreamsConfiguration() {
  Map<String, Object> streamsConfigMap = new HashMap<>();
  streamsConfigMap.put(StreamsConfig.APPLICATION_ID_CONFIG,
                  "loan-processing-app");
  streamsConfigMap.put(StreamsConfig.BOOTSTRAP_SERVERS_CONFIG,
                  bootstrapServers);
  return new KafkaStreamsConfiguration(streamsConfigMap);
}
```

Bean annotation;
provides the name

Creates the
HashMap of
required
configurations

Returns the KafkaStreamsConfiguration instance

When starting a Spring Boot application with the @EnableKafkaStreams annotation, the application will expect to find a Spring Bean named defaultKafkaStreamsConfig providing the configurations used to create the StreamBuilderFactoryBean, which creates the StreamsBuilder instance and will also control the starting and stopping of the Kafka Streams application. If you don't require any modifications on the

`KafkaStreams` instance itself, then at this point, you've done everything needed to run a Kafka Streams application. All that is left is to create the application itself.

But before we build the application, let's quickly discuss what steps you can take when you need access to the underlying `KafkaStreams` instance. For those times you require access, Spring provides the `StreamsBuilderFactoryBeanCustomizer` interface, a functional or single abstract method interface. The functional interface is ideal to work with as you can use a Java lambda to represent it versus a concrete object instance. So when would you need to access the `KafkaStreams` object? Consider the case where you'd like a `StateListener` to notify you of when `KafkaStreams` transitions to a running state so you can log the active tasks' topic-partition assignment information.

So, to set the `StateListener`, you'd create two new bean definitions in your configuration class. The first is the `KafkaStreamsCustomizer`, which gives you access to `KafkaStreams` before it starts. The second bean is the `StreamsBuilderFactoryBeanCustomizer`, which accepts the `KafkaStreamsCustomizer` for applying your desired changes.

Listing 12.19   Kafka Streams customizer for setting a `StateListener`

```
@Bean
KafkaStreamsCustomizer getKafkaStreamsCustomizer() {
 return  kafkaStreams ->                                             Sets the state
   kafkaStreams.setStateListener((newState, oldState) -> {          listener on
   if (newState == KafkaStreams.State.RUNNING) {                    Kafka Streams
      LOG.info("Streams now in running state");
    kafkaStreams.metadataForLocalThreads()                          Logs the
      .forEach(tm -> LOG.info("{} active task info: {}",            currently active
                    tm.threadName(), tm.activeTasks()));            tasks for each
   }                                                                StreamThread
   });
}
```

So with the `KafkaStreamsCustomizer`, we can access the `KafkaStreams` instance—in this case, to set the `StateListener`. To get this bean into the `StreamsBuilderFactoryBean`, you create the second bean definition as shown in the following listing.

Listing 12.20   Creating a customizer to apply `KafkaStreams` object settings

```
                                            Creates the builder factory customizer
@Bean
StreamsBuilderFactoryBeanCustomizer kafkaStreamsCustomizer() {
  return  streamsFactoryBean ->
 streamsFactoryBean
    .setKafkaStreamsCustomizer(getKafkaStreamsCustomizer());
}
                                            Sets the KafkaStreamsCustomizer
```

With these two bean definitions added to the configuration class, you can access the `KafkaStreams` object.

Now that we've covered how to access the `KafkaStreams` object in the Spring Boot application, let's build the application itself. Building the Kafka Streams application with the `@EnableKafkaStreams` annotation differs from what we've seen before. To show the differences, let's dive right into an example. Let's say you've taken your loan application and converted it from using Kafka producer and consumer clients to a Kafka Streams application.

Listing 12.21   Converted Kafka producer and consumer app to Kafka Streams

```
@Component                                          ◁——  Declares the class
public class LoanApplicationProcessor {                   as a component

  @Value("${loan.app.input.topic}")
  private String loanAppInputTopic;
                                                          Method constructing
                                                          the topology for Kafka
                                                          Streams
@Autowired
public void loanProcessingTopology(StreamsBuilder builder) {    ◁——

KStream<String, LoanApplication> processedLoanStream =    ◁——  Constructs
   builder.stream(loanAppInputTopic,                            the KStream
               Consumed.with(stringSerde,
                     loanApplicationSerde))
               .mapValues(loanApp -> {
double monthlyIncome = loanApp.getReportedIncome() / 12;
double monthlyDebt = loanApp.getReportedDebt() / 12;
double monthlyLoanPayment =
       loanApp.getAmountRequested() / (loanApp.getTerm() * 12);
double debtRatio =
       (monthlyDebt + monthlyLoanPayment) / monthlyIncome;

boolean loanApproved =
       debtRatio <= 0.33 && loanApp.getCreditRating() > 650;

return LoanApplication.Builder.newBuilder(loanApp)
                          .withApproved(loanApproved)
                          .build();
  });
}
```

From looking at this code listing, it's standard Spring configuration, declaring a class as a `@Component` and injecting the desired objects with the `@Autowired` annotation, this time on a method, not a constructor. I mentioned some differences previously, and if you look closely, you'll notice the `void` return type for the `loanProcessing-Topology` method. The method doesn't return the `StreamsBuilder` instance since the Spring container manages it.

When starting the container to build a stream application, Spring will pass around a singleton `StreamsBuilder` to all methods that reference it and have the `@Autowired` annotation on it. Then, when `StreamsBuilderFactoryBean` begins starting the `Kafka-Streams` instance, it will execute the `StreamsBuilder.build` method.

The effect of having Spring control the `StreamsBuilder` instance in this way means you can potentially have the classes making up your topology spread out among different classes. While it's possible to take this approach, I'd still recommend having the entire topology in a single class, as when it comes time to debug any potential issues, it will be much easier to track down what's wrong by viewing the entire topology in one place.

You've just seen how to build a Kafka Streams application with Spring Boot and the `@EnableKafkaStreams` annotation. Taking this approach does make things easier, as Spring takes care of most of the configuration details for you. But everything is a tradeoff. Here, while you gain the ease of starting up and managing Kafka Streams, you also lose some visibility into what's happening with the application.

But there's another tradeoff you can make: you can give up some of the convenience in exchange for more control and visibility into how Kafka Streams is put together, and we're going to cover that next. You'll still use Spring to wire the application together. Instead of the Spring container managing the lifecycle of the `KafkaStreams` object and how the topology gets put together, you'll take on that responsibility. We will take the same Kafka Streams application and make minor changes, starting with the class that builds the topology (some details are omitted for clarity).

Listing 12.22   Renaming the class to reflect its role

```
@Component
public class LoanApplicationTopology {          ◁——— New name of the class


}
```

The first change is to rename the `LoanApplicationProcessor` to `LoanApplication-Topology`. By changing the name, you're declaring the intent that the class will contain the entire topology for the Kafka Streams application, not just a single processor or section of it. Next, you'll update the signature of the `loanProcessingTopology` method, which itself will require an additional change.

Listing 12.23   Updating the method and explicitly creating a `StreamsBuilder`

```
private final KafkaStreamsConfiguration streamsConfigs;

@Autowired
public LoanApplicationTopology(                       ◁  Injects configurations
    KafkaStreamsConfiguration streamsConfigs) {           into the class via
  this.streamsConfigs = streamsConfigs;                   constructor injection
}

public Topology topology() {                          ◁  Changes the method
  StreamsBuilder builder = new StreamsBuilder();          name and removes the
                                                          StreamsBuilder parameter
// Build topology
                                                      ◁  Directly creates a
                                                         StreamsBuilder
return builder.build(streamsConfigs.asProperties());  ◁  Returns a Topology
```

So, the first change to support the method change is to update the constructor to auto-wire the `KafkaStreamsConfiguration`. You'll see next how this is related to refactoring the method. The difference at annotation two is that you've changed the method's name to `topology` to reflect its role and removed the `StreamsBuilder` parameter, creating it directly instead at annotation three. You've also updated the return type from `void` to `Topology`, reflecting the change at the last line of the method where you execute `StreamBuilder.build` returning the `Topology` instance.

The next change is an addition—creating a class to support making the Kafka-Streams instance and controlling the lifecycle of the streams application (some details are omitted for clarity).

Listing 12.24   Creating a class to support Kafka Streams

```
@Component
public class KafkaStreamsContainer {

@Autowired
public KafkaStreamsContainer(
        final LoanApplicationTopology loanApplicationTopology,     ⟵   Injects the LoanApplicationTopology object
        final KafkaStreamsConfiguration appConfiguration) {     ⟵   Injects the application configurations
    this.loanApplicationStream = loanApplicationTopology;
    this.appConfiguration = appConfiguration;
}
```

You've created the `KafkaStreamsContainer` class to handle the tasks required to build and run the Kafka Streams application. Notice the constructor has the `@Autowired` annotation and two parameters, the `LoanApplicationTopology` and the `KafkaStreams-Configuration` class that Spring will inject for us. Next, let's see how you'll use these two objects to get Kafka Streams up and running.

Listing 12.25   Creating and running Kafka Streams

```
@PostConstruct     ⟵   The PostConstruct annotation
public void init() {
    Properties properties = appConfiguration.asProperties();
    Topology topology = loanApplicationStream.topology();     ⟵   Creates the topology
    kafkaStreams = new KafkaStreams(topology, properties);     ⟵   Builds the KafkaStreams instance

    kafkaStreams.setStateListener((newState, oldState) -> {
      if (newState == KafkaStreams.State.RUNNING) {
        LOG.info("Streams now in running state");
        kafkaStreams.metadataForLocalThreads().forEach(tm ->
          LOG.info("{} assignments {}", tm.threadName(), tm.activeTasks()));
      }
    });
    kafkaStreams.start();     ⟵   Starts KafkaStreams
}
```

Next, you add the `init` method that creates the `KafkaStreams` instance and starts it running. But the question is when to call the `init` method. This part is handled by

Spring for you via the `PostConstruct` you've decorated the method with. When you add a `PostConstruct` to a container component class, the Spring container will execute the method when the object is fully constructed, so in this case, after the required dependencies are injected, Spring will call the `init` method, automatically creating the `KafkaStreams` instance and starting it up. But we did say you would handle the full lifecycle, so what about stopping a `KafkaStreams` application? For that, you'll add one more method.

Listing 12.26 Stopping the Kafka Streams application

```
@PreDestroy                                        ⟵─── The PreDestroy annotation
public void tearDown(){
    kafkaStreams.close(Duration.ofSeconds(10));    ⟵─── Stops KafkaStreams
}
```

To stop the application, you'll use a similar approach, placing the `PreDestroy` annotation on a method that the Spring container will call before it tears down the managed component when it begins shutting down.

Now, the changes are complete, giving you more visibility into the construction of the Kafka Streams application and how it's started and shut down. There isn't a right or wrong decision here with the approach you can take, and it comes down to personal preference and your different requirements.

Summary

- Spring Kafka provides abstractions, making it easier to work with Kafka. You can use the standard approach of classic Spring, but you'll have to add the required configuration for the `KafkaTemplate` and `KafkaListener` supporting classes. By contrast, by using Spring Boot, the configuration level you need is significantly reduced by adding `@SpringBoot` and `@EnableKafka` annotations at the class declaration level. Typically, you'd also want to place this on a configuration class denoted with a `@Configuration` annotation. Since Spring Boot follows the practice of convention over configuration, if you're happy with the defaults, there is little configuration for you to do.
- When setting the concurrency level for a `KafkaListener`, the Spring Container will create a corresponding number of threads running a `KafkaConsumer` for the listener. You must ensure the method decorated with the `@KafkaListener` annotation is thread safe when setting the concurrency level to a value of more than one.
- Spring Boot also provides an `@EnableKafkaStreams` annotation that covers the lifecycle of working with Kafka Streams. But this is not required, and if you prefer to have more control over your stream application, you can use Spring just for the dependency injection capabilities only.

Kafka Streams Interactive Queries

This chapter covers

- Learning about querying the state of a Kafka Streams application
- Discovering what's required to enable Interactive Queries
- Building an Interactive Queries app with Spring Boot

Earlier in this book, you learned how to build stateful applications in Kafka Streams. When you enable a stateful operation like an aggregation, Kafka Streams creates a state store to hold the calculation results.

The point of running an aggregation or something similar is you need to gain information or insight from the combined values that it contains. For example, your application keeps track of failed login attempts. Too many indicate someone is trying to gain illegal entry into an account or machine. Another example of a less nefarious nature is tracking page views of a website to determine the most advantageous time to run ads. I could continue listing examples, but you understand. But you have to be able to view this information promptly for it to be helpful. Otherwise, it's not worthwhile.

In this chapter, you'll learn how to enable Interactive Queries (IQ) from your Kafka Streams applications. You'll also learn how to build your web-based dashboard application with Spring Boot that will render the stateful results of a Kafka Streams example you previously created in the book.

13.1 Kafka Streams and information sharing

Typically, to view the results of a calculation performed by an application, you need first to export the results to a relational database. Then, connect a dashboard application to the database to pull the information for display. Figure 13.1 shows how this could look.

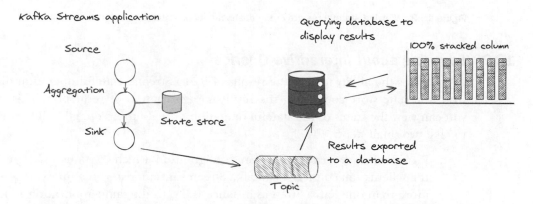

Figure 13.1 Exporting stateful results to a database for viewing from a dashboard application

From this picture, you can easily see the flow of information. An application performs a stateful computation and exports the results to a database, and the dashboard queries the database to display the results. This approach is an acceptable way to access information generated from an application. But Kafka Streams offers the ability to simplify this information-sharing process. This ability comes from IQ. IQ provides the ability to query the information contained in the state store directly. This direct access will simplify your architecture because you can now view the data as Kafka Streams calculates it, eliminating a couple of moving parts from your application infrastructure. Figure 13.2 provides a picture to help demonstrate what I'm talking about here.

So, as you can see here, by directly accessing the information from the state store, you get to comprehend the events as they're happening. You've also simplified your architecture by removing the database and a data "transfer."

> **NOTE** I'm not saying that IQ will eliminate the need for a relational database. You'll most certainly always require them in your applications. But with IQ, you now have another option that, in some situations, helps you view the information you need more efficiently.

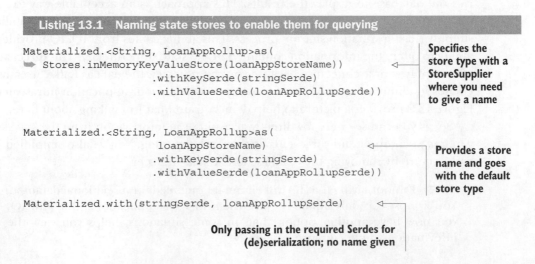

Figure 13.2   Viewing stateful results in the dashboard application directly from the source

13.2   *Learning about Interactive Queries*

IQ provides the ability to view the results of Kafka Streams stateful operation directly from the state store containing the results of aggregation or reduction. This means you can view the status of the stateful operation in real time. To enable IQ, you need to take two small steps:

1   Provide the `application.server` configuration, which contains a host and port that clients can connect to a Kafka Streams instance to run a query. If you have more than one Kafka Streams instance (all with the same application ID), then each instance must provide a unique host:port combination.

2   Give each state store a unique name via the `Materialized` configuration object. While it's true that Kafka Streams will give each state store a name if none exists, the store is unavailable for querying if you don't explicitly provide one. The following listing provides a code example demonstrating what I mean.

Listing 13.1   Naming state stores to enable them for querying

```
Materialized.<String, LoanAppRollup>as(
➥  Stores.inMemoryKeyValueStore(loanAppStoreName))          ◁
               .withKeySerde(stringSerde)
               .withValueSerde(loanAppRollupSerde))


Materialized.<String, LoanAppRollup>as(
               loanAppStoreName)                             ◁
          .withKeySerde(stringSerde)
          .withValueSerde(loanAppRollupSerde))


Materialized.with(stringSerde, loanAppRollupSerde)           ◁
```

Specifies the store type with a StoreSupplier where you need to give a name

Provides a store name and goes with the default store type

Only passing in the required Serdes for (de)serialization; no name given

You've indicated you want to use an in-memory store with the given `StoreSupplier`, and the name you provide is the store name. You're only giving the name and accepting the default store type, which is persistent. Either way, both of these approaches enable the store for queries. The choice of the `StoreSupplier` doesn't matter. With any of them, you must give a name, which becomes the store's name. You don't use a name or a supplier, so this store will not be eligible for querying.

To refresh your memory, Kafka Streams ends up with a state store per task, and since each task represents a single partition, you end up with a store per partition. What makes this critical to our discussion here is that when you deploy multiple Kafka Streams application instances, each instance is only responsible for a subset of the total number of tasks (partitions), as Kafka Streams spreads the processing load across other applications. For example, if the source topic has six partitions and you've started three application instances, each one will be responsible for two.

Since the state stores for Kafka Streams are local for each instance, the effect of spreading the processing load is that you're distributing state across multiple machines. Figure 13.3 is an illustration that should help make this concept clear to you.

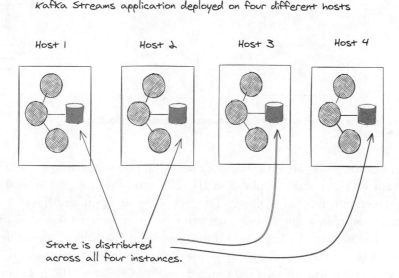

Figure 13.3   State in Kafka Streams distributed across multiple machines

Given that you have distributed state and the state stores in Kafka Streams are key-value pairs, how do you know which instance to query? The excellent news is that Kafka Streams provides the infrastructure, so you don't have to know which instance is responsible for the partition where a given key is. You pick one to query, and if its state doesn't have the key in question, it knows which one does. It will forward the query on your behalf and return the results to you.

This routing of queries is possible because the rebalance protocol allows for encoding arbitrary data in the payload. So when Kafka Streams application instances rebalance, in addition to providing all the partitions each one is responsible for, they also include the individual `application.server` configurations. Let's look at figure 13.4, which is a graphic illustrating this process.

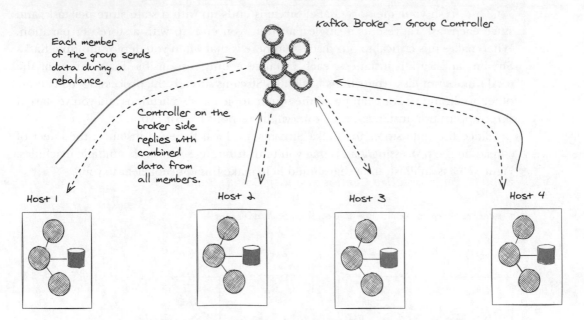

Figure 13.4   Rebalance distributes metadata to all instances with the same application ID.

So, each application ends up with metadata knowing all the partitions each other instance (again with the same application ID) is responsible for. Additionally, the application server information is known for each one, so when you query a Kafka Streams application with a given key, it determines the partition the key would fall into by taking the hash of the key modulo the number of partitions (assuming hash partitioning). If the current Kafka Steams app doesn't own that partition, it knows which one does, and it will forward the request since it also knows the host and port for the queries on that machine. This metadata sharing and request forwarding is a somewhat complex process, so figure 13.5 shows this process in action.

While Kafka Streams provides the internal plumbing for application instances with the same application ID to share metadata, the communication layer between them does not. You must implement the query serving layer and the internal communication parts, which you will do in the next section.

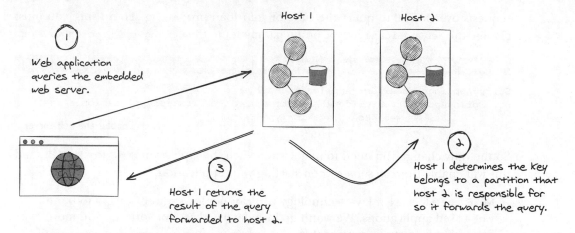

Figure 13.5 Determining the correct host for a given query

13.2.1 Building an Interactive Queries app with Spring Boot

We will take a step-by-step process to build the IQ layer in your Kafka Streams application. The good news is that you've already laid the groundwork with the example application, `bbejeck.spring.streams.container.KafkaStreamsLoanApplication-Application`. The steps you're going to take next are as follows:

1 Enable the Spring Boot application to have an embedded web server when starting up.
2 Add a `RestControler` to handle incoming query requests either directly from users or sibling applications.
3 Inside the controller, the class adds logic to process queries.
4 Build an HTML index page to make REST API calls and continually display loan application results on a single page.

Along the way with building this, you'll also learn about the latest version of IQ (IQv2), a significant improvement over the first version.

So, let's tackle each item on our list in order, starting with enabling a web server on start-up. With our previous Spring Boot application in the main method, we explicitly turned off the web server, as shown in the following listing.

Listing 13.2 Disabling the web-server

```
SpringApplicationBuilder applicationBuilder =
new SpringApplicationBuilder(LoanApplicationProcessingApplication.class)
        .web(WebApplicationType.NONE);
```
Explicitly turns off the web server

You turned off the web server as you did not need it. This application ran with no need to accept external requests. But now you want to expose your app to receive requests over HTTP to query the state of your loan processing. To do this, you must change the `WebApplicationType` enum parameter to the `SpringApplicationBuilder`.

Listing 13.3   Enabling a web server for a Spring Boot application

```
SpringApplicationBuilder applicationBuilder =
new SpringApplicationBuilder(KafkaStreamsLoanApplicationApplication.class)
              .web(WebApplicationType.SERVLET);          Enables the web-server
```

So the only change you need to make is to switch the type from NONE to SERVLET, and this will enable the web server when starting your application.

> **NOTE**   Servlets are a Java technology you can deploy on web servers to build web-based applications. We won't go into details here, but you can find more information at http://mng.bz/0laJ.

The next step is creating a class to handle the incoming query requests, as in the following listing (some details are omitted for clarity).

Listing 13.4   Creating the controller for responding to HTTP requests

```
@RestController                              Adds a RestController annotation
@RequestMapping("/loan-app-iq")
public class LoanApplicationController {      Specifies the base URL
                                              this controller handles

}
```

The first step is to create the class and add two new annotations at the class declaration level. The `@RestController` annotation does two main things for us: it marks this class as a web controller and enables the automatic conversion to JSON of the object you return in response to a request. The `@RequestMapping` annotation maps incoming HTTP requests to request handling classes and methods. The annotation at the class level specifies that this controller handles all requests with the specified base URL.

Before we go any further, let's take a minute to define what the term *controller* means. A web controller is part of an MVC (Model–View–Controller) design pattern for web applications. In this section, you will build a Spring MVC application. The model represents the data, the view is responsible for presenting a visual component, and the controller is responsible for accepting and processing incoming requests and responding to them displayed via the view component. Figure 13.6 illustrates how the MVC components fit together in a web application.

Now, with the brief description of the MVC pattern for web applications, let's get back to building the controller. So far, we've covered declaring the class and providing the required annotations for the controller to receive HTTP requests. Now let's look further into how it will work by looking at listing 13.5.

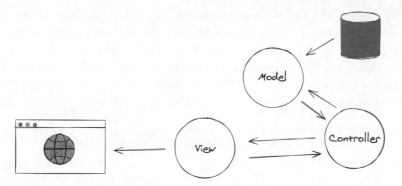

Figure 13.6   MVC components in a web application

Listing 13.5   Dependencies needed for the controller

```
@RestController
@RequestMapping("/loan-app-iq")
public class LoanApplicationController {

  @Value("${store.name}")                        ◁──┐ Field for the state
  private String storeName;                           store name

  @Value("${application.server}")                     Injects the
  private String applicationServer;              ◁──  application.server
                                                      configuration

  private final KafkaStreams kafkaStreams;       ◁──┐ KafkaStreams instance field
  private final RestTemplate restTemplate;       ◁──  RestTemplate used for
                                                      handling HTTP REST calls
  @Autowired
  public LoanApplicationController(KafkaStreams kafkaStreams,
                                   RestTemplate restTemplate) {
      this.kafkaStreams = kafkaStreams;
      this.restTemplate = restTemplate;
  }
}
```

Here you see the injected configuration fields that you've seen before and the
@Autowired annotation on the constructor injecting the KafkaStreams instance built
into KafkaStreamsContainer and RestTemplate from Spring. You'll use these to communicate with the other Kafka Steams applications when handling requests with keys the
current application instance is not responsible for. To inject the KafkaStreams instance,
we will need to make a quick modification to the KafkaStreamsContainer class.

Listing 13.6   Exposing a KafkaStreams object as a Spring bean

```
@Bean                                          ◁──  Adds the @Bean annotation
public KafkaStreams kafkaStreams() {                to expose the method as a
    return kafkaStreams;                            provider of a Spring bean
}
```

By adding this method to the `KafkaStreamsContainer` class with the `@Bean` annotation, when any other class in the container has a reference to a `KafkaStreams` object as a dependency, the Spring container executes this `kafkaStreams` method to inject into the class. When the Spring container calls the `LoanApplicationController` constructor, it follows this exact process. When we look at a request handling method, you'll see how the controller uses the `KafkaStreams` object next.

Listing 13.7   Method annotated to handle incoming requests for a given loan category

**Maps the method to handle requests
for individual loan rollups**

**The URL path variable
represents the loan category.**

```
@GetMapping(value = "/loantype/{category}")
 public QueryResponse<LoanAppRollup> getCategoryRollup(@PathVariable
                              String category) {
    KeyQueryMetadata keyMetadata =
            getKeyMetadata(symbol,
                Serdes.String().serializer());
  if (keyMetadata == null) {
        return QueryResponse.withError(String.format(
    "ERROR: Unable to get key metadata after %d retries", MAX_RETRIES));
  }


}
```

**Retrieves the
KeyQueryMetadata**

The `@GetMapping` annotation on the method means web requests with the URL `http://loan-app-iq/loantype/<category>` are routed to the `LoanApplication-Controller.getCategoryRollup` method. The `KeyQueryMetadata` is an essential object as it will contain the information required to determine whether the current instance queried is responsible for the partition the key in the query belongs to. The `getKeyMetadata` is simply an internal method, and it's worth taking a quick look at what's going on there (some details are omitted for clarity).

Listing 13.8   Implementation of the getKeyMetadata method retrieving KeyQueryMetadata

```
private <K> KeyQueryMetadata getKeyMetadata(K key,
                    Serializer<K> keySerializer) {
  int currentRetries = 0;
  KeyQueryMetadata keyMetadata =
      kafkaStreams.queryMetadataForKey(storeName,
                          key,
                          keySerializer);

  return keyMetadata;
```

**Executes the
KafkaStreams
method for
extracting key
metadata**

In this code block, you can now see one of the reasons we need a reference to the `KafkaStreams` object: the ability to extract essential metadata for the key involved in the query. One of the critical fields of the `KeyQueryMetadata` is the `HostInfo` field, which contains the host and port of the server instance, with the Kafka Streams instance containing the data in question. To enable this comparison of host information, when building the controller, you'll use the `application.server` configuration to build a `HostInfo` object for comparison.

Listing 13.9 Building a HostInfo object from the application server configuration

```
@PostConstruct
public void init() {
    String[] parts = applicationServer.split(":");
    thisHostInfo = new HostInfo(parts[0],
                    Integer.parseInt(parts[1]));
}
```

> Constructs a HostInfo object from the configuration

In the `init` method (executed by the Spring container after the controller constructor call completes), you build a `HostInfo` object. You'll use this to compare it to the one returned in the metadata, as shown in the following listing (some details are omitted for clarity).

Listing 13.10 Comparing HostInfo from key metadata to `HostInfo` for the current host

Creates a StateQueryRequest object
for executing the query

```
if (targetHostInfo.equals(thisHostInfo)) {

  Set<Integer> partitionSet =
        Collections.singleton(keyMetadata.partition());

  StateQueryResult<LoanAppRollup> keyQueryResult =
      kafkaStreams.query(StateQueryRequest.inStore(storeName)
          .withQuery(query)
          .withPartitions(partitionSet));

  QueryResult<LoanAppRollup> queryResult =
        keyQueryResult.getOnlyPartitionResult();

} else {
  String path = "/loantype/" + type;
  String host = targetHostInfo.host();
  int port = targetHostInfo.port();
  queryResponse = doRemoteRequest(host, port, path);
}
```

> Compares the target host with the current host

> Extracts the results into the QueryResult object

> Else block if the target host doesn't match the current one

> Executes a query remotely on the correct host

To determine whether you've queried the correct host, you compare the `HostInfo` object retrieved from the metadata with the one created by the controller. If they are equal, you run the query with the `KafkaStreams.query` method and extract and return the results. Otherwise, you extract the host and port information and execute the search remotely on the correct host with a REST API call. I'm skipping over several details in this code listing, so let's take a minute to go over those details now.

NOTE   These code examples create a few internal objects to support running examples and are not part of Spring or Kafka Streams. I won't describe those as they don't add anything to what you're learning, but you can view them by looking at the source code for the book.

The critical component of IQ is the `Query<R>` interface that all queries must implement. There are currently four implementations:

- `KeyQuery`
- `RangeQuery`
- `WindowKeyQuery`
- `WindowRangeQuery`

You'll use a `KeyQuery` when looking for results of an individual key. The `RangeQuery` is useful when seeing results from a range of keys. You'd use a `WindowKeyQuery` with windowed state stores as it allows you to specify a time from and to for the query, and the `WindowRangeQuery` is also for windowed stores and enables you to identify a time range for querying. Other than the specific parameters you provide to the individual `Query` object, you'll use them similarly when executing the query. You create a `Key-Query` by using the static factory method provided by the class.

Listing 13.11   Creating a `KeyQuery`

```
KeyQuery<String, LoanAppRollup>
        keyQuery = KeyQuery.withKey(loanType);
```

With the `KeyQuery.withKey` method, you provide the key, and the method returns a `KeyQuery` object. The types on the `KeyQuery` are the expected query results, the key, and the value.

The next step is to create a `StateQueryRequest` object that you'll pass to the `Kafka-Streams.query` method. It uses the builder pattern, as shown in the following listing.

Listing 13.12   Building the `StateQueryRequest`

```
StateQueryRequest.inStore(storeName)        ◁————| Provides the store name
        .withQuery(query)                   ◁————| Supplies the Query object
        .withPartitions(partitionSet))      ◁————| Specifies the partition(s) involved
```

The `StateQueryRequest` object contains essential information like the name of the state store, the query object, and the partition for the key (you got this earlier from the `KeyMetadata` object). Other than the required parameters of the store name and the query itself, the different parameters of the `StateQueryRequest` are optional. They include the partitions requiring active tasks and the amount of acceptable lag when querying standby tasks. We'll cover queries with standby tasks a little later in this chapter.

After executing the `KafkaStreams.query` method, you're set to capture the resulting `StateQueryResult` in a variable.

Listing 13.13   Setting the query result in a variable

```
StateQueryResult<LoanAppRollup> stateQueryResult =
        kafkaStreams.query(StateQueryRequest.inStore(storeName)
                    .withQuery(query)
                    .withPartitions(partitionSet));
```

Stores the query result in a variable

Once you have the `StateQueryResult`, you can extract the underlying result from the state store. The overall query result(s) are stored in a `Map` with the partition as the key and a `QueryResult` object as the value. In this case, we know the result only relates to a single partition since it's a key query, so we can use the convenience method `StateQueryResult.getOnlyPartitionResult`.

Listing 13.14   Extracting the query result

```
QueryResult<LoanAppRollup> queryResult =
    stateQueryResult.getOnlyPartitionResult();

LoanAppRollup loanAppRollup = queryResult.getResult()
```

Gets the single partition result

Retrieves the LoanAppRollup result

Now that you've extracted the result, you're ready to return it to the web application, as shown in the following listing (some details are omitted for clarity).

Listing 13.15   Returning the result to the web view

```
queryResponse =
    QueryResponse.withResult(
    queryResult.getResult().value());
// possibly add some metadata from the query
return queryResponse;
```

Gets the raw query result

Returns the result to the view

After getting the raw result, you store it as a custom object capable of holding metadata for display in the view. The web controller automatically converts everything to JSON for rendering in the web application. For the multipartition results, you'll extract the `Map` and iterate over the contents, as in the following listing (some details are omitted for clarity).

Listing 13.16   Extracting multiple partition results

```
Map<Integer, QueryResult<KeyValueIterator<String,
    LoanAppRollup>>> allPartitionsResult =
    result.getPartitionResults();

    allPartitionsResult.forEach((key, queryResult) -> {
        // Do something with the results
    });
```

Retrieves a Map of results

Iterates over the contents of the map

There are a couple of essential notes here to keep in mind. First, when retrieving the results of a query, if you use the `getOnlyPartition` method but there are multiple partition results, the `QueryResult` object will throw an exception. But the rule is evident in this case. With a query involving a single key (i.e., `KeyQuery`), you can safely use the single partition extraction approach, and for all others, you'll want to iterate over the resulting map.

So far, you've seen running the query on the current host, but how would you execute it on a remote one when the current one does not have the key? Earlier in this section, we saw an internal method, `doRemoteQuery`, used by the controller for that case. The following listing shows the implementation of that method (some details are omitted for clarity).

Listing 13.17   Executing a query on the remote host

```
private <V> QueryResponse<V> doRemoteRequest(String host,
                                             int port,
                                             String path) {
    QueryResponse<V> remoteResponse;
    try {
        remoteResponse = restTemplate.getForObject(BASE_IQ_URL + path,
                                             QueryResponse.class,
                                             host,
                                             port);
    } catch (RestClientException exception) {
        remoteResponse = QueryResponse.withError(exception.getMessage());
    }
    return remoteResponse;
```

Catches a possible exception (annotation pointing to the `catch` block)

Executes request on a remote host via RestTemplate call (annotation pointing to the `getForObject` call)

For the case where another Kafka Streams application instance is responsible for the key, you'll use the `RestTemplate` to issue a REST API call to the remote host, where it will follow the exact execution path and return the result to the current host which will return it to view for rendering.

Before we wrap up this chapter, we have some additional items to discuss. First is the different approach you'll take with a range query. Unlike the `KeyQuery`, where you can specifically query a single host, the range query needs to execute against all Kafka Streams instances due to the distributed partition assignment. To run the query across all application instances, you'll want to know which ones contain the state store in question and its partition assignment.

Listing 13.18   Setting up to query all instances for a range query

```
Collection<StreamsMetadata> streamsMetadata =
                kafkaStreams.streamsMetadataForStore(storeName);
List<LoanAppRollup> aggregations = new ArrayList<>();
```

Gets metadata for all Kafka Streams instances (annotation pointing to the `streamsMetadataForStore` call)

```
streamsMetadata.forEach(streamsClient -> {          ◄─┐  Iterates over each
  Set<Integer> partitions =                             │  StreamsMetadata
             getPartitions(streamsClient.topicPartitions());  object
  QueryResponse<List<LoanAppRollup>> queryResponse =
                  doRangeQuery(streamsClient.hostInfo(),
Executes a ┌─▷    Optional.of(partitions),
range query │     Optional.empty(),
remotely    │     lower,
                  upper);
```

Since you need to execute a range query across all hosts, you'll need to use the `Kafka-Streams.streamsMetadataForStore`, which returns metadata from each Kafka Streams application that contains the store in its topology. You can then iterate over the metadata for each one and execute the query for the current instance and remotely for the rest.

The second item to discuss is the ability to query a standby task. When a Kafka Streams instance goes offline, there will be a rebalance to reassign its partitions to the remaining active members. But this can take some time to complete, so Kafka Streams allows us to make the tradeoff of availability over consistency by allowing for querying a standby task. The tradeoff is that the standby will only partially catch up to the active members before it goes offline, but you can still serve query results.

To query a standby, you'll look for any error when executing a query on the active host by catching a `RestClientException`. Then, in your query processing, if you find an error, you'll issue it to the standby, as shown in the following listing.

Listing 13.19 Issuing a standby query in the event of an error querying the active host

```
if (queryResponse.hasError() && !standbyHosts.isEmpty()) {   ◄─┐  Detects errors
  Optional<QueryResponse<LoanAppRollup>> standbyResponse =      │  and enabled
           standbyHosts.stream()                                │  standbys
Iterates over ┌─▷ .map(standbyHost -> doKeyQuery(standbyHost,
the standby(s) │                                 keyQuery,
and queries    │                                 keyMetadata,
               │                                 symbol,
               │                                 HostStatus.STANDBY))
               .filter(resp -> resp != null && !resp.hasError())
               .findFirst();                               ◄─┐
    if (standbyResponse.isPresent()) {                       │  Short-circuits the
       queryResponse = standbyResponse.get();                │  loop when finding
    }                                                        │  a result
}

    return queryResponse;         ◄──── Returns the result
```

When detecting an error and standby tasks are enabled, you can elect to query the standby task when you receive an error querying the active task. You must have standby tasks enabled via configuration to use this feature.

Summary

- IQ provides the ability to query the information contained in the state store directly.
- Direct access to stateful results simplifies architecture and, in some cases, may eliminate the need for a relational database to display results.
- Providing a name to the `Materialized` object either directly or via a `Store-Supplier` enables the store for queries.
- Spring Boot, by default, starts a web server automatically with the application. This embedded web server is excellent for Kafka Streams applications as you'll automatically have the serving layer for IQ available.
- By using the `@RestController` annotation on a controller class, methods handling web requests will automatically have the return objects converted to JSON for the web response.

14

Testing

This chapter covers

- Understanding the difference between unit and integration testing
- Testing Kafka producers and consumers
- Creating tests for Kafka Streams operators
- Writing tests for a Kafka Streams topology
- Developing effective integration tests

So far, we've covered the components for building an event streaming application: Kafka producers and consumers, Kafka Connect, and Kafka Streams. But I've left out another crucial part of this development until now: how to test your application. One of the critical concepts we'll focus on is placing your business logic in standalone classes entirely independent of your event streaming application because that makes your code much more accessible to test. I expect you're aware of the importance of testing, but I'd like to cover the top two reasons testing is just as necessary as the development process itself.

First, as you develop your code, you're creating an implicit contract of what you and others can expect about how the code performs. The only way to prove that the application works is by testing it thoroughly. You'll use testing to provide a good breadth of possible inputs and scenarios to ensure everything works appropriately under reasonable circumstances. The second reason you need comprehensive testing is that it helps you deal with the inevitable changes that occur with software. A rigorous set of tests gives you immediate feedback when the new code breaks the expected behaviors.

Additionally, when you do a major refactor of the application, having your tests pass gives you confidence about releasing the updated software. Once you understand the importance of testing, writing tests for distributed applications like Kafka Streams can be challenging. You can still run these applications with simple inputs and observe the results, but this approach has a severe drawback. You aren't using a suite of repeatable tests that you can run at any time, and that's also part of your continuous integration build. Another component of testing is the need to run the tests as fast as possible. Thus, you'll likely want to run a large segment of your tests without a Kafka broker. Testing without a Kafka broker is one of the most essential points in this chapter. But there will be times when you need a live broker for effective testing. This tension between using a broker or not in testing is the boundary between unit and integration testing.

14.1 Understanding the difference between unit and integration testing

This section will provide an opinionated definition of unit and integration testing. I define unit testing as a test that exercises a specific subcomponent of an application, a particular point of the logic. For example, let's say you have an international sales application, and when you receive an order, you immediately convert the transaction amount into US dollars. A unit test would validate only the currency conversion part, with separate tests or a parameterized test for each type of expected currency.

These types of tests are usually at the method level of a class and run very fast. The problem with unit testing is that often there are external dependencies—for example, a Kafka broker may be required to run the application. So, going with our immediate example here, does that mean you need to run a Kafka broker to feed the different currency types? The answer to that is no, and I'll explain why you wouldn't by way of an analogy.

Let's say you're in a play at your local theater and must learn your part. You don't need other actors on stage to do this, just someone, a stand-in, to feed you the required lines to speak your parts. The person helping can also tell you if you got everything correct.

The same is true with a unit test. You don't need the whole system to test a specific part; you just need a mechanism to provide the input and the ability to validate the results. This mechanism is called a mock object. A mock is an object that has the same interface as the remote connecting component but no actual behavior. You explicitly tell the mock object what it should do, such as supplying specific values, and then you validate the results.

Going back to our analogy, there will be times when you need to get everyone together to rehearse to ensure all the actors know their lines and interact with each other as expected. This type of test in code is an integration test, not the live production, but there's no mocking. All external components are the real thing.

Integration tests are also essential, but you will have a different number of them than you will have unit tests. One of the big reasons you need more unit tests versus integration tests is the execution speed. A typical unit test will run in a quarter to a half a second, but an integration test can run for up to several seconds and as high as 30 seconds to a couple of minutes. Those numbers for a single test are OK, but once you get tests numbering in the hundreds or thousands, you can see the need to have your test suite run as quickly as possible.

So, what's a good indicator for writing a unit test versus an integration test? As I stated earlier, you'll want to use unit tests to validate the behavior of individual methods. An indicator of needing an integration test is when you need the actual behavior of a remote component. For example, you want to see how a Kafka client application behaves when a rebalance occurs. In that case, you'll want an integration test where you can trigger an actual rebalance and validate the behavior. Continuing with our Kafka application example, you'll run the integration test with a Kafka broker in a Docker image, so you still have a live broker. Still, it's contained entirely in your local (i.e., on your laptop) development environment. Let's wrap up this section with table 14.1, which summarizes the difference between unit and integration testing:

Table 14.1 Unit testing compared to integration testing

Type	Purpose	Speed	Percentage of use
Unit	Method level, finer-grained logic, objects in isolation	Subsecond	Majority
Integration	Holistic, integrated components, course grained	Seconds to minutes	Minority

14.1.1 Testing Kafka producers and consumers

Let's say you have an application that runs a simple currency exchange operation, as shown in figure 14.1.

The results of non-US currency transactions are produced to a Kafka topic named `exchange-input`. Your application consumes from the `exchange-input` topic, converts the currency amount to US currency, and then produces the converted amount to another Kafka topic, `exchange-output`. Listing 14.1 shows an abbreviated look at the code (some details are omitted for clarity).

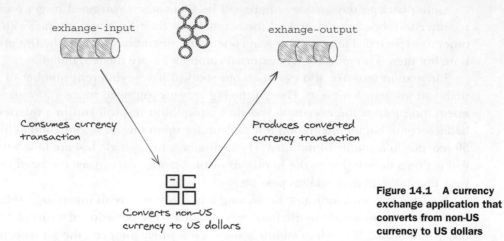

Figure 14.1 A currency exchange application that converts from non-US currency to US dollars

Listing 14.1 Currency exchange class `CurrencyExchangeClient`

```
public void runExchange() {
  while (keepRunningExchange) {
  ConsumerRecords<String, CurrencyExchangeTransaction> consumerRecords=
              exchangeConsumer.poll(Duration.ofSeconds(5));
  consumerRecords.forEach(exchangeTxn -> {
      CurrencyExchangeTransaction tx = exchangeTxn.value();
      double convertedAmount =
            tx.currency().exchangeToDollars(tx.amount());
  CurrencyExchangeTransaction converted =
            new CurrencyExchangeTransaction(convertedAmount,
                  CurrencyExchangeTransaction.Currency.USD);
  ProducerRecord<String, CurrencyExchangeTransaction> producerRecord =
              new ProducerRecord<>(outputTopic, converted);
  exchangeProducer.send(producerRecord...)
  }
  }
```

It's a straightforward application, and you'd like to validate the currency exchange process. You need to write a test for this application and want it to run as a unit test, meaning there is no need for a live Kafka broker. You've designed the class in such a way that only the `Consumer` and `Producer` interface are the types expected.

Listing 14.2 Specifying interfaces in the constructor

```
public CurrencyExchangeClient(
  final Consumer<String, CurrencyExchangeTransaction>
  exchangeConsumer,                               ◁── Consumer interface constructor parameter
  final Producer<String, CurrencyExchangeTransaction>
  exchangeProducer,                               ◁── Producer interface constructor parameter
  final String inputTopic,
  final String outputTopic) {
```

```
        this.exchangeConsumer = exchangeConsumer;
        this.exchangeProducer = exchangeProducer;
        this.inputTopic = inputTopic;
        this.outputTopic = outputTopic;
    }
```

By only specifying the interface used by your application (a proper design decision at all times), you have set yourself up for quickly testing the `CurrencyExchangeClient` class by using the `MockConsumer` and `MockProducer` classes. The `MockConsumer` implements the `Consumer` interface, and the `MockProducer` does likewise with the `Producer` interface, so we can substitute these in a test, allowing you to fully execute the application without needing a live broker. Additionally, you can verify the interactions of the consumer and producer and the final output.

The tradeoff for using mock objects is that you have to specify all the interactions and steps it needs to take. Let's start with the consumer. There are a couple of things to consider when we construct the test. The currency exchange client runs in a loop indefinitely until the `CurrencyExchangeClient.close` method executes. But since it runs in a loop, once we call `CurrencyExchnageClient.runExchange` in the test, the control won't return to the test until the loop terminates, giving us a chicken-and-egg situation.

So, we need a clean way to stop the loop without starting an additional thread in the test. Working with a loop is common when dealing with code using a `KafkaConsumer`. Ideally, you want the consumer to run indefinitely, as event streams never stop. Fortunately, the `MockConsumer` provides this capability with the `schedulePollTask` method, which allows you to supply a task added to a queue the consumer will execute for each `poll(Duration)` call. Figure 14.2 is a picture of this process.

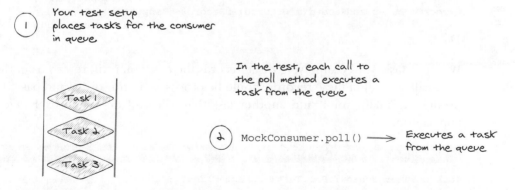

Figure 14.2   Describing consumer behavior with a task queue

So, for each `MockConsumer.schedulePollTask` expected, you'll add a task for the consumer (as a `Runnable` instance) that it will execute, in order, for each `poll()` execution. Let's take a look at this in action.

Listing 14.3   Using the schedule poll task

```
@Test
void runExchangeApplicationTest()
CurrencyExchangeClient exchangeClient = new CurrencyExchangeClient(          ◄─┐   Creates the
            mockConsumer,                                                        client instance
            mockProducer,                                                        for the test
            "input",
            "output");                                        Adds a task to
                                                              the queue
mockConsumer.schedulePollTask(() -> {                 ◄─┐
    final Map<TopicPartition, Long> beginningOffsets = new HashMap<>();
    TopicPartition topicPartition = new TopicPartition("input", 0);
    beginningOffsets.put(topicPartition, 0L);
    mockConsumer.rebalance(Collections.singletonList(topicPartition));
    mockConsumer.updateBeginningOffsets(beginningOffsets);
});
```

Here, you create an instance of the class under test; next, you add the first task to the queue for the consumer. In this case, it's all the required setup for getting the Mock-Consumer in an initial state. Note that the tasks you supply here don't have to provide records the consumer will return from the poll call. They are arbitrary code you need to run. However, at some point, we want to provide records so the consumer can return them, exercising the code in the loop, which is what you'll do next.

Listing 14.4   Adding a task for returning records

```
mockConsumer.schedulePollTask(() -> {
  mockConsumer.addRecord(new ConsumerRecord<>("input", 0, 0, null,
                                                  euroTransaction));
  mockConsumer.addRecord(new ConsumerRecord<>("input", 0, 1, null,
                                                  gbpTransaction));
  mockConsumer.addRecord(new ConsumerRecord<>("input", 0, 2, null,
                                                  jpyTransaction));
});
```

With this task, you're supplying three records the consumer will return from the next poll call, and all of these records should be processed in this order and passed to the producer. Finally, you'll add another task that will shut down the exchange client application.

Listing 14.5   Adding a final task for shutting down the exchange loop

```
mockConsumer.schedulePollTask(exchangeClient::close);     ◄─┐
exchangeClient.runExchange();                               │
                                                  Adds a method handle
┌─►                                               that will close down
Starts the application                            the application
in the test
```

So, the final task you provide here executes the CurrencyExchangeClient.close method, which will shut down the application. The following command is where you start the `CurrencyExchangeClient` in the test. The loop will run three times because you've provided three tasks, and the application shuts down cleanly. But we have another part to test: Has the producer received the expected records with the currency exchanged to US dollars?

Listing 14.6   Validating the producer in the test

```
List<CurrencyExchangeTransaction> actualTransactionList = mockProducer
                                   .history()
                                   .stream()
                                   .map((ProducerRecord::value))
                                   .toList();

assertThat(actualTransactionList.get(0), equalTo(expectedEUROToUS));
assertThat(actualTransactionList.get(1), equalTo(expectedGBDToUS));
assertThat(actualTransactionList.get(2), equalTo(expectedJPYToUS));
```

Uses the history method of the MockProducer

To validate that the producer received records with the correct currency translation and in the proper order, you use the `MockProducer.history` method. You map the resulting `ProducerRecord` list to a list of the value objects and then compare them one at a time to the expected value constructed earlier in the test. You can find the entire test at streams/src/test/java/bbejeck/chapter_14/CurrencyExchangeClientTest.java in the book's source code. There are also tests for the Kafka Connector you developed in chapter 5 located at custom-connector/src/test/java/bbejeck/chapter_5. I didn't include examples of the code in this chapter as it would be repetitive since it also uses mock objects.

You have tested the `CurrencyExchangeClient` application in a unit test that runs very fast. You can and should add additional tests for different error conditions that occur, but I won't cover those here. Now, let's continue the discussion of testing with mocks for testing Kafka Streams operators.

14.1.2 Creating tests for Kafka Streams operators

A Kafka Streams topology will contain one or more operations that take a Single Abstract Method (SAM) interface. This development style allows for quickly creating an application without needing any concrete classes for these operations, and you can supply a lambda to satisfy the behavior requirements. But the tradeoff by doing so makes it next to impossible to provide a test for just that operation as you provide the implementation in line. The other option is to create a concrete class implementing the expected SAM interface. Still, some of these interfaces may need to work with other Kafka Streams objects internal to the application, making testing challenging.

For example, consider the `Punctuator` interface. Since it only has one method, `punctuate` it qualifies as a SAM interface; the `Punctuator` typically doesn't work by itself and requires collaboration with other objects.

In this section, I will show you how to mock arbitrary interfaces and objects with Mockito (https://site.mockito.org/) to make testing with external collaborating objects

a breeze. For this example, you will create a test for the `bbejeck.chapter_9.punctuator`
`.StockPerformancePunctuator` class (figure 14.3).

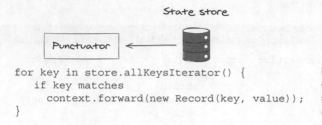

```
for key in store.allKeysIterator() {
    if key matches
      context.forward(new Record(key, value));
}
```

Figure 14.3   The `Punctuator`
examines all records in the state
store and forwards those matching
a condition.

When executed, this punctuator instance will examine the contents of a state store
and forward any records that meet specific criteria. I won't discuss how Kafka Streams
use punctuators here; instead, I'll focus exclusively on the test. To get an idea of the
collaborators for the Punctuator, let's take a look at the constructor.

Listing 14.7   The constructor for the `Punctuator` shows us the required collaborators

```
public StockPerformancePunctuator(double differentialThreshold,
        ProcessorContext<String, StockPerformance> context,
        KeyValueStore<String, StockPerformance> keyValueStore) {

    this.differentialThreshold = differentialThreshold;
    this.context = context;
    this.keyValueStore = keyValueStore;
}
```

As you can see, the first constructor parameter is a Java primitive type, which poses no
issue with testing. Still, the other two are Kafka Streams interfaces and have expected
behavior during the punctuation call. Namely, you'll iterate over everything from the
state store and forward records matching a performance metric. By using mock
objects, however, writing this test will be straightforward. First, let's look at creating
the mock objects for the test.

Listing 14.8   Creating the required mock objects

```
@BeforeEach
 public void setUp() {
    context = mock(ProcessorContext.class);          ◁── Creates a mock for
    keyValueStore = mock(KeyValueStore.class);       ◁──    the ProcessorContext
    stockPerformancePunctuator =                            Creates a mock for
      new StockPerformancePunctuator(                ◁──    the KeyValueStore
                    differentialThreshold,
                    context,                         ◁── Passes all parameters to the
                    keyValueStore);                     StockPerformancePunctuator
}                                                       constructor
```

Creating a mock object is as simple as calling the `Mockito.mock` (shown here as a static import) and passing the object's class or interface to mock. In our case, Mockito returns an object implementing the expected interface, so you pass the returned objects to the `StockPerformancePunctuator` constructor to satisfy the parameter requirements.

While mock objects satisfy the interface requirements, they lack any behavior, so the next step is to give the mock objects instructions on how they will behave, which you'll do in the test method (some details are omitted for clarity).

Listing 14.9 **The test for the** `StockPerformancePunctuator`

```
@Test
void shouldPunctuateRecordsTest() {                      Creates the
  StockPerformance stockPerformance =                    Protobuf
      getStockPerformance();                    ◁        object needed

  Iterator<KeyValue<String, StockPerformance>>
    storeKeyValues =                                     Builds an Iterator to
        List.of(KeyValue.pair("CLFT",                    supply to the mock
  stockPerformance)).iterator();              ◁          KeyValueStore
  long timestamp = Instant.now().toEpochMilli();

  Record<String, StockPerformance> record =              The record
      new Record<>("CFLT", stockPerformance, timestamp);  ◁   to return

  when(keyValueStore.all())                      ◁       Tells the mock
        .thenReturn(                                     key-value store
        TestUtils.kvIterator(storeKeyValues.iterator())); what to do when
  context.forward(record);                               the all method is
                                                         called

  stockPerformancePunctuator.punctuate(timestamp);  ◁
  verify(context, times(1)).forward(record);      ◁
                                                         Executes the
  Sets an expected call for the      Validates the mock   method under test
  mock ProcessorContext            ProcessorContext actions
```

Some of these steps are creating the necessary objects to work with. You create the `StockPerformance` object with the correct properties to pass the performance criteria. Then, you create an `Iterator` by first building an `ArrayList` containing the `KeyValue` object we want the store to return. Next, we create a `Record` instance to give to the `ProcessorContext` to forward.

Next, you tell the mock `KeyValueStore` that when the `all` method gets called, return this stubbed-out instance of a `KeyValueIterator` (created by a testing utility method in the source code by wrapping an `Iterator` instance with a `KeyValueIterator`), which will use the actual iterator you created before. Then, you set the mock `Processor-Context` behavior. Specifically, it should expect to execute the `forward` method with the `Record` object created before.

You run the `Punctuator.punctuate` method, which will exercise all the code inside the method, including the mock objects. Finally, you validate that the mock

`ProcessorContext` did what it was expected to do. You'll find the `TestUtils` class in the source at streams/src/main/java/bbejeck/utils/TestUtils.java.

In this section, you've learned how to unit-test Kafka Streams operators using mock objects to stand in for the collaborating ones. Next, we'll test a Kafka Streams application without mock objects or a live broker.

14.1.3 *Writing tests for a Kafka Streams topology*

When testing a Kafka Streams application, there should be two levels of testing. It's usually a best practice to write the different operations, filter, mapping, aggregations, etc., as concrete classes so you can test them individually. But the Kafka Streams DSL takes most of these as lambda functions, making it easy to write a complete application without much effort. But even with individual unit tests for these various functions, it's essential to have tests that exercise the entire topology to ensure everything works together as expected.

But how can we develop fast tests since a Kafka Streams application is designed to work against a Kafka broker? Enter the `TopologyTestDriver` class, designed to thoroughly test a Kafka Streams topology (complete with state stores) but without needing a live broker, a unit test for your entire Kafka Streams topology. The best way to learn the `TopologyTestDriver` is to dive in with examples. Let's start with our first example of a Kafka Streams application—the Yelling application.

Since this is our first time using the `TopologyTestDriver`, we'll step through each part of setting the test up, but for future examples, we'll only show the central part of what's being covered.

> **Listing 14.10   Setting up to test the Yelling Kafka Streams application**

```
@Test
@DisplayName("Should Yell At Everyone")
void yellingTopologyTest() {
 KafkaStreamsYellingApp yellingApp =          Creates an instance of the
 ➥ new KafkaStreamsYellingApp();            ◁──  KafkaStreamsYellingApp
 Topology yellingTopology =                       Gets the Topology
 ➥ yellingApp.topology(new Properties());  ◁──   object of the app
 Serializer<String> stringSerializer =            Creates a serializer
 ➥ Serdes.String().serializer();           ◁──   for String objects
 Deserializer<String> stringDeserializer =
         Serdes.String().deserializer();    ◁──  Creates a deserializer
                                                  for Strings
```

So, you start like you usually would for any test method by decorating the method with the `@Test` and `@DisplayName` annotations. Inside the method, you first create an instance of the `KafkaStreamsYellingApp`, which you'll need to extract the topology on the following line with the `KafkaStreamsYellingApp.topology` method. This way, we extract the same topology used when running the application. You also create a serializer and deserializer, which you'll see in action in the next step when we get into the crucial parts of building the test.

Listing 14.11   Building the `TopologyTestDriver` and input and output topics

```
try (TopologyTestDriver driver =
              new TopologyTestDriver(yellingTopology)) {

    TestInputTopic<String, String> inputTopic =
          driver.createInputTopic("src-topic",
                                  stringSerializer,
                                  stringSerializer);

    TestOutputTopic<String, String> outputTopic =
          driver.createOutputTopic("out-topic",
                                   stringDeserializer,
                                   stringDeserializer);
```

Creates the
TopologyTestDriver instance

Builds a
TestInputTopic

Builds a
TestOutputTopic

In this next step, you construct the `TopologyTestDriver` instance, which will be the harness for running the topology under test, which you pass as a constructor parameter. There are a few overloaded constructors for the `TopologyTestDriver`; we'll cover them later with other examples.

Next, you need to create a `TestInputTopic` that you'll use to pipe records into the topology for running the test. When you make the `TestInputTopic`, the name for the topic is the first parameter, and this name must match the topic name you use when building the topology. The other parameters here are the serializers for the keys and the values you'll provide for running the test. `TestInputTopic` will serialize each key and value you provide so that your Kafka Streams application will receive the expected byte arrays as it would when running for real.

You'll also need an output topic for the topology to write results to, which is what you're doing when you create the `TestOutputTopic`. It has the exact requirement for the topic name parameter to match the name in the actual application. For the `TestOutputTopic`, you supply key and value deserializers since the topology serializes the output, so you'll need a way to convert them back to concrete objects to validate the test result.

The next step in our test is to send input records through the topology and then validate the output, which looks something like figure 14.4.

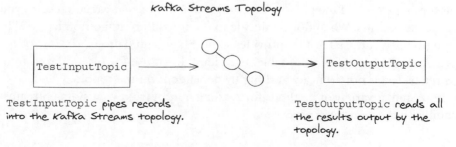

Figure 14.4   Piping records into the topology and then reading the output

So, the process is calling the appropriate `TestInputTopic` method to send value(s) through the topology and then using the `TestOutputTopic` to capture the results.

Listing 14.12 Sending records through the topology and validating results

```
List<String> inputValues = List.of("if you don't eat your meat",          ◁
                "you can't have any pudding!",
                "How can you have any pudding",               Puts together a
                "if you don't eat your meat!" );              list of input values

inputTopic.pipeValueList(inputValues);          ◁───┐  Sends all input
List<String> expectedOutput = inputValues.stream()        into the topology
                                    .map(String::toUpperCase)
                                    .toList();              ◁

                                                   Creates a list of
  ┌─▷ List<String> actualOutput = outputTopic.readValuesToList();    the expected
  │                                                                  values
  │   assertThat(actualOutput, equalTo(expectedOutput));     ◁
  │
Reads the output of the           Asserts that actual results
topology into a list              equal the expected ones
```

In this final section of the example, you first create some sample input (bonus points if you can name the artist and song lyrics we're using). You'll then take the sample input and send it through the topology using the `TestInputTopic.pipeValueList` method. In this case, since our Kafka Streams application only works on the values of the key-value pair, it's perfectly acceptable to provide all the values in a list. You're not limited to sending in a list of values, although the `TestInputTopic` offers additional methods for piping input into it for testing. I'll list a few of them in the following listing.

Listing 14.13 Other methods for piping input from the `TestInputTopic`

```
      pipeInput(K key, V value)                 ◁───┐  Sends a key and value
  ┌─▷ pipeInput(K key, V value, Instant)
  │   pipeInput(TestRecord<K, V> testRecord)         ◁

Provides a key and value with              Sends a
a timestamp of the event                   TestRecord object
```

I want to note that there are also overloaded versions of the methods listed here, accepting a `List` of the parameters. In the case of the `key`, `value` variants, you'd use a `List<KeyValue>`. When do you decide to use the different method types? While there are no fixed rules, table 14.2 provides some general guidance.

As you can see from the table, the method you use for piping input into a Kafka Streams test is not arbitrary and highly dependent on the topology's actions. There's also a similar variety of methods for reading output from your Kafka Streams application with the `TestOutputTopic`.

Table 14.2   Method parameters general guidance

Method parameters	Reason to use
single value or list of values	Simple topology; input topic has no keys, no stateful operations or stateful ones where you extract the key entirely from the value.
single key-value or list of key-values	Stateful operations; input topic does have keys.
Single `TestRecord` or list of `TestRecord`	Topologies using timestamps and headers; the test will advance stream time based on the timestamps in the records.
`Instant` and `Duration`	Using an `Instant` provides the timestamp for that record. Overloads with an `Instant` and `Duration` will use the `Instant` for the starting timestamp and the `Duration` for the advance of each record.

Listing 14.14   Methods for reading output from the test

```
readKeyValue()
readKeyValuesToList()
readRecord()
readRecordsToList()
```

The method you'll use to verify the output could mirror how you feed the records into the test. While not a strict rule, it's something that I do myself, dumping an entire list of input. Then you can read a list out and compare the expected to the actual output you've received. Other times, you may want to pipe in a record, assert the outcome, and then pipe another record, assert the result, and so on.

So far in this section, I've introduced you to setting up a `TopologyTestDriver` for a basic Kafka Streams application, but it's not limited to testing simple topologies. You can use the `TopologyTestDriver` to test highly complex topologies, and we'll see some examples of that in the next section.

But before we move on to more advanced examples, I'd like to point out another use for the `TopologyTestDriver` beyond testing your application for correctness. You can also use the `TopologyTestDriver` to build quick Kafka Streams application prototypes. For the most part, using `TopologyTestDriver` will provide most of the functionality you'll need to observe the behavior of a Kafka Streams application. Some parts that `TopologyTestDriver` can't provide include task assignments, rebalancing, repartitions, etc. Still, you can speed up your development by building a Kafka Streams application without needing a Kafka broker. Of course, you'll always need tests with a live broker, and I'll get to those a bit later in the chapter.

14.1.4 *Testing more complex Kafka Streams applications*

This section will explore using the `TopologyTestDriver` for more advanced Kafka Streams applications. Since we covered creating a test from start to finish previously, I'm only going to show the specific sections of the test you'll need to know for testing more advanced Kafka Streams applications. For our first venture into advanced testing, we'll look into a stateful application that performs a `reduce` operation. I should mention that for stateful topologies, `TopologyTestDriver` does not buffer any records; each input generates an output record.

In chapter 7, we discussed building stateful Kafka Streams applications, and one of the operations covered is a `reduce` operation, so let's look at a test for one of the examples (bbejeck.chapter_7.StreamsPokerGameInMemoryStoreReducer) now.

Listing 14.15   Setting up the test for a `reduce` operation

```
try (TopologyTestDriver driver = new TopologyTestDriver(topology)) {
  TestInputTopic<String, Double> inputTopic = driver.createInputTopic..
  TestOutputTopic<String, Double> outputTopic = driver.createOutputTopic...

  inputTopic.pipeInput("Anna", 65.75);          ◁── Executes the pipeInput
  inputTopic.pipeInput("Matthias", 55.8);            method three times
  inputTopic.pipeInput("Neil", 47.43);
```

The poker game Kafka Streams application has input from players of an online poker game where the key is the username, and the value is their current score in the game. So here, to start the test, you input three scores, and you first want to verify the reduction happens in order, so you'll want to read the output for the following three records and assert the order matches the input order.

Listing 14.16   Asserting the order of records in a reduce

```
KeyValue<String, Double> actualKeyValue =
➥   outputTopic.readKeyValue();              ◁── Reads the first record
assertThat(actualKeyValue,                        from the output topic
➥   equalTo(KeyValue.pair("Anna", 65.75)));   ◁── Asserts the key-value pair
                                                  matches the expected order
actualKeyValue = outputTopic.readKeyValue();
assertThat(actualKeyValue, equalTo(KeyValue.pair("Matthias", 55.8)));

actualKeyValue = outputTopic.readKeyValue();<
assertThat(actualKeyValue, equalTo(KeyValue.pair("Neil", 47.43)));
```

In this block, you read three records and assert the order matches the order you piped into the application. So far, so good, but you can perform another level of testing. This current application is stateful, so a state store keeps the state of the latest `reduce` operation. While you just verified the output, you can also inspect the state store and validate that its content matches the newest output.

Listing 14.17   Validating the state store contents match the latest output

```
KeyValueStore<String, Double> kvStore =                    Retrieves the
     driver.getKeyValueStore("memory-poker-score-store");  ◁── state store from
                                                               the topology
assertThat(kvStore.get("Anna"), is(65.75));     ◁── Validaties the store contents
assertThat(kvStore.get("Matthias"), is(55.8));      match the latest output
assertThat(kvStore.get("Neil"), is(47.43));
```

To validate the contents of a state store, `TopologyTestDriver` provides a `getKeyValue-Store` method allowing you to retrieve a `KeyValueStore` by name, which you do here, and validate the contents match the latest records output.

> **TIP** The `TopologyTestDriver` provides several methods for retrieving store types, session, window, and timestamped stores. For the cases where you haven't named the store, Kafka Streams generates its name; the `getAllState-Stores` method returns a `Map` of the stores. From there, you can iterate over the entries and extract the store.

To conclude the test, you pipe in many records in random order. Since you've already validated that everything gets processed in order, you want to verify that the total final output matches your expectations. But since you just input a bunch of random records, you'll need a way to get the final result for each key.

Listing 14.18 Retrieving the last output for inputs

```
Map<String, Double> allOutput = outputTopic.readKeyValuesToMap();        ◁──────┐
    assertThat(allOutput.get("Neil"), is (252.43));                             │
    assertThat(allOutput.get("Anna"), is (185.75));                    Gets the latest
    assertThat(allOutput.get("Matthias"), is (180.8));                 record for each key

    assertThat(kvStore.get("Anna"), is(185.75));            ◁──────   Validates the final
    assertThat(kvStore.get("Matthias"), is(180.8));                   state of the store
    assertThat(kvStore.get("Neil"), is(252.43));
```

To get the last output per key, you use `TestOutputTopic.readKeyValuesToMap`, which presents a final table view of the results where more recent entries update and replace previous ones. If you want to inspect each result separately, you will use one of the `TestOutputTopic.readXXXToList` methods.

> **NOTE** With the built-in Kafka Steams aggregations, don't feel compelled to test the contents of a store. However, I recommend validating the store's contents for Kafka Streams applications involving a state store with the Processor API. I've provided this example to show how you can do it.

You just learned how to test a stateful application. Let's move on to testing an application where the timestamps of the records drive the behavior with a test for the `Stock-PerformanceApplication` from chapter 10. For a quick review, `StockPerformance-Application` only emits records after a `punctuation` call scheduled to run every 10 seconds based on stream time. That means the punctuation only executes when stream time advances due to the timestamps of the records. To drive this punctuation behavior, you'll need to provide the corresponding timestamps, as in the following listing from the test bbejeck.chapter_9.StockPerformanceApplicationTest (some details are omitted for clarity).

Listing 14.19   Providing timestamps to drive stream-time behavior

```
try (TopologyTestDriver driver = new TopologyTestDriver(topology)) {

inputTopic.pipeInput("ABC", transactionOne, instant);           ◁─── Passes in the first record
inputTopic.pipeInput("ABC", transactionTwo,                          with the current time
                            instant.plus(15, ChronoUnit.SECONDS));
inputTopic.pipeInput("ABC", transactionThree,
                            instant.plus(25, ChronoUnit.SECONDS));  ◁───

// punctuation should fire three times                          Pipes in a final
assertThat(outputTopic.getQueueSize(), is(3L));                 record advancing
                                                                time by 25 seconds
```

Adds 15 seconds for the
second record passed in

So, to drive the timestamp behavior, you use the `TestInputTopic.pipeInput` method that accepts the following parameters: a key, value, and a timestamp for the record represented as a `java.time.Instant`. Providing these timestamps with a forward time setting will move the stream time appropriately for Kafka Streams to perform the expected correct number of punctuations. Here, we validate the current number of times Kafka Streams executed a `punctuate` call by verifying the number of records in the internal queue of the `TestOutputTopic`.

> **TIP** For Kafka Streams applications with punctuation based on wallclock time, you'll have to explicitly move the internal wallclock time of `Topology-TestDriver` with the `advanceWallClockTime(Duration advanceAmount)` method. We discussed wallclock time in a previous chapter.

You should note that the `pipeInput` method accepting a timestamp parameter advances stream time tracked by the `TopologyTestDriver` and does not advance the internal event time of the `TestInputTopic`. To understand what this means is that if you provide another record to `TestInputTopic` without explicitly providing a timestamp, that record's event time will be the initial timestamp of `TestInputTopic` when you created it in this case, as when you provide a record without an explicit timestamp, it uses the current event time of the topic. To advance the event time of the input topic, you'd take an approach like in the following listing.

Listing 14.20   Providing timestamps by advancing the event time of the input topic

```
inputTopic.pipeInput("ABC", transactionOne);
inputTopic.advanceTime(Duration.ofSeconds(15));        ◁─── Advances the event time of
inputTopic.pipeInput("ABC", transactionTwo);                the topic by 15 seconds
inputTopic.advanceTime(Duration.ofSeconds(25));        ◁───
inputTopic.pipeInput("ABC", transactionThree);              Another advance of event
                                                            time by 25 seconds
```

This example provides the same behavior for the test and validates Kafka Streams performing three punctuations. Which approach should you use? The answer to that depends on how you structure your test. If you provide a small number of records, manually setting each timestamp is ideal, as you can see the timestamp for each record. However, if you have a test where you want to generate a large number of inputs, it could be cumbersome to set a timestamp for each record, so advancing the event time of `TestInputTopic` at various intervals would be more effective in that case.

Before we wrap up our coverage of the `TopologyTestDriver`, we have one more example to cover regarding timestamps on records—when you have windowed Kafka Streams applications. For this example, you'll look at a test for the `bbejeck.chapter_9``.window.StreamsCountTumblingWindowSuppressedEager` application, which does a windowed count of incoming records where the window size is 1 minute with no grace period.

When you have a windowed operation with suppression, Kafka Streams will not emit a result until the window closes. You must input records with timestamps moving stream time forward to validate the results. To do window stream time advancement, you'll take a similar approach to what you did with the punctuation test found in `bbejeck.chapter_9.window.StreamsCountTumblingWindowSuppressedEagerTest`.

Listing 14.21   Setting timestamps to emit suppressed windowed results

```
Stream.generate(() -> "Foo").limit(10)          Generates 10
         .forEach(item ->                        input records
  inputTopic.pipeInput(item, item));
assertThat(outputTopic.getQueueSize(), is(0L));   Validates no
inputTopic.pipeInput("Foo", "Foo",                emitted records
    instant.plus(75, ChronoUnit.SECONDS));
assertThat(outputTopic.readValue(), is(10L));     Adds a new input record
                                                  and advances stream time
              Asserts the window emitted
              the correct count
```

For this test, you first generate 10 records, input them into the topology, and assert that Kafka Streams has not emitted anything by validating that the queue of the output topic is empty. You then add another record and explicitly set its timestamp to 1 minute 15 seconds in the future. Since you have set the application window to 1 minute, you should observe Kafka Streams forward the count of all events by key from the previous window with a count of 10.

> **NOTE**   This approach of setting forward timestamps to move stream time is the way to test any windowed Kafka Streams application.

So far, you've learned about unit testing for Kafka producer and consumer clients and testing a Kafka Streams application without needing a live broker. But unit testing and not using a broker is only part of the picture. Some of your tests should include a live Kafka broker; we'll cover that next in our final section.

14.1.5   Developing effective integration tests

For developing an integration test for Kafka Streams, the main difference you'll notice is the extra code needed to interact with a live broker. Additionally, you'll have to account for how Kafka Streams works as a live application, especially when you have a stateful operation due to the caching behavior. You'll still write a JUnit 5 test, but it will operate slightly differently, mainly due to the time it takes to start the broker, which is where we'll start.

You will use `TestContainers` to provide a Kafka broker for the integration tests. Testcontainers (https://www.testcontainers.org/) is a Java library allowing access to external components running in Docker containers directly in your JUnit tests. I assume you are familiar with Docker, but for more information, go to the Docker website: https://www.docker.com/.

> **NOTE**   Testcontainers for Kafka in JUnit 5 tests require the following dependencies: `org.testcontainers:junit-jupiter:1.17.1` and `org.testcontainers:kafka:1.17.1`. The source code for the book already does this, but I'm adding this here for your information.

Let's begin your integration testing by creating a test for the `bbejeck.chapter_9.window.StreamsCountHoppingWindow` application. The Testcontainers library provides annotations that you'll use to annotate your code that will handle the lifecycle of your Kafka Docker container. Let's start building the test now.

Listing 14.22   Integration test with Testcontainers annotations

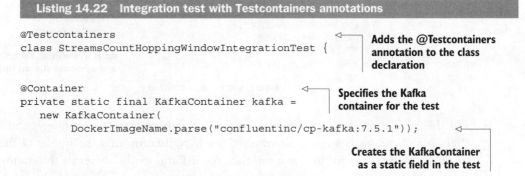

```
@Testcontainers
class StreamsCountHoppingWindowIntegrationTest {        ◁──  Adds the @Testcontainers
                                                             annotation to the class
                                                             declaration

@Container                                              ◁──  Specifies the Kafka
private static final KafkaContainer kafka =                  container for the test
   new KafkaContainer(
        DockerImageName.parse("confluentinc/cp-kafka:7.5.1"));   ◁──

                                                          Creates the KafkaContainer
                                                          as a static field in the test
```

You first create the test class `StreamsCountHoppingWindowIntegrationTest` and add `@Testcontainers` at the class level. The `@Testcontainers` annotation is a JUnit Jupiter extension, and it automatically manages the lifecycle for any containers in the test by finding any fields with a `@Container` annotation. You can see the `@Container` here annotating the `KafkaContainer` field. When you define the field as static, the container is shared with all test methods, meaning it starts before the first test and shuts down after the last test completes.

If you define the container field as nonstatic, the container starts and stops for each test. Unless you have a specific reason for stopping and starting a container for each test, I'd recommend using the static field approach, as this will save a few CPU cycles by only starting and stopping the container once.

You create the `KafkaContainer` instance by passing a `DockerImageName` object, which you make by passing in a string in the standard Docker format (registry/name:tag).

NOTE   I'm not going to cover Docker, but you can learn more about it in the Manning books *Docker in Practice*, by Ian Miell and Aidan Hobson Sayers (2019), and *Docker in Action*, by Jeff Nickoloff and Stephen Kuenzli (2019).

You've now set up the Kafka Docker container for the test! What's left is for us to discuss the elements you'll need for running the Kafka Streams application in the test. Since Kafka Streams gets its input from topics and produces the final results to a Kafka topic, you'll need a `KafkaProducer` to feed a topic for the test and a `KafkaConsumer` to help validate the test results. Let's take a look at the details you need to put together (some details are omitted for clarity).

Listing 14.23   Setting the test up with producer and consumer configurations

```
@BeforeEach
 public void setUp() {                              ← Setup method executed
                                                       before each test

    streamsCountHoppingWindow = new StreamsCountHoppingWindow();
    kafkaStreamsProps.put("bootstrap.servers",
            kafka.getBootstrapServers());                      ←

    kafkaStreamsProps.put(StreamsConfig.APPLICATION_ID_CONFIG,
            "hopping-windows-integration-test");

    //Producer configs
    producerProps.put("bootstrap.servers",
            kafka.getBootstrapServers());                      ←
    producerProps.put(ProducerConfig.KEY_SERIALIZER_CLASS_CONFIG,
                    StringSerializer.class);
    producerProps.put(ProducerConfig.VALUE_SERIALIZER_CLASS_CONFIG,
                    StringSerializer.class);
                                            Sets the bootstrap.servers
                                              config for the clients
    //Consumer configs
    consumerProps.put("bootstrap.servers",
            kafka.getBootstrapServers());                      ←
    consumerProps.put(ConsumerConfig.KEY_DESERIALIZER_CLASS_CONFIG,
                        StringDeserializer.class);
    consumerProps.put(ConsumerConfig.VALUE_DESERIALIZER_CLASS_CONFIG,
                        LongDeserializer.class);
    consumerProps.put(ConsumerConfig.GROUP_ID_CONFIG,
                        "integration-consumer");
    consumerProps.put(ConsumerConfig.AUTO_OFFSET_RESET_CONFIG,
                        "earliest");

    Topics.create(kafkaStreamsProps,                   Creates the
            streamsCountHoppingWindow.inputTopic,      required
            streamsCountHoppingWindow.outputTopic);  ← topics
```

In the test `setUp` method, you set all the required configurations for Kafka Streams, producer, and consumer clients. You won't directly create a `KafkaProducer` and `KafkaConsumer` in the test, relying instead on some static helper methods to produce and consume records as needed in the test, and I'll cover those methods soon. The last bit of code in the setUp method creates the required topics. There's a corresponding method, tearDown, executed after each test completes as well.

Listing 14.24 The teardown method that runs after each test

```
@AfterEach
public void tearDown() {
    Topics.delete(kafkaStreamsProps,
                  streamsCountHoppingWindow.inputTopic,
                  streamsCountHoppingWindow.outputTopic);
}
```

The tearDown method deletes all the topics created before the test runs. It's essential to follow this practice of creating and deleting topics for each test, ensuring we have a clean starting point so the results aren't affected by the running of a previous test. To create and delete topics, we use the utility class bbejeck.utils.Topics provided in the book's source code.

Now, let's get to the heart of the integration test. Since the Kafka Streams application under test, StreamsCountHoppingWindow, uses hopping windows (windows that have an advance smaller than the window size) with a length of 1 minute and an advance of 10 seconds, you'd like to provide values in such a way that you can see the overlap of each window advance. That means each time the window moves forward by 10 seconds, the count includes previous results until the window size is reached. You'll see the count start to decline. Figure 14.5 shows what we want our test to validate.

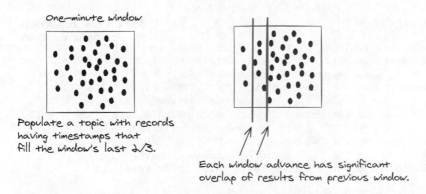

Figure 14.5 The integration test for a hopping window count should increase up to the window size and then start declining.

NOTE Since I've covered Kafka Streams windowing in a previous chapter, I will only focus on details relevant to the test.

Now that you have all the setup details complete, let's get to writing the test itself.

Listing 14.25   Writing the test for a hopping window Kafka Stream application

```
@Test
@DisplayName("Integration test for hopping window")
void shouldHaveHoppingWindowsTest() {
    final Topology topology =                                    ⟵  Creates the
        streamsCountHoppingWindow.topology(kafkaStreamsProps);       topology
    AtomicBoolean streamsStarted = new AtomicBoolean(false);
    try (KafkaStreams kafkaStreams =
            new KafkaStreams(topology, kafkaStreamsProps)) {
        kafkaStreams.cleanUp();
        kafkaStreams.setStateListener((newState, oldState) -> {
            if (newState == KafkaStreams.State.RUNNING) {
                streamsStarted.set(true);
            }
        });
        kafkaStreams.start();                          ⟵  Starts the KafkaStreams
        while (!streamsStarted.get()) {                    instance
            time.sleep(250);                  ⟵  Waits for the RUNNING
        }                                         state for the test to
    }                                             proceed
```

Sets a StateListener for the RUNNING state

In the test method, you create the `KafkaStreams` instance, and before starting, you set a `StateListener` for when Kafka Streams goes into a RUNNING state; that way, we can wait to start the test until Kafka Streams is ready for work. While this step is not required, I prefer to start the test once that application is in a known state.

Now let's move on to producing input for Kafka Streams to work with.

Listing 14.26   Producing records for the Kafka Streams application

```
long startTimestamp = Instant.now().toEpochMilli();       ⟵  First timestamp set
for (int i = 1; i <= 6; i++) {                                to the current time
    List<KeyValue<String, String>> list =
    Stream.generate(() ->
            (KeyValue.pair("Foo", "Bar"))).limit(i).toList();  ⟵  Generates a list
    TestUtils.produceKeyValuesWithTimestamp(          ⟵            of KeyValue pair
            streamsCountHoppingWindow.inputTopic,
            list,                                     Uses a utility method
            producerProps,                            for producing records
            startTimestamp,
            Duration.ofMillis(100L));
    startTimestamp += 10_000;            ⟵  Increases the timestamps
}                                           by the window boundary
```

Here, you're using the helper method `TestUtils.produceKeyValuesWithTimestamp` from the book's source code that takes care of all the details of producing a message to Kafka for the `StreamsCountHoppingWindow` to process. For each iteration, we send the number of records matching the loop index and increment the timestamp by 10 seconds to ensure the next batch of records produced will end up in the next window advance. We expect each advance to include the sum of the previous records up to the window size boundary, and then we should see the count start declining as records age off. To test our expectations, we add the following code to the test.

Listing 14.27   Setting an expected list of records we expect Kafka Streams to produce

```
List<KeyValue<String, Long>> expectedKeyValues =
    List.of(KeyValue.pair("Foo", 1L),
    KeyValue.pair("Foo", 3L),
    KeyValue.pair("Foo", 6L),
    KeyValue.pair("Foo", 10L),
    KeyValue.pair("Foo", 15L),
    KeyValue.pair("Foo", 21L),
    KeyValue.pair("Foo", 20L),
    KeyValue.pair("Foo", 18L),
    KeyValue.pair("Foo", 15L),
    KeyValue.pair("Foo", 11L),
    KeyValue.pair("Foo", 6L));
```

The max window size of 1 minute is reached.

Utility method to consume results

```
List<KeyValue<String, Long>> actualConsumed =
    TestUtils.readKeyValues(streamsCountHoppingWindow.outputTopic,
            consumerProps,
            45_000,
            12);
    assertThat(actualConsumed, equalTo(expectedKeyValues));
```

The maximum time the test is willing to wait for results.

The max number of records to retrieve; once consuming this number of results, the method returns.

Asserts the actual results match the expected ones

So, we create an expected list where the count increments by the number of records sent plus the previous count $(1 + 2 + 3 + 4 \ldots)$. Then, we use another utility method to consume from the topic and return the results to the test to assert the results match our expectations of the application output. When running this test, it should pass every time. The main points of this section are to first create an expected result for comparison and use utility helper methods that facilitate reuse across the different integration tests. I haven't included examples of integration testing for the producer and consumer classes. Still, you'd follow the same pattern: create a test class with the `@Testcontainer` annotation and a `@Container` field in the class and use the helper methods for producing and consuming to drive the test. Examples of these integration tests are in the source code for the book.

Summary

- Testing is a critical part of development, and tests should correspond to each piece of the application to a reasonable level.
- Unit tests should make up most of your testing strategy, but you'll still need to include integration tests to ensure everything works together as expected.
- Using the MockProducer and MockConsumer is practical for performing testing of client applications without requiring the use of a live Kafka broker.
- It's best to write your Kafka Streams operators map, filter, punctuators, etc., as concrete instances that allow you to write isolated tests for each component.

- To validate their expected behavior, you should perform unit tests for the various mappers, aggregators, and punctuators for Kafka Streams applications. Still, you can also use the `TopologyTestDriver` to test your entire topology without needing a live Kafka broker.

- The `TopologyTestDriver` provides for creating `TestInputTopic` and `TestOutputTopic` instances, which you'll use to run records through the topology and then capture the results, respectively.

- When testing a Kafka Streams application with the `TopologyTestDriver`, you may need to provide timestamps to ensure windows advance.

- For your integration tests, you should use Testcontainers to provide the Kafka broker for the test, as it takes care of getting the container and managing the container lifecycle for you.

- By default, using the JUnit 5 Testcontainers extension will spin up a new container for each test, which can be expensive timewise. Consider making the Testcontainers field static, and you'll only start one container for all tests in the test class.

appendix A
Schema compatibility workshop

In this appendix, you'll take a guided walkthrough of updating schemas in different compatibility modes. You'll change the schemas compatibility mode, make changes, test those changes, and finally run updated producers and consumers to see the different compatibility modes in action. I've already made all the changes. You only need to read along and run the provided commands. There are three subprojects: `sr-backward`, `sr-forward`, and `sr-full`. Each subproject contains producers, consumers, and schemas updated and configured for the representative compatibility mode.

> **NOTE** There is a lot of overlap between the code and the build.gradle files between the subprojects. This overlap is intentional, as I wanted each module isolated. The focus of using these modules is learning about evolving schemas in Schema Registry and the related changes you need to make to Kafka producers and consumers, not how to set up the ideal Gradle project!

In this section, I'll discuss how Schema Registry ensures client compatibility. For schema compatibility rules of the serialization frameworks, you'll want to look at each one specifically. Avro schema resolution rules are available at http://mng.bz/ngE2. Protobuf provides backward compatibility rules in the language specification found at http://mng.bz/v8r4.

Let's go over the different compatibility modes now. For each compatibility mode, you'll see the changes made to the schema and run a few steps needed to migrate a schema successfully.

For clarity, each schema migration for the different compatibility modes has its own Gradle submodule in the book source. I did this as each changing Avro schema file results in different Java class structures when you build the code. Instead of having you rename files, I opted for a structure where each migration type can stand independently. In a typical development environment, you will not follow this practice. You'll modify the schema file, generate the new Java code, and update the producers and consumers in the same project.

All schema migration examples will modify the original `avenger.avsc` schema file. The following listing provides the original schema file for reference so it's easier to see the changes made for each schema migration.

Listing A.1   The original Avenger Avro schema

```
{
  "namespace": "bbejeck.chapter_3.avro",
  "type": "record",
  "name": "AvengerAvro",
  "fields": [
    {"name": "name", "type": "string"},
    {"name": "real_name", "type": "string"},
    {"name": "movies", "type":
                        {"type": "array", "items": "string"},
      "default": []
    }
  ]
}
```

NOTE For working through schema evolution and the compatibility types, I've created three submodules in the source code, `sr-backward`, `sr-forward`, and `sr-full`. These submodules are self-contained and intentionally contain duplicated code and setup. The modules have updated schemas, producers, and consumers for each type of compatibility mode. I did this to make the learning process easier, as you can look at the changes and run new examples without stepping on the previous ones.

A.1   *Backward compatibility*

Backward compatibility is the default migration setting. With backward compatibility, you update the consumer code first to support the new schema. The updated consumers can read records serialized with the new or immediate previous schema (figure A.1).

As shown in figure A.1, the consumer can work with the previous and the new schemas. The allowed changes with backward compatibility are deleting fields or adding optional fields—an optional field is when the schema provides a default value. If the serialized bytes don't contain the optional field, then the deserializer uses the specified default value when deserializing the bytes back into an object.

Before we get started, let's run the producer with the original schema. That way, after the next step, you'll have records using both the old schema and the new one,

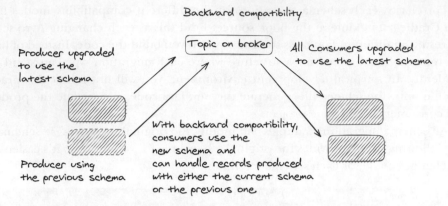

Figure A.1   Backward compatibility updates consumers first to use the new schema. Then, they can handle records from producers using either the new schema or the previous one.

and you'll be able to see backward compatibility in action. Make sure you've started docker with `docker-compose up -d` and then run the following commands.

Listing A.2   Producing records with the original schema

```
./gradlew streams:registerSchemasTask
./gradlew streams:runAvroProducer
```
Makes sure you've registered the original avengers.avsc schema

Runs a producer with the original schema

Now, you'll have records with the original schema in the topic. When you complete the next step, having these records available will clarify how backward compatibility works, as the consumer can accept records using the old and the updated schema. So, let's update the original schema by deleting the `real_name` field and adding a `powers` field with a default value.

> **NOTE**   You'll find the schema file and code in the `sr-backward` sub-module in the `src/main/avro` directory in the source code.

Listing A.3   Backwards compatible updated schema

```
{
  "namespace": "bbejeck.chapter_3.avro",
  "type": "record",
  "name": "AvengerAvro",
  "fields": [
    {"name": "name", "type": "string"},
    {"name": "powers", "type":
                {"type": "array", "items": "string"},
       "default": []
    },
```
The powers field replaces the deleted real_name field.

Provides a default value of an empty powers list for backward compatibility

```
    {"name": "movies", "type":
                        {"type": "array", "items": "string"},
      "default": []
    }
    ]
}
```

Now that you have updated the schema, you'll want to test its compatibility before uploading it to Schema Registry. Fortunately, testing a new schema is a simple process. You'll use the testSchemasTask in the sr-backward module from the Gradle plugin for testing compatibility. So, let's test the compatibility first by running this command from the root of the project.

Listing A.4 Testing whether a new schema is backward compatible

```
./gradlew :sr-backward:testSchemasTask
```

> **WARNING** To run the example successfully, you need to run the command exactly as it's displayed here, including the leading : character.

The result of running the testSchemasTask should be BUILD SUCCESSFUL, which means that the new schema is backward compatible with the existing one. The testSchemas-Task calls Schema Registry to compare the proposed new schema against the current one to ensure it's compatible. Now that we know the new schema is valid, let's go ahead and register it with the following command.

Listing A.5 Registering the new schema

```
./gradlew :sr-backward:registerSchemasTask
```

Running the register command prints a BUILD SUCCESSFUL on the console. Before we move on to the next step, let's run a REST API command to view the latest schema for the avro-avengers-value:

```
curl -s "http://localhost:8081/subjects/avro-avengers-value/
  versions/latest" | jq '.'
```

Running this command should yield results resembling the following:

```
{
  "name" : "avro-avengers-value",
  "version" : 2,
  "schema" : "{\"type\": \"record\", \"namespace\": \"bbejeck.chapter_3\",
    \"name\": \"AvengerAvro\"..."
}
```

The results show the increase in the version from 1 to 2 as you've loaded a new schema. With these changes in place, you'll need to update your clients, starting with the consumer. With the compatibility of BACKWARD, you want to update the consumer first to handle any records produced using the new schema.

For example, you initially expected to work with the real_name field, but you deleted it in the schema, so you want to remove references to it in the new schema. You also added the powers field, so you'll want to be able to work with that field. That also implies you've generated new model objects.

Earlier, when you ran the clean, build command, it generated the correct objects for all modules. So, you should not have to do that now.

Note that since we are in BACKWARD compatibility mode, it wouldn't blow up if your updated consumer were to consume in the previous format. The update ignores the real_name field, and the powers field uses the default value.

After you have updated the consumer, you'll want to update your producer applications to use the new schema. The AvroProducer in the sr-backward submodule has had the updates applied already. Now run the following command to producer records using the new schema.

Listing A.6 Producing records with the new schema

```
./gradlew :sr-backward:runAvroProducer
```

You'll see some text scroll followed by the familiar BUILD SUCCESSFUL text. If you remember, you ran the produce command from the original submodule just a few minutes ago, adding records in the previous schema. So now that you've run the producer using the new schema, you have a mix of old and new schema records in the topic. But our consumer example should be able to handle both types since we are in the BACKWARDS compatibility mode.

Now, when you run the consumer, you should be able to see the records produced with the previous schema and those produced with the new schema. Run the following command to execute the updated consumer.

Listing A.7 Consuming records against the new schema

```
./gradlew :sr-backward:runAvroConsumer
```

You should see the first results printing with powers [] in the console. The empty value indicates those older records using the default value since the original records did not have a powers field on the object.

> **NOTE** For this first compatibility example, your consumer reads all the records in the topic. This happened because we used a new group.id for the consumer in the sr-backward module, and we've configured it to read from the earliest available offset if none were found. We'll use the same group.id for the rest of the compatibility examples and modules, and the consumer will only read newly produced records. Chapter 4 contains the details for the group.id configuration.

A.2   Forward compatibility

Forward compatibility is a mirror image of backward compatibility regarding field changes. With forward compatibility, you can add fields and delete optional ones. Let's go ahead and update the schema again, creating `avenger_v3.avsc`, which you can find in the `sr-forward/src/main/avro` directory.

Listing A.8   Foward compatible Avenger schema

```
{
  "namespace": "bbejeck.chapter_3.avro",
  "type": "record",
  "name": "AvengerAvro",
  "fields": [
    { "name": "name", "type": "string" },
    { "name": "powers", "type": {
            "type": "array", "items": "string"},
        "default": []
    },
    {"name": "nemeses","type": {                    Adds a new
            "type": "array", "items": "string"       field, nemeses
      }
    }
  ]
}
```

In this new schema version, you've removed the `movies` field, which defaults to an empty list, and added a new field `nemeses`. In forward compatibility, you would upgrade the producer client code first (figure A.2).

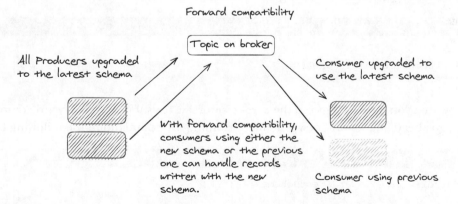

Figure A.2   Forward compatibility updates producers first to use the new schema, and consumers can handle the records either the new schema or the previous one.

Upgrading the producer code first ensures the new fields are correctly populated, and only records in the new format are available. Consumers you need to upgrade can still work with the new schema as they will simply ignore the new fields, and the deleted fields have default values.

Now you must change the compatibility mode from BACKWARD to FORWARD. In the `sr-forward` sub-module, the configuration for the Schema Registry plugin has the following code section, setting the compatibility.

Listing A.9   Compatibility in `build.gradle` for `sr-forward` submodule

```
config {
        subject('avro-avengers-value', 'FORWARD')
    }
```

Now, with the configuration set, to change the compatibility mode, run the following command.

Listing A.10   Changing the compatibility mode to FORWARD

```
./gradlew :sr-forward:configSubjectsTask
```

As we've seen before, the result of this command produces a BUILD SUCCESSFUL result on the console. If you want to confirm the compatibility mode for your subject, you can use this REST API command.

Listing A.11   REST API to view configured compatibility mode for a subject

```
curl -s "http://localhost:8081/config/avro-avengers-value" | jq '.'
```

The jq at the end of the curl command formats the returned JSON, and you should see something like the following code listing.

Listing A.12   Formatted configuration response

```
{
    compatibility: FORWARD
}
```

Now that you have configured the avro-avengers-value subject with forward compatibility, go ahead and test the new schema by running the command in the following listing.

Listing A.13   Testing a new schema is forward-compatible

```
./gradlew :sr-forward:testSchemasTask
```

This command should print a BUILD SUCCESSFUL on the console, and then you can register the new schema.

Listing A.14   Registering the new forward-compatible schema

```
./gradlew :sr-forward:registerSchemasTask
```

Then run a producer already updated to send records in the new format with the command.

Listing A.15   Running the producer updated for records in the new schema format

```
./gradlew :sr-forward:runAvroProducer
```

Now that you've run the producer with an updated schema, let's run the consumer that is *not* updated.

Listing A.16   Running a consumer not yet updated for the new schema changes

```
./gradlew :sr-backward:runAvroConsumer
```

The command results show how, with forward compatibility, even if the consumer is *not* updated, it can still handle records written using the new schema. Now we need to produce some records again for the *updated* consumer.

Listing A.17   Running the producer again

```
./gradlew :sr-forward:runAvroProducer
```

Now run the updated consumer for the new schema.

Listing A.18   Running the consumer updated for the new schema

```
./gradlew :sr-forward:runAvroConsumer
```

In both cases, the consumer runs successfully, but the details in the console are different due to having upgraded the consumer to handle the new schema.

Now, you've seen two compatibility types: backward and forward. As the compatibility name implies, you must consider record changes in one direction. In backward compatibility, you updated the consumers first, as records could arrive in either the new or old format. In forward compatibility, you first updated the producers to ensure the records from that time are only in the new format. The last compatibility strategy to explore is the FULL compatibility mode.

A.3   *Full compatibility*

You can add or remove fields in full compatibility mode, but there is one catch. Any changes you make must be to *optional* fields only (figure A.3).

Since the fields in the updated schema are optional, these changes are compatible with existing producer and consumer clients. This means that the upgrade order, in this case, is up to you. Consumers will continue to work with records produced with the new or old schema. Let's take a look at a schema to work with FULL compatibility.

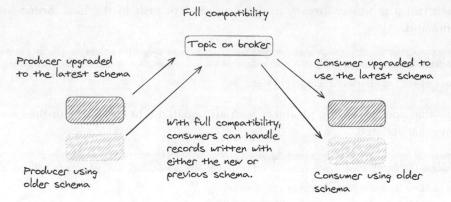

Figure A.3 Full compatibility allows producers to send with the previous or new schema, and consumers can handle the records with either the new schema or the previous one.

Listing A.19 Full compatibility `schema avengers_v4.avsc`

```
{
  "namespace": "bbejeck.chapter_3.avro",
  "type": "record",
  "name": "AvengerAvro",
  "fields": [
    { "name": "name", "type": "string" },
    { "name": "yearPublished", "type": "int", "default": 1960 },
    { "name": "realName", "type": "string", "default": "unknown" },
    { "name": "partners", "type": {
                          "type": "array","items": "string"},
          "default": []
    },
    {"name": "nemeses" "type": {
                       "type": "array", "items": "string"},
          "default": []
    }
  ]
}
```

Adds new optional field yearPublished

Adds back optional field realName

Adds new field partners

Before you update the schema, let's produce a set of records one more time to have a batch of records in the format before our next schema change.

Listing A.20 Creating a batch of records before we migrate the schema.

```
./gradlew :sr-forward:runAvroProducer
```

This will give us a batch of records to read with an updated consumer. But first, let's change the compatibility, this time to `FULL`.

Listing A.21 Changing the compatibility to `FULL`

```
./gradlew :sr-full:configSubjectsTask
```

To keep consistent with our process, let's test the compatibility of the schema before we migrate it.

Listing A.22 Testing the schema for full compatibility

```
./gradlew :sr-full:testSchemasTask
```

With the migrated schema compatibility tested, let's go ahead and register it.

Listing A.23 Registering the FULL compatibility schema

```
./gradlew :sr-full:registerSchemasTask
```

With the new schema version registered, let's have some fun with the order of records we produce and consume. Since all schema updates involve optional fields, the order in which we update the producers and consumers doesn't matter.

A few minutes ago, I had you create a batch of records in the previous schema format. I did that to demonstrate that we can use an updated consumer in FULL compatibility mode to read older records. Remember, before, with FORWARD compatibility, it was essential to ensure the updated consumers would only see records in the new format.

Let's run an updated consumer to read records using the previous schema. But watch what happens after running the following command. Now run the updated consumer, as in the following listing.

Listing A.24 Consuming with the updated consumer

```
./gradlew :sr-full:runAvroConsumer
```

And it runs just fine! Now, let's flip the order of operations and run the updated producer.

Listing A.25 Producing records with the new schema

```
./gradlew :sr-full:runAvroProducer
```

And now you can run the consumer that we haven't updated yet for the new record format.

Listing A.26 Consuming new records with a consumer not updated

```
./gradlew :sr-forward:runAvroConsumer
```

As you can see from playing with the different versions of producers and consumers with FULL compatibility when you update the producer and consumer, it is up to you; the order doesn't matter.

appendix B
Confluent resources

B.1 Confluent Cloud

Confluent Cloud is a resilient, scalable, streaming data service based on Apache Kafka®, delivered as a fully managed service. Confluent Cloud allows you to quickly scale up a Kafka cluster without consuming any resources on your laptop. Combined with the Confluent command-line interface (CLI), you can quickly and easily spin up a cluster, create and delete topics, and enable Schema Registry. You can also take the latest product from Confluent, Flink SQL, for a spin.

Readers of this book can try Confluent Cloud out for free with the code "KAFKASTREAMSBOOK", which gives you $100 in free credit. You can find Confluent Cloud at https://www.confluent.io/confluent-cloud/.

B.2 Confluent command-line interface

The Confluent CLI interface is a powerful tool that makes working with Confluent Cloud very easy. You can create and delete clusters and consume from and produce to topics from the command line. To install the CLI tool, users of macOS or Linux can use the command `brew install confluentinc/tap/cli`, but all major operating systems are supported. Go to http://mng.bz/KZMZ for complete instructions on the different options available for installing the CLI.

You also can extend the behavior of the CLI with plugins. There's an official Confluent plugin repository at https://github.com/confluentinc/cli-plugins. You can find available plugins to install by running `confluent plugin search`. For example, you can use the kickstart plugin `confluent cloud kickstart`, and it will automatically provision a Kafka cluster, enable Schema Registry, create API keys, and generate the required client-connection properties file for you. You can run `confluent cloud kickstart -h` to see all available options.

B.3   *Confluent local*

The CLI also has the option to run a Docker-based Kafka broker locally (http://mng .bz/9dMo) with the command `confluent local kafka start`. This command will pull a Kafka Docker image and start it. You must have Docker and Docker Desktop installed to do so.

appendix C
Working with Avro, Protobuf, and JSON Schema

C.1 Apache Avro

An Avro schema is composed of two main types: primitive and complex. The supported primitive types are `null`, `boolean`, `int`, `long`, `float`, `double`, `bytes`, and `string`. Complex types are `records`, `enums`, `arrays`, `maps`, `unions`, and `fixed`. For a complete description of primitive and complex types, see the Avro documentation at http://mng.bz/Y7Oo and http://mng.bz/GZ2M, respectively.

In this book, you'll mostly work with the complex type of `record` as this corresponds to an object. You may also encounter a few examples using primitive values, but for the most part, we'll stick with records. You'll also use the `union` type, especially when working with the `RecordNameStrategy` and `TopicRecordName` strategies. The `union` type is represented as an array and is very useful. `unions` allow you to specify that a field may be of one or more types. You'll see `union` types in the following examples.

The `record` type contains the elements `name`, `doc`, `aliases`, and `fields`. The `fields` element can have the properties `name`, `type`, `doc`, `default`, `order`, and `aliases`. Figure C.1 shows an Avro schema with some descriptions of what each of these properties is.

By default, when serializing records, Avro encodes fields in the listed order of the schema. Next, I want to discuss the `default` and `aliases` from the previously mentioned properties.

```
{
    "type": "record",
    "namespace": "bbejeck.chapter_3.nested"
    "name": "Transaction",
    "fields" : [
        {"name": "txn_type", "type": [
            {
                "type": "record",
                "namespace": "bbejeck.chapter_3.nested",
                "name":"Purchase",
                "fields": [
                    {"name": "item", "type":"string"},
                    {"name": "amount", "type": "double"}
                ]
            },
            {
                "type":"record",
                "namespace": "bbejeck.chapter_3.nested",
                "name":"Return",
                "fields": [
                    {"name": "item", "type":"string"},
                    {"name": "amount", "type": "double"}
                ]
            },
            {
                "type":"record",
                "namespace": "bbejeck.chapter_3.nested",
                "name":"Exchange",
                "fields": [
                    {"name": "item", "type": "string"},
                    {"name": "amount", "type": "double"},
                    {"name": "new_item", "type": "string"}
                ]
            }
        ]},
        {"name": "identifier", "type": "long"}
    ]
}
```

```
{
    "type": "record",
    "namespace": "bbejeck.chapter_3",
    "name": "Transaction",
    "fields" : [
        {"name": "txn_type", "type": [
                                        "bbejeck.chapter_3.Purchase",
                                        "bbejeck.chapter_3.Return",
                                        "bbejeck.chapter_3.Exchange"
        ]},
        {"name": "identifier", "type": "long"}
    ]
}
```

Schema using references is easier to read and will pick up changes to original schema.

Schema with full definition of nested object fields inline.

Figure C.1 Comparing a schema with nested records vs. using schema references

C.1.1   Default and alias

The `default` property is essential as it provides a field value if it's missing when deserializing an Avro encoded object. In other words, supplying a `default` value in Avro is the same as saying the field is optional. With regard to compatibility, optional values factor heavily in whether a schema is backward- or forward-compatible.

An `alias` refers to the name of a previous schema you want to use for the current schema. This means that if you evolve a schema named `event` to `event_v2`, but you specified an alias of `event` for the new schema, that means readers of the new schema will read it as if it were named `event`. The same is true of a field with an alias. Specifying aliases of the expected name in the old schema will cause the reader to use the alias name. An alias is useful if you've updated a field name but want the new schema to remain backward-compatible for existing downstream users expecting the previous field name.

C.1.2   Union

I mentioned the `union` type as an important concept with Avro schemas. A `union` allows you to specify that a field can be of one of more than one type. First, consider the following basic example:

```
{
    "type": "record",
    "namespace": "bbejeck.chapter_3",
    "name": "transaction",
```

```
    "fields" : [
      {"name": "txn_type", "type": "string"},
      {"name": "identifier", "type": ["string", "long"]}    ◁——
    ]
}
```

> Union type for
> the identifier
> field; it can be a
> string or a long.

This sample schema shows an example of a simple union type. The identifier could be either a string or a long. But the union type can be more complex and is integral to working with the RecordNameStrategy or TopicRecordName strategies.

For example, let's extend the example schema to have a record type for the transaction. Instead of a string, it could be one of three record types—Purchase, Return or Exchange. Our schema would look something like this:

```
{
  "type": "record",
  "namespace": "bbejeck.chapter_3.nested"
  "name": "transaction",

  "fields" : [
    {"name": "txn_type", "type": [    ◁——
```

> Start of union
> type as an array

```
        {
          "type":"record",
          "namespace": "bbejeck.chapter_3.nested",
          "name":"Purchase",            ◁——
          "fields": [
            {"name": "item", "type":"string"},
            {"name": "amount", "type": "double"}
          ]
        },
```

> The Purchase
> record type

```
        {
          "type":"record",
          "namespace": "bbejeck.chapter_3.nested",
          "name":"Return",              ◁——
          "fields": [
            {"name": "item", "type":"string"},
            {"name": "amount", "type": "double"}
          ]
        },
```

> The Return
> record type

```
        {
          "type":"record",
          "namespace": "bbejeck.chapter_3.nested",
          "name":"Exchange",            ◁——
          "fields": [
            {"name": "item", "type": "string"},
            {"name": "amount", "type": "double"},
            {"name": "new_item", "type": "string"}
          ]
        }
```

> The Exchange
> record type

```
    ]},

    {"name": "identifier", "type": "long"}
  ]
}
```

The union type can be one of three `record` types from this example schema. If the nested types have several fields and possibly contain other union types, you can see how a union type can quickly get unwieldy. Luckily, Schema Registry now supports schema references.

C.2 *Protocol Buffers*

For cross-language communication, Google developed Protocol Buffers (https:// developers.google.com/protocol-buffers). What exactly are Protocol Buffers? Well, let's get the definition straight from the Protocol Buffers home page:

> *Protocol buffers are Google's language-neutral, platform-neutral, extensible mechanism for serializing structured data—think XML, but smaller, faster, and more straightforward. You define how you want your data to be structured once. You can use generated source code to quickly write and read your structured data to and from various data streams and using a variety of languages.*

>   —Protocol Buffers definition from the home page

Since Protocol Buffers aim to be in a language-neutral serialization format, they are perfect for use with Kafka and Schema Registry. As you can imagine, protocol buffers are a deep subject, so I will only go into specific details. But in this section, I'll review the information you'll need to start using Protocol Buffers with Schema Registry.

NOTE Currently, versions 2 and 3 of Protocol Buffers are available, with version 3 being the latest. In this book, I'm going to focus on version 3. Also, while Protocol Buffers support different languages, I will focus on using Java.

You'll need to install the Protobuf compiler on your machine to use Protocol Buffers. If you are using macOS as I am, you can use `homebrew` (https://brew.sh/) to install Protobuf with one line: `brew install Protobuf`.

NOTE While command-line tools are available for working with Protobuf, the source code project for the book is gradle-based. As a result, you'll use the `Protobuf-gradle-plugin` provided by Google (https://github.com/google/ Protobuf-gradle-plugin).

Protocol Buffers have a concept of a `message`, the outermost part of the record hierarchy. A `message` will have one or more fields defined inside the message. You'll write the schema for a Protocol Buffer message in a file ending with `.proto`, compared to Avro where the record schemas come in files ending with `.avsc`.

The fields contained within a `message` can take the form of a scalar type—`long`, `int`, `float`, `double`, `boolean`,—`String` and for the Java API, a `ByteString`, which is analogous to a byte array in Java. Protocol bBuffers support generating code from the `.proto` files. Protobuf also supports complex structures so that messages can contain nested message types. Let's look at a basic proto file to cement your understanding of what we just covered in listing C.1.

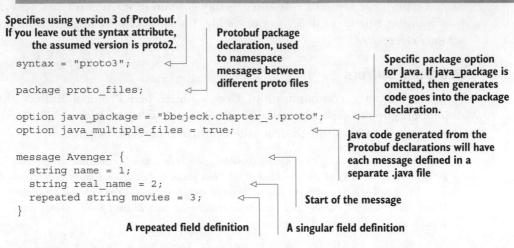

Listing C.1   Simple Protobuf schema file `avenger_v1.proto`

```
syntax = "proto3";

package proto_files;

option java_package = "bbejeck.chapter_3.proto";
option java_multiple_files = true;

message Avenger {
  string name = 1;
  string real_name = 2;
  repeated string movies = 3;
}
```

Specifies using version 3 of Protobuf. If you leave out the syntax attribute, the assumed version is proto2.

Protobuf package declaration, used to namespace messages between different proto files

Specific package option for Java. If java_package is omitted, then generates code goes into the package declaration.

Java code generated from the Protobuf declarations will have each message defined in a separate .java file

Start of the message

A repeated field definition

A singular field definition

The format of a Protobuf schema file is relatively straightforward. There is one thing I want to point out in this example. I want to emphasize that with Protobuf version 3, all fields are considered optional.

When deserializing a message, if a field is missing, then Protobuf uses the default for the fields' type. For example, numeric types use 0; a Boolean will use `false`, and strings will use an empty string. The default value for missing message fields is language dependent, so in the case of Java, it will be `null`.

You'll notice you've set the field name equal to a number. Protobuf uses that number to uniquely identify the field in the message binary format, which means you can't reuse numbers you assign to fields once you start using the message.

The numbers 1 to 15 take up only 1 byte in the encoded format, so using these numbers for the frequently appearing message fields is a good idea. For more information on numbering fields, please refer to the "Assigning Number Fields" in the Protobuf documentation (http://mng.bz/RZGK).

Fields in a message can take two forms: singular fields (the default) or repeated fields. The schema file we just reviewed shows an example of this. The first three fields in the schema are singular. Singular fields are the default, and you don't need to add any additional information. `repeated string movies = 3;` in the previous schema is an example of a repeated field, which means there can be exactly 0 or more values. Repeated fields are analogous to resizable arrays, and in the Java API, they are represented as the `List<T>` type.

C.2.1   Complex messages

Protobuf allows you to define a message type within a message. Also, a field in a message can refer to another message. Consider the previous `Avenger` proto example. Let's say you want to expand the information for the movie beyond just the title.

You've decided to add a new message type, AvengerMovie, to the proto file from the previous example.

> **Listing C.2 Adding new message movies to avenger_v1.proto**

```
syntax = "proto3";

package proto_files;

option java_package = "bbejeck.chapter_3.proto";
option java_multiple_files = true;

message Avenger {                         ◄───┤ Adds a new nested
                                               message type
    message AvengerMovie {          ◄──┐
        string title = 1;              Uses the message
        string year = 2;              type as a field
        string producer = 3;
    }

  string name = 1;
  string real_name = 2;
  repeated AvengerMovie = 3;
}
```

In this example, I've used a nested type and used it as a field, but I did this for demonstration purposes. There's no requirement to declare a nested type as a field if you don't need to.

> **NOTE** At the risk of stating the obvious, the numbering of fields for each nested message is independent of any outer message types. In other words, when creating a nested message type, you should restart the numbering of the fields.

With Protobuf, you can nest message types to any level. But it's essential to remember how manageable the nesting is to maintain. At some point, files with deeply nested types are hard to maintain. Sometimes, it's better to break out complex nested types into their own proto files. Fortunately, Protobuf provides an easy way to work with separate proto files.

C.2.2 *Importing*

While Protobuf allows you to define multiple message types in a file, that is not always possible. You could want to use a proto from another code module. Whatever the reason, the ability to import is valuable as you can use existing proto definitions you need to reuse throughout your code base. To import a proto file, you add an import statement to the top of the proto file. Let's revisit our avenger_v2.proto file example and extract the message AvengerMovie into its own proto file.

Listing C.3   Separate proto file `avenger_movie.proto`

```
syntax = "proto3";

package proto_files;

option java_package = "bbejeck.chapter_3.proto";
option java_multiple_files = true;

message AvengerMovie {
        string title = 1;
        string year = 2;
        string producer = 3;
    }
```

Now, to use the `AvengerMovie` message, copy the `avengerV1.proto` file into the `avengerV2.proto` file and add an import statement at the top of the file like this:

```
syntax = "proto3";

package proto_files;

import "avenger_movie.proto";

option java_package = "bbejeck.chapter_3.proto";
option java_multiple_files = true;

message AvengerV2 {
  string name = 1;
  string real_name = 2;
  repeated AvengerMovie = 3;
}
```

Whether to combine several message types in one proto file or break things out into separate files probably lies along the lines of functionality. But ultimately, it will depend on what works best for your use case.

Before we conclude the coverage of Protobuf, I want to discuss one more aspect that allows for using multiple event types per topic.

C.2.3   *Oneof type*

Protobuf has a `oneOf` type that allows a message to have a field consisting of multiple types, but only one at a time. The Protobuf `oneOf` is similar to Avro's `union` type. It's important to note that you can't define a `oneof` as a top-level element. You must wrap a `oneof` in a message; you can't have it defined alone in a proto file.

Let's look at an example of a proto using the `oneOf` type. We'll use the same purchase example from the Avro union example so you can make a clear comparison.

Listing C.4 Protobuf with `oneof` type

```
syntax = "proto3";

package proto_files;

option java_package = "bbejeck.chapter_3.proto";
option java_multiple_files = true;

message TransactionType {

        message Purchase {              ⟵─┤ Defines the
            string item = 1;                 purchase message
            double amount = 2;
        }

        message Return {                ⟵─┤ Defines the
            string item = 1;                 return message
            double amount = 2;
        }

        message Exchange {              ⟵─┤ Defines the
            string item = 1;                 exchange message
            double amount = 2;
            string new_item = 3;
        }

    string identifier = 1;

    oneof txt_type {                    ⟵─┤ The oneof
        Purchase purchase = 2;               type field
        Return return = 3;
        Exchange exchange = 4;
    }

}
```

As you can see here, you define the possible message types in the proto file. Then, when you declare the `oneoftxn_type` field, you list the possible types the field can contain. This ability to have a field that can take multiple types is essential to using the `RecordNameStrategy` or `TopicRecordNameStrategy` since, with those strategies, the topics may have more than one record type.

In this example, defining all the message types in one file makes sense since they are closely related. But if you need to have several record types in one topic, but the proto files are in different locations, you can also use schema references with Protobuf. Let's look at how you would take this schema and use schema references.

I won't show the individual proto files for the example here of schema references with Protobuf. But you can safely assume we've created separate proto files for each nested message type. So the updated proto schema file now looks like the following listing.

```
Listing C.5  Protobuf with one of type transaction_type.proto
```

```
syntax = "proto3";

package proto_files;

import "purchase.proto";        ◄─┐  Adds the import
import "return.proto";             │  statements
import "exchange.proto";

option java_package = "bbejeck.chapter_3.proto";
option java_multiple_files = true;

message TransactionType {

    string identifier = 1;

    oneof txt_type {            ◄─┐  References the imported
        Purchase purchase = 2;     │  messages for the oneof field
        Return return = 3;
        Exchange exchange = 4;
    }
}
```

Here, the proto file looks much cleaner with the other message types extracted from separate files. Now that you've set up the proto file for schema references, let's look at how you would register a proto schema for using different event types.

NOTE When using schema references, you should use the `TopicNameStrategy`, which uses the topic name to determine the subject used for schema lookup. Using this strategy is important because it enforces subject-topic constraints for all the types used in the schema refs to *only* the topic using the schema references.

```
Listing C.6  Protobuf schema reference
```

```
{
    "schema" :"syntax = \"proto3\";   ◄─┐  The portion of the request with the
                                          schema containing the references
        package proto_files;

        import \"purchase.proto\";
        import \"return.proto\";
        import \"exchange.proto\";

        option java_package = \"bbejeck.chapter_3.proto\";
        option java_multiple_files = \"true\";

        message TransactionType {

            string identifier = 1;
```

```
                    oneof txt_type {
                        Purchase purchase = 2;        ◁──┐  Schema
                        Return return = 3;               │  references by
                        Exchange exchange = 4;           │  message name
                    }

                ",                                       Specifies the schema type;
                                                         if this is left off, the format
    "schemaType": "PROTOBUF",          ◁────┐            defaults to AVRO.

    "references": [                         ┌──          The references to the already
        {                                   │            registered schemas for these
            "name": "purchase.proto",       │            event objects
            "subject": "purchase",
            "version": 1
        },
        {

            "name": "return.proto",
            "subject": "return",
            "version": 1
        },
        {
            "name": "exchange.proto,
            "subject": "exchange",
            "version": 1
        }
    ]

}
```

TIP When using schema references with Protobuf, you do not need to turn
off `auto.shema.registration` because Protobuf recursively autoloads all referenced schemas.

So far, you have learned how to register single Protobuf schemas and use multiple schemas via schema references. Next, let's move on to using Protobuf code generation.

C.2.4 Code generation

Protobuf, like Avro, provides a mechanism for generating code from a schema file. I
mentioned the importance of generating code in the Avro section, but I'll repeat it
here. You must generate the source code for the object types you want to work within
your applications.

The great thing about the code generation tools is that they remove the burden of
creating the model objects you'll use. In addition to removing the tedious boilerplate
work, the generated objects are guaranteed to match the schema specifications.

Protobuf supports code generation in several languages, but in this book, you will
use the Protobuf Java code generation. Now, let's take a minute to walk through generating the Java source code for this example schema.

Protobuf provides the `protoc` tool, which generates Java source code from proto
files. But you'll use a `gradle` plugin for the Protobuf code generation. The plugin uses

protoc, but the code generation is part of the project build process, so it's automatic and one less step for you.

To do this, go to the root directory where you cloned the source code for the book. From the root of the project, run the command ./gradlew clean build and then go and look at the build/generated-main-proto-java/main/java/bbejeck.chapter_3/proto directory, and you'll see the AvengerProto java file.

As with the Avro-generated source files, the Java objects generated by Protobuf are immutable. As a result of this immutability, Protobuf provides Builders that allow you to construct the message objects. Each message gets its own builder, taking the form of MessageName.Builder. Nested messages still get their own builder, too, referenced by OuterMessage.NestedMessage.Builder.

> **NOTE** What's the builder pattern? The builder pattern provides a flexible way for object construction. Instead of passing parameters to a constructor, you add the required fields through methods on a builder. Each method on the builder represents a field on the object. It's one of the original Gang of Four Design patterns (https://martinfowler.com/bliki/GangOfFour.html).

The example proto files have two option fields specific to the Protobuf Java API, java_package and java_multiple_files. Let's take a quick minute to discuss those fields now:

- java_package—When specified, the java_package specifies the package for the generated Java code. Otherwise, the Protobuf uses the package field. It would be best to use the Protobuf package field to namespace proto files and the java_package to specify a proper Java package name.
- java_multiple_files—Setting this configuration to true tells Protobuf to write separate .java files for all top level messages and enums in the proto file at the package level. Otherwise, proto writes the messages and enums as members of the containing outer class.

So now you know how to use the included tools with Protobuf to generate source code files. Remember, Protobuf supports many languages, but we focused on using Java, which I used in the book. You may have noticed that while the tools are different, the code Avro and Protobuf generate are reasonably similar. Some minor differences exist with the helper methods Protobuf provides when you use the oneof field.

C.2.5   *Specific and dynamic types*

When the Schema Registry Protobuf deserializers construct an object from the serialized format, the resulting object can be a specific or dynamic type. This typing is just like Avro. Depending on the information available when deserializing, the object can be either a particular type of class or a GenericRecord.

Usually, you'll want to use the specific type class because you'll know the object's structure. For instance, when you deserialize and work with the AvengerProto object, you'll know exactly how to handle it in your code. But just like the Avro generic type, working with the Protobuf dynamic type means you'll have to query the object to discover the fields it contains and what data the object ultimately represents.

Now you've learned the basics of working with Protobuf. We'll move on to the third schema format supported by Schema Registry, JSON.

C.3 JSON Schema

Now we'll get to the third schema type supported by Schema Registry, JSON Schema. For JSON Schema, I will use version draft-07 (http://json-schema.org/draft-07/schema#).

> **NOTE** As with the Avro and Protobuf sections, I won't go into every detail of using a JSON schema. I'll cover enough here to get you started. For the complete information, you should go to https://json-schema.org/.

When working with the JSON schemas, you'll notice some similarities with Avro schemas, as they are both defined in JSON. However, the JSON Schema format is different, so you can't reuse an Avro schema for a JSON schema.

Let's jump in and look at JSON Schema. To help with the comparison to the other formats, we'll use the `Avenger` object again here.

Listing C.7 JSON Schema for an Avenger object

```
{
  "$schema" : "http://json-schema.org/draft-07/schema#",     ←  Declares the schema version
  "title" : "Avenger",                                            The title is not required but is good for metadata.
  "description" : "A JSON schema of Avenger object",         ←  The description, again, is optional but good for informational purposes.
  "javaType": "bbejeck.chapter_3.SimpleAvenger",
  "type" : "object",                                         ←  Declaring the type, which is object here. Analogous to record in Avro or message in Protobuf.
  "properties" : {                                           ←  The properties, which are the fields on the object
  
    "name" : {
      "type": "string"
    },
    "movies" : {
      "type": "array",
      "items": {
        "type": "string"
      },
      "default": []
    },
    "realName": {
      "type": "string"
    }
  },
  "required" : ["name", "realName"]                           ←  Declares the required fields
}
```

As you can see, the JSON schema is relatively close to what you've seen with Avro. The type is `object` and it's one of the `simpleTypes`. The `object` type is essentially a map with the `properties` attribute representing the keys and values contained in the object.

The other simple types are `array`, `boolean`, `integer`, `null`, `number`, and `string`. Properties on an object are not required by default. You must add the property name to the `required` array to enforce required fields.

C.3.1   Nested objects

JSON Schema supports having nested object types as well. Let's expand the example to add a `movies` field as we've done before.

Listing C.8   JSON Schema with nested object type

```
{
  "$schema" : "http://json-schema.org/draft-07/schema#",
  "title" : "Avenger",
  "description" : "A JSON Avenger object with a nested object",
  "javaType" : "bbejeck.chapter_3.SimpleAvenger"
  "type" : "object",
  "properties" : {

    "name" : {
      "type" : "string"
    },
    "realName": {
      "type" : "string"
    },
    "movies" : {
      "type" : "array",
        "items" : {
          "type" : "object",
          "javaType": "bbejeck.chapter_3.AvengerMovie"
            "properties" : {

              "title" : {
                "type" : "string"
              },
              "year" : {
                "type" : "string"
              },
              "producer" : {
                "type" : "string"
              }
            },
            "required" : ["title", "year", "producer"]
        }
    }
  },
  "required" : ["name", "realName"]
}
```

Adds the movies property

The movies property is an array, and you need to define the types in the array.

Specifies the class name of AvengerMovie for the array type

All three properties of the nested movie object are required.

This example shows that all the properties on the nested object in the array are required, but the array `movies` itself is not. However, the nesting of object structures can become hard to manage very quickly. Fortunately, JSON Schema also supports references, which can help keep the main object structure more manageable to view.

C.3.2   JSON references

So, let's update this schema to use references for the object definitions.

Listing C.9   JSON Schema with references

```
{
  "$schema" : "http://json-schema.org/draft-07/schema#",
  "title" : "Avenger",
  "description" : "A JSON schema of Avenger object with an
    internal schema ref",

  "definitions" : {

        "avenger_movie" : {          ←——| Creates the object under
            "type" : "object" ,          | the definitions keyword
            "properties" : {
              "title" : { "type" : "string" },
              "year" : { "type" : "string" },
              "producer" : { "type" : "string" }
            },
            "required" : ["title", "year", "producer"]
        }

  },

  "description" : "A JSON schema of Avenger object",
  "type" : "object",
  "javaType": "bbejeck.chapter_3.SimpleAvenger",
  "properties" : {

    "name" : {
      "type" : "string"
    },
    "realName": {
      "type" : "string"
    },
        "movies" : {                                    Refers to the
            "type" : "array",                         avenger_movie
            "items": { "$ref" : "#definitions/avenger_movie" }  definition
        }                                                    ←——

  }
}
```

The way the $ref key works is to replace where you are using it with the entire contents of what the ref points to. The updated schema is just as long as the previous schema, but it's easier to read and understand the structure. But you can also use the $ref attribute to refer to separate files containing a JSON schema. So, let's revisit the previous JSON schema you just reviewed. Assume you've extracted avenger_movie to avenger_movie .schema.json files. Now, you can update the schema to refer to those files, but the $ref behavior is the same: it replaces the content where you place the reference. The following listing provides the updated schema with external file references.

Listing C.10   JSON Schema with external file references

```
{
  "$schema": "http://json-schema.org/draft-07/schema#",
  "title": "Avenger with an external schema reference",
  "description": "A JSON schema with refs to external schemas",
  "javaType": "bbejeck.chapter_3.ComplexAvenger",        ◁─── Changes the Java
  "type": "object",                                             class name to reflect
  "properties": {                                               the changes to the
    "name": {                                                   schema
      "type": "string"
    },
    "realName": {
      "type": "string"
    },
    "movies": {
      "type": "array",
      "items": {
        "$ref": "avenger_movie.schema.json"
      }
    }
  }
}
```

Now, you've simplified the schema even more by using the external references. Please note that for this example, I'm assuming you placed the referenced JSON files in the same directory as the referring.

> **NOTE**   When working with JSON schemas, it might be helpful to have a valida-
> tor handy to ensure you are building schemas correctly. As of the writing of this
> book, there is an online validator located at https://www.jsonschemavalidator
> .net/, which I found helpful.

The JSON schema provides some keywords that allow for combining schemas. Those keywords are allOf, anyOf, oneOf, and not. In particular, we want to look at the oneOf keyword for use in Schema Registry's multiple event types per topic. The oneOf key-word works similarly to the same in Protobuf.

Next, look at an example of using Schema Registry schema imports with JSON.

C.3.3   *JSON Schema Registry schema references*

Since I've already covered complex schemas and references with JSON, we'll jump into how you would submit a JSON schema with schema imports. Again, the JSON shown here is a more readable version of what you'll submit to Schema Registry. The top "schema" information will be on one line with no line breaks.

Listing C.11   JSON Schema reference

```
{

  "schema" :                    ◁─┘  The JSON schema
```

```
"{
    \"schema\" : \"http://json-schema.org/draft-07/schema#\",
    \"title\" : \"JSON schema example\",
    \"description\" : \"Schema imports with JSON\",
    \"javaType" = "transactionInfo\"              ← Specifies the className

        \"oneOf\": [
                { \"$ref\" : \"purchase.schema.json\" },
                { \"$ref\" : \"return.schema.json\" },
                { \"$ref\" : \"exchange.schema.json\" }
        ]
}",
"schemaType": "JSON",                ← Specifies the schema
                                        type as JSON

"references" : [                     ← The references of the already
    {                                   regisered JSON schemas
        "name": "purchase.schema.json",
        "subject": "purchase",
        "version": 1
    },
    {
        "name": "return.schema.json",
        "subject": "return",
            "version": 1
    },
    {
        "name": "exchange.schema.json,
        "subject": "exchange",
        "version": 1
    }
  ]

}
```

JSON schema refs using oneOf

Other than the differences between the different formats, the format for submitting
schema references looks similar to what you have seen for Avro and Protobuf. When
using schema references with JSON schemas, the same configuration changes you
must take with Avro apply here. If you remember, you must set the properties `auto
.register.schema=false` and `use.latest.version=true` when configuring a producer to
work with JSON schemas. Chapter 4 covers configuring the various schema types with
clients (producers and consumers).

There's another essential point to working with JSON schemas, especially with stat-
ically typed languages like Java. That is the specification of the class name. If you compare the JSON schemas to Avro or Protobuf, you'll notice a lack of information you
can use for the package or class name. But JSON schemas allow adding fields not
explicitly listed in the schema specification, known as the open content model.

The open content model allows adding a top-level field to a schema file and desig-
nating the class name for deserializing. You can turn off adding additional fields by
setting `"additionalProperties": false` in the schema for an object, but otherwise, it's

true by default. When using JSON schemas, adding a top-level field specifying the class name is essential for deserialization when you can have multiple types in a topic.

At this point, we've covered enough information for you to understand JSON schemas well. But the schema is a specification, and you need to work with the objects that follow the schema specification. The next section will discuss how to go from schema to objects.

C.3.4   JSON Schema code generation

As I said before, working with JSON schemas differs from working with Avro or Protocol Buffers regarding source code generation. The difference is a JSON schema is a *specification*, and Avro and Protocol Buffers are frameworks complete with tooling. So, with Avro and Protobuf, code generation is a first-class citizen. But code generation with JSON schemas requires using third-party tools.

There are some popular tools for working with Java objects and JSON. While there are many good choices, in this book's source code, I will use the FasterXML/Jackson utilities (https://github.com/FasterXML/jackson). Jackson provides serialization support and annotations you can use on a Java object. But I will use an additional open source library for the annotation support—the Jackson jsonSchema Generator (https://github.com/mbknor/mbknor-jackson-jsonSchema). While Jackson also offers schema generation, the mbknor-jackson-jsonSchema offers functionality beyond Jackson's.

Before we get into the annotation extensions, let's take a look at a Java class using the basic annotations (some details are omitted for clarity).

Listing C.12   Java code with Jackson annotations

```
public class Customer {
    @JsonProperty
    private String name;

    @JsonProperty
    private int id;

    @JsonProperty
    private String email;

}
```

Using the annotations gives you a couple of options. First, you don't have to generate a schema physically. You can rely on the Schema Registry JSON serializers to parse and register the schema during the serialization process. But if you have multiple event types in a single topic, you must add a top-level field specifying the class name, as discussed in the previous section. So, you'll need to generate a physical schema in that case.

The need to add the top-level field specifying a class name is where using the mbknor-jackson-jsonSchema project comes into play. The mbknor-jackson-jsonSchema library provides a `@JsonSchemaInject` annotation that supports injecting JSON Schema fragments. Let's update the previous example and add the `@JsonSchemaInject` annotation (some details are omitted for clarity).

Listing C.13 Java code with annotations to inject class name

```
@JsonSchemaInject(strings=
        {@JsonSchemaString(
                path="javaType",
                value="bbejeck.chapter_3.codegen.Customer")})
public class Customer {
    @JsonProperty
    @JsonProperty
    private String name;

    @JsonProperty
    private int id;

    @JsonProperty
    private String email;

}
```

So the code here is the same; you've just added the `@JsonSchemaInject` annotation, which places a top-level field `javaClassName` and enables multiple event types in a single topic. Now, to generate the physical schema with the mbknor-jackson-jsonSchema schema generator, you could do something like the following listing.

Listing C.14 Using the schema generator

```
theClass = Class.forName(args[0]);
ObjectMapper mapper = new ObjectMapper();
JsonSchemaGenerator schemaGenerator = new JsonSchemaGenerator(mapper);
JsonNode jsonNode = schemaGenerator.generateJsonSchema(theClass);
String schema = mapper.writerWithDefaultPrettyPrinter()
        .writeValueAsString(jsonNode);
```

The resulting schema will look like the following listing.

Listing C.15 Generated JSON Schema

```
{
  "$schema" : "http://json-schema.org/draft-04/schema#",
  "title" : "Customer",
  "type" : "object",
  "additionalProperties" : false,
  "javaType" : "bbejeck.chapter_3.codegen.Customer",
  "properties" : {
    "name" : {
      "type" : "string"
    },
    "id" : {
      "type" : "integer"
    },
```

```
    "email" : {
      "type" : "string"
    }
  },
  "required" : [ "id" ]
}
```

There is an example of schema generation in the source code for the book in the bbe-jeck.chapter_3.codegen package.

The approach you've seen in this section is "code first, schema second". You started with existing Java objects or wrote the objects first, applied annotations, and then generated a schema. However, some may find hand-rolling the code awkward and prefer to write a schema first and then generate the code. We'll cover code generation from a schema next.

Another tool, https://github.com/joelittlejohn/jsonschema2pojo, offers Java code generation from a JSON schema. The jsonschema2pojo library optionally supports generating with annotations. The supported annotation styles are for the following Java-JSON tools: Jackson 1.x, Jackson 2.x, Gson, and Moshi.

> **NOTE**   I've already mentioned Jackson, but here I'll point out two new Java-JSON tools. Gson (https://github.com/google/gson) comes from Google, and Moshi (https://github.com/square/moshi) comes from Square. Both are libraries for converting Java objects to and from their JSON representation. I won't go into details about either of them in this book. But if you want to learn more about either, follow up using the provided links.

One unique point about jsonschema2pojo is that it offers an online solution, http://www.jsonschema2pojo.org/, that can generate Java source code directly from the browser. It's also available as a command-line tool, Maven, or Gradle plugin. Since the source code project for this book is Gradle-based, I will use the `gradle` plugin.

> **NOTE**   On the jsonschema2pojo GitHub site, the `gradle` plugin is deprecated and will only work with Gradle versions <7.0. But there is a fork of the `gradle` plugin: plugin, https://github.com/eirnym/js2p-gradle, which will continue new development, so I'll use the forked version in this book.

To generate your Java code from the schema, run the command `./gradlew build`. Running the Gradle build generates the source code for Avro, Protobuf, and JSON Schema simultaneously. I won't go into details about the configuration of the various build options, but you can look at the `build.gradle` file for more information.

So, that concludes our coverage of code generation with JSON schemas. Next, we'll discuss how to configure the different serializers provided by Schema Registry to work with Avro, Protobuf, and JSON.

> **NOTE**   For those readers using macOS (or Linux), jsonschema2pojo is available for installation with Homebrew: `brew install jsonschema2pojo`. If you are unfamiliar with Homebrew, it's a package manager for installing software. You can find out more about Homebrew at its home page: https://brew.sh/.

C.3.5 Specific and generic types

Like Avro and Protobuf, two categories of objects are available using a JSON schema when deserializing. When configured with the specific class name, the Schema Registry deserializer returns an object of that type. If you don't configure the exact class for deserialization or the deserializer can't determine the class, then the deserializer returns the object of `JsonNode`. You'll work with the `JsonNode` similarly to working with a DOM (Document Object Model) from an XML structure.

To use schema references, you'll first need to register the referred schemas. Then, from within the schema you want to use a reference from, you'll need to specify three things for the referred-to schema:

1 The reference name
2 The subject you registered the schema under
3 The exact version of the schema under the subject (the one in point 2 here)

In Avro, the reference name is the fully qualified name of the schema file. For Protobuf, it's the name of the proto file, and with JSON, it's a URL.

Schema references can be convenient when you have a topic expecting one type of object and another where the object uses the same type as one of its fields. However, you don't need to use schema references strictly because you have more than one topic requiring the same schema. You can use schema references to keep a schema with a union of types from getting too difficult to manage. It's much cleaner to have a reference to the schema for a complete field versus having a complex schema in the field definition.

> **NOTE** Schema Registry supports schema references with Protobuf and JSON as well. We'll talk about schema references as well in those sections. Still, we'll only cover the syntactic details specific to Protobuf and JSON since we've covered the more general discussion in this section.

Let's look at figure C.2 to clarify what I mean here.

Now that you've learned how schema references work, let's revisit our example from section C.1.2 on Avro unions to see how this will look.

```
{
  "type": "record",
  "namespace": "bbejeck.chapter_3"
  "name": "Transaction",

  "fields" : [                                   Uses a union type to
    {"name": "txn_type", "type": [            represent one of three
                                                        event types
                       "bbejeck.chapter_3.Purchase",     ◄─┐
                       "bbejeck.chapter_3.Return",
                       "bbejeck.chapter_3.Exchange"
                 ]},

    {"name": "identifier", "type": "long"}
  ]

}
```

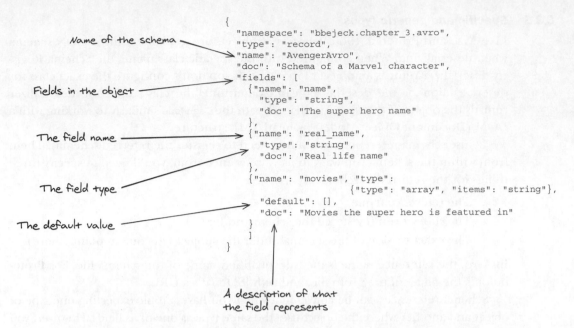

Name of the schema

Fields in the object

The field name

The field type

The default value

A description of what
the field represents

Figure C.2 Schema references help when you already have a schema for another topic you need to reuse.

So in this schema, you are using a union for the txn_type field. Compared with the schema directly above this one is much cleaner as you have a small array of types versus showing full schemas.

Now let's walk through another example, and this time we'll have an example of how you can register a schema with references. First here is a readable example of the JSON you'll use to register your schema containing references:

```
{
                                    The portion of the request
  "schema" :                        with the schema containing
      "{                            the references
        \"type\": \"record\",
        \"namespace\": \"bbejeck.chapter_3\"
        \"name\": \"transaction\",

        \"fields\" : [
                {\"name\": \"txn_type\", \"type\": [
              \"bbejeck.chapter_3.Purchase\",
              \"bbejeck.chapter_3.Return\",        Schema
              \"bbejeck.chapter_3.Exchange\"       references by
            ]},                                     class name

                {\"name\": \"identifier\", \"type\": \"long\"}
              ]

      }",
```

```
"schemaType": "AVRO",

"references" : [
      {
          "name": "bbejeck.chapter_3.Purchase",
          "subject": "purchase",
          "version": 1
      },
      {
          "name": "bbejeck.chapter_3.Return",
          "subject": "return",
          "version": 1
      },
      {

          "name": "bbejeck.chapter_3.Exchange",
          "subject": "exchange",
          "version": 1
      }
  ]

}
```

◁—— **The references to the already registered schemas for these event objects**

As you can see from the JSON here, you post the schema along with the references needed. This format is required; otherwise, by just sending a list of references, Schema Registry wouldn't know which schema to associate the references.

> **NOTE** The JSON presented here is formatted for readability. The schema property needs to be a string value without any line breaks when submitting the schema reference JSON.

appendix D
Understanding Kafka
Streams architecture

In this book, you've learned that Kafka Streams is a directed, acyclic graph of processing nodes called a topology. You've seen how to add processing nodes to a topology for processing events in a Kafka Topic. But we still need to discuss how Kafka Streams get events into a topology, how the processing occurs, and how processed events are written back to a Kafka topic. We'll take a deeper look into these questions in this appendix.

D.1 High-level view

Figure D.1 shows a high-level view of what we're going to discuss.

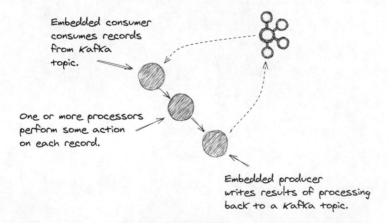

Embedded consumer consumes records from Kafka topic.

One or more processors perform some action on each record.

Embedded producer writes results of processing back to a Kafka topic.

Figure D.1 **Componetized view of a Kafka Streams application. There are three sections: consuming, processing, and producing.**

446

As you can see from the figure, at a high level, we can break up how a Kafka Streams application works into three categories:

- Consuming events from a Kafka topic
- Assigning, distributing, and processing events
- Producing processed events results to a Kafka topic

Given that we've already covered the Kafka clients in a previous chapter and that Kafka Streams is an abstraction over them, we won't get into those details here. Instead, I'll combine consuming and producing into a more general discussion on clients and then go deeper into Kafka Streams architecture for assigning, distributing, and processing events.

D.2 *Consumer and producer clients in Kafka Streams*

Our examples show that a Kafka Streams application creates a KStream instance that starts with a source node and ends with a sink node. Kafka Streams uses embedded KafkaConsumer and KafkaProducer clients to consume events from a source node and produce transformed results back to Kafka.

So, where do the clients come from? Kafka Streams uses an internal class, Default-KafkaClientSupplier, which implements the KafkaClientSupplier interface that provides all the clients Kafka Streams needs.

> **NOTE** The KafkaClientSupplier has methods that provide an admin client, a restore consumer, a global consumer, and a producer client. We've covered the admin client in a previous chapter. The restore consumer and global consumer are plain KafkaConsumer objects with specific roles within Kafka Streams, and we'll discuss their roles later.

Occasionally, a user will ask about providing their own clients. While most of the time it's not necessary, it's possible. For example, you may have some custom observability code you've added by extending the KafkaConsumer or KafkaProducer classes.

The KafkaStreams object has a few constructor overloads; a few accept the Kafka-ClientSupplier interface as a parameter. So, to provide Kafka Streams with your own clients, you would implement an instance of the KafkaClientSupplier and pass it in as one of the constructor parameters like in the following listing (see bbejeck.chapter_6.client_supplier.CustomKafkaStreamsClientSupplier.java).

Listing D.1 Supplying Kafka Streams with your version of clients example

```
KafkaStreams kafkaStreams =
new KafkaStreams(topology,
  properties,
  new CustomKafkaStreamsClientSupplier());
```

The CustomKafkaStreamsClientSupplier in the source doesn't do anything special; it's there simply to show how you would provide your implementations of the client interfaces. When would you provide your client implementations? I've seen users who wanted to use some company-specific auditing logic in the past. But again, in most cases, you'll want to stick with the clients Kafka Streams builds internally.

So, a natural follow-up would be that you don't want to provide any custom clients, but you'd like to supply some client configurations to meet your needs. Usually, the default configurations Kafka Streams uses are sufficient, but there are cases where you need to provide some custom configuration.

It's straightforward to provide your client configurations by adding them to the properties you supply when building the `KafkaStreams` object. Kafka Streams passes the properties you provide to every internal object requiring configuration. Any configurations that do not match the expected ones are ignored.

For example, let's say you have some processors that do some time-intensive work, and there's a reasonable probability that the processing won't be complete in time before the internal consumer needs to make another `poll` call before getting kicked out of the consumer group. So you decide to increase the maximum time the `Kafka-Consumer` can take between `poll` calls. The following listing shows how you would set the configuration.

Listing D.2   Adjusting a configuration for the embedded `KafkaConsumer`

```
props.put(StreamsConfig.APPLICATION_ID_CONFIG, "yelling_app_id");
props.put(StreamsConfig.BOOTSTRAP_SERVERS_CONFIG, "localhost:9092");
props.put(ConsumerConfig.MAX_POLL_INTERVAL_MS_CONFIG, 45_000);
```
Adds a custom value for the max poll interval

So, all you need to do here is set the `max.poll.interval.ms` configuration along with the stream-specific properties. But you should note that when passing in configurations, Kafka Streams applies it in all cases where applicable. From this example, the main consumer (the consumer that gets records for the topology), restore consumer, and global consumer all have increased the time allowed between `poll` calls.

To avoid the issue of property conflicts, Kafka Streams provides prefixes to target the specific client you wish to configure. So, to make sure you only adjust the main consumer, you would update how you set the configuration like in the following listing.

Listing D.3   Adjusting max poll interval for the main consumer only

```
props.put(
StreamsConfig.mainConsumerPrefix(ConsumerConfig.MAX_POLL_INTERVAL_MS_CONFIG)
, 45_000);
```

Only the configuration for the main consumer is affected by using the `StreamsConfig .mainConsumerPrefix` method to set the parameter. The other prefix methods offered are `consumerPrefix`, `producerPrefix`, `topicPrefix` (used for internal topics created by

Kafka Streams), `restoreConsumerPrefix`, `globalConsumerPrefix`, and `adminClientPrefix`. I recommend using the prefix methods when setting any client configurations.

One final note on client configurations within Kafka Streams: you can't update all client configurations with a Kafka Streams application. For example, `auto.commit` is set to `false`, and changing it to `true` is impossible. This is because Kafka Streams may not fully process all records from the previous `poll` call by the time it makes another. So, by not using auto-commit, you are guaranteed to reprocess any partially processed records should a failure occur. For a full explanation of when not to use auto-commit, see section 4.3.5.

We've completed our discussion of using the embedded clients in Kafka Streams, so let's move on to the architecture of Kafka Streams and how it affects the processing of records in the topology.

D.3 Assigning, distributing, and processing events

This section will discuss the structure of a Kafka Streams application and how it organizes and distributes work. In section 6.6.3, we discussed the structure of a Kafka Streams application and how each source node creates a sub-topology (figure D.2).

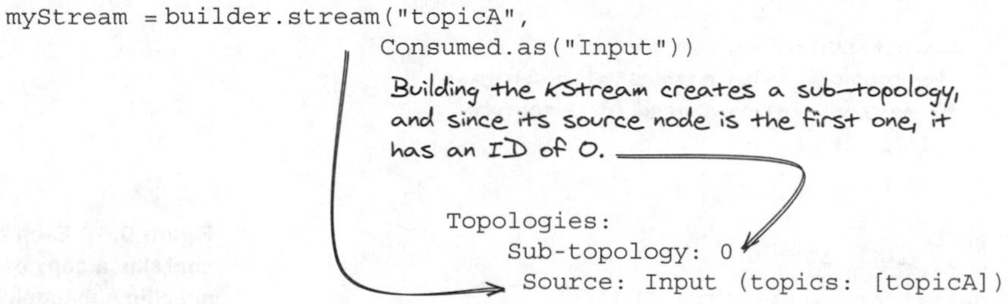

```
myStream = builder.stream("topicA",
                 Consumed.as("Input"))
```

Building the KStream creates a sub-topology, and since its source node is the first one, it has an ID of 0.

```
            Topologies:
               Sub-topology: 0
                  Source: Input (topics: [topicA])
```

Figure D.2 Creating a KStream instance with a source creates a sub-topology

A sub-topology is a discrete section of the overall application with its own source node and processing node(s), which could include a terminal node (most likely a sink for writing results back to a Kafka topic). Several of the diagrams we've looked at in this chapter illustrate what a sub-topology is. But first, we need to discuss how Kafka Streams determines how it will process records.

Kafka Streams uses the concept of a `Task` as its basic unit of work. A `Task` processes records from a specific sub-topology and partition number. Figure D.3 helps describe this assignment mapping.

From looking at figure D.3, each `Task` is tied to a specific sub-topology and a partition number from which it will process records. Note that I'm using the term *partition number* instead of *partition*; we'll discuss what that means later in this section.

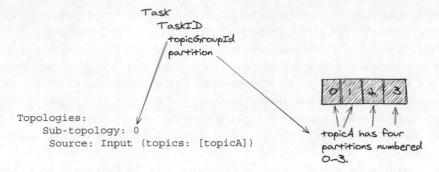

Figure D.3   Task assignment in Kafka Streams

NOTE   I'm not going too much into specific details of classes and field names, but providing more of a conceptual view as the implementation details may change over time.

Additionally, each `Task` has a full copy of the specific sub-topology it's responsible for. Let's look at figure D.4 to demonstrate this point.

Figure D.4   Each task contains a copy of the specific sub-topology it's responsible for processing records.

Since each `Task` has a copy of its assigned sub-topology, we can see that Kafka Streams has a shared-nothing architecture, meaning that each task processes records independently of other tasks. This principle of not sharing and working on its copy of the topology is fundamental when we talked about state in chapter 7.

So far, we've learned that Kafka Streams uses tasks for the unit of work in the application. But what determines how many tasks there are? It's the number of partitions in the underlying topic. Let's start with the simplest case of a single source node with a single topic. Figure D.5 will help make this clear.

From the illustration here, the topic with four partitions results in four tasks. Each task is represented by the format of sub-topology partition number, which is also how Kafka Streams prints a task's ID, which you may see in the logs. So, looking at the first

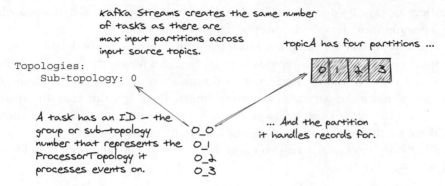

Figure D.5 With a single topic, the number of tasks is equal to the total number of partitions in that topic.

task, `0_0`, it's responsible for processing records on the first sub-topology, 0, from partition 0 of topic A.

But as we've noted before, a source node can subscribe to more than one topic, so let's take a look figure D.6 to see how having multiple topics may or may not affect the total number of tasks.

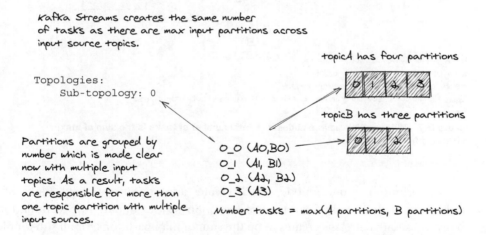

Figure D.6 With multiple topics in a single source node, the total number of tasks is the maximum number of partitions across the topics.

By adding another topic with three partitions, we can see that the total number of tasks remains unchanged at four. However, the number of topic partitions has increased for three tasks. This is because Kafka Streams groups topic partitions for a task by partition number. So, task 0_0 processes records in the first sub-topology and any records from the 0 partition for all topics in that particular source node.

The same is true for the other tasks. Until there are no more partitions to assign, the fourth task only handles records from topicA. This means that when you add more than one topic to a source node, the total number of tasks created is the maximum number of partitions across all the input topics. If we put this into an equation, it would look like tasks = max(topicA partitions, topicB partitions). Since the new topic has three partitions, the total number of tasks remains unchanged.

Finally, let's consider when you add a new KStream instance, which adds its own source (figure D.7).

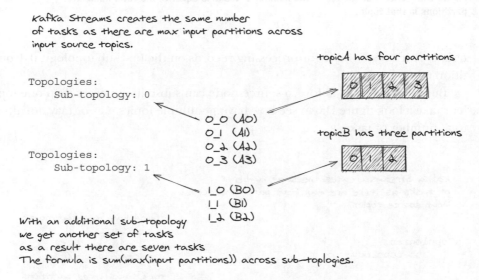

Figure D.7   **With multiple source nodes, the total number of tasks is the sum of max partitions for all source nodes.**

By adding another source node, which creates another sub-topology, Kafka Streams adds another three tasks for a total of seven. So we can update our formula for how many tasks Kafka Streams creates to be the sum of max partitions for all sub-topologies. Understanding the number of tasks created is essential to get maximum performance out of your applications. In the next section, we'll discuss this concept in more detail as we continue exploring Kafka Streams architecture.

NOTE   Each task operates independently and does not share information or state with other tasks.

Now that you've seen the unit of work for the Kafka Streams application, we need to talk about execution.

D.4   Threads in Kafka Streams: StreamThread

Kafka Streams uses `StreamThread` for running tasks. Note that `StreamThread` extends the `java.lang.Thread` class.

The default number of threads in a Kafka Streams application is one. To increase the number of threads, you set the `StreamsConfig.NUM_STREAM_THREADS_CONFIG` configuration to a value greater than 1. The question is, how many threads should you allocate? To answer that question, we need to consider the number of cores on your machine, but for now, let's stick to considerations for Kafka Streams only.

From our discussion of `KafkaConsumer` behavior from the client chapter, the maximum number of active consumer clients corresponds to the number of partitions; any beyond that will be idle. So, the same rules apply here as well, but the maximum number of stream threads to create depends on the number of tasks (figure D.8).

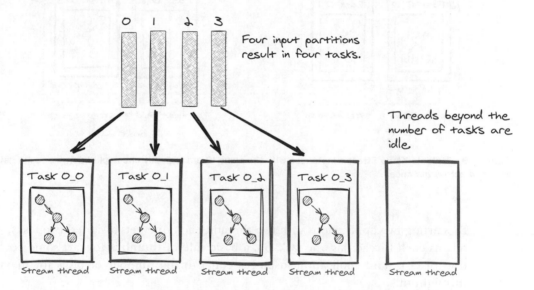

Figure D.8   You can have as many threads as tasks; any additional threads will be idle.

Going with our first task example, you could allocate up to four stream threads. Kafka Streams will evenly distribute the tasks to the running threads, so in our illustration here, adding threads over four is idle as there are no more tasks to assign.

But when determining the correct number of threads to use and throughput considerations, you're not limited to thinking about stream threads alone. Kafka Streams instances with the same `application.id` are logically considered one application. So that means that multiple applications started with the same app ID go through a similar task assignment process. Let's look at figure D.9 to demonstrate.

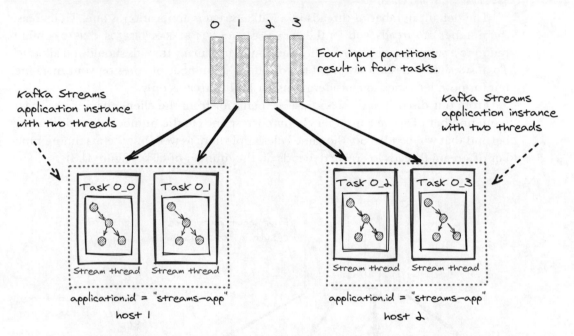

Figure D.9   Multiple Kafka Streams instances with the same ID are logically one application, so tasks get assigned across instances.

By starting two applications with two threads, each one gets assigned two tasks, which means each thread has one task. So, a basic rule of thumb is to start some combination of tasks and application instances equal to the number of tasks for maximum throughput.

This flexibility of application instances and threads leads to a powerful concept of Kafka Streams: dynamic membership. In chapter 4 on clients, we discussed how the consumer rebalance protocol worked. When group membership changes, a rebalance occurs, assigning resources. Since Kafka Streams uses embedded Kafka consumers, it uses the same rebalance functionality. Figure D.10 demonstrates this process.

When a Kafka Streams application comes online, current members give up some tasks assigned to the newly running application. The reverse process is true as well. When an instance goes down, Kafka Streams reassigns the tasks it holds to the other

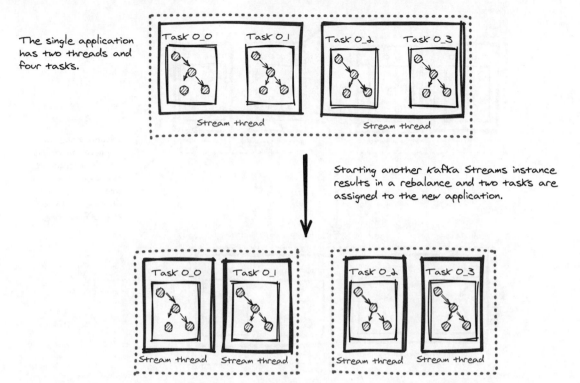

The single application has two threads and four tasks.

Starting another Kafka Streams instance results in a rebalance and two tasks are assigned to the new application.

Figure D.10   Starting a second Kafka Streams instance causes a rebalance and distribution of tasks to the new instance.

active applications (figure D.11). This dynamic task assignment is compelling because it means you can respond to changes in demand for processing by either starting up or shutting down applications on the fly, and Kafka Streams will automatically handle it.

From the illustration here, when demand is low, you can run one Kafka Streams instance with one thread processing all six tasks. But when the number of events increases and you need more processing power, you can start two more additional streams apps, and after rebalancing, the original application gives away four tasks. The newly started applications pick up two tasks each.

So far, we've covered how to maximize throughput by providing a combination of threads in a Kafka Streams instance and the total number of Kafka Streams applications equal to the number of tasks. There's one more situation I'd like to cover that is more nuanced: when you want to maximize throughput, and you have a source node with multiple topics. If you remember, the number of tasks for a source node with multiple topics is the maximum partition count across all the input topics.

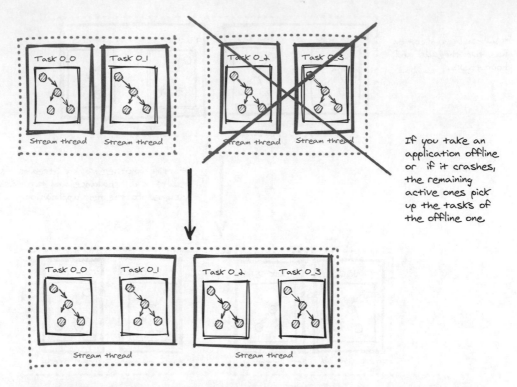

If you take an application offline or if it crashes, the remaining active ones pick up the tasks of the offline one

Figure D.11   Kafka Streams also dynamically handles applications dropping out and reassigns tasks to the active members left.

For example, consider you're building a Kafka Streams application that starts like in the following listing.

Listing D.4   Defining a stream with one source and multiple topics

```
StreamsBuilder builder = new StreamsBuilder();
// topicA has 4 partitions and topicB has 3
KStream<String, String> myStream = builder.stream(List.of("topicA",
  "topicB"));
myStream.filter(..).mapValues(..).to(..);
```

We know from our previous conversation that we'll end up with four tasks even though there are seven input partitions. While getting the same number of stream threads per task is no problem, you can access multiple servers, so spreading the application to a few servers is not an issue.

Ideally, you would like to use a thread per task to maximize your processing throughput. But since you've combined topics, you have three tasks doubled up with two topic-partitions to process. In this case, we can make some minor changes to the application, and you can increase the task count to equal the number of input partitions, giving you maximum processing with a thread per task.

What you're going to do is to create a `KStream` for each input topic. Then, you take each `KStream` object and pass it to the `buildStream` method, where we add the required operations for the topology.

Listing D.5 Creating a KStream per topic to maximize task count

```
StreamsBuilder builder = new StreamsBuilder();
KStream<String, String> streamA = builder.stream("topicA");     Creates a stream
KStream<String, String> streamB = builder.stream("topicB");     per topic

buildStream(KStream<String, String> sourceStream) {          ◁── Method to build
  return sourceStream.filter(..).mapValues(..);                 streaming topology
}

buildStream(streamA).to("output");     Call method for
buildStream(streamB).to("output");     each source stream
```

By making this small change, we've gone from one sub-topology to having two, and the significance of this is we now have seven tasks instead of four, allowing us to run a stream thread per task for maximum throughput.

While it's true that we have some code duplication, we've tried to keep it to a minimum, and you'll need to consider the tradeoff of having maximum throughput for the application. To be clear, I presented this case to demonstrate the relationship between tasks and a sub-topology. It's not meant to represent a best practice; instead, it's something to consider when looking at the performance factors for a Kafka Streams application.

D.5 *Processing records*

So far, we've talked about how a task is the unit of work and how threads are executed in a Kafka Streams application. To complete the discussion, let's go into some details on the Kafka Streams tasks in action. We'll look into the lifecycle of a single `Stream-Thread` in a Kafka Steams application. For our tour of the stream thread lifecycle, we assume you've already built and deployed your Kafka Streams application, and it's up and running successfully.

Starting your application with the `KafkaStreams.start()` method will start all of the `StreamThread` instances you have configured.

> **NOTE** For clarity, we will only discuss one `StreamThread` in a single Kafka Stream application, but there could be several of both in practice.

A `StreamThread` has several states it can transition to, as shown in figure D.12.

What you see here is the full state graph, but for a successful *initial* start, the states will progress in the manner shown in figure D.13.

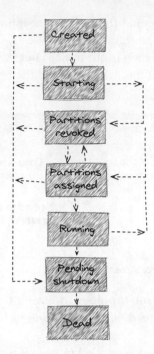

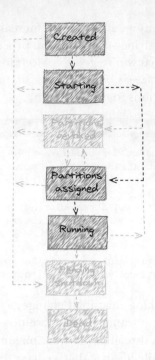

Figure D.12   A `StreamThread` has several states that it can go through processing events.

Figure D.13   The `StreamThread` state progression on an initial, successful startup

From this illustration, you can see the stream thread starts and then goes through a progression of starting, receiving its task assignment, and then finally running, which is the state where it's processing events. We will jump directly into the running state since we discussed task assignments in the previous section. I'm not going to describe every detail of the lifecycle while a thread is running, but it's more of an overview of the main points you'll want to be aware of when running a Kafka Streams application.

For a reference point, we're going to follow along with figure D.14, illustrating what is going on with the stream thread while it's doing its job.

So, the first step is the embedded consumer subscribes to the topics you provided to the `StreamBuilder#stream()` method, whether you provided full names or a regular expression.

After subscribing to topics, the next step is to poll for records for processing through the topology. With records returned from the `poll` call, they need to be distributed to the tasks. The distribution process involves retrieving the appropriate task for a batch of records by topic partition. Once located, the task adds the records to its processing queue. There's one point we should discuss briefly here, and we'll use figure D.15 to help.

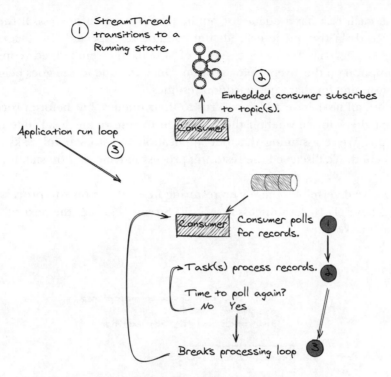

Figure D.14 `StreamThread` lifecycle running in an application

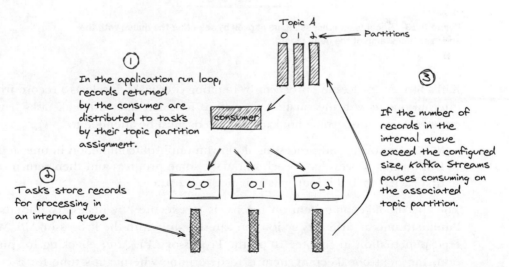

Figure D.15 Pausing a topic partition if the number of records exceeds the configured max buffer size

As we showed, each task has a queue for holding records it will process. If the total number of records in the queue exceeds the configured expected size (`StreamsCo-nifg.BUFFERED_RECORDS_PER_PARTITION_CONFIG`) of 1,000, the embedded consumer pauses consumption on the given topic partition. Once the queue size goes below the threshold, consuming for that topic partition resumes.

At this point, all tasks have added records to their queues, but before processing starts, there's a check to see whether there's a need to restore any local state from a changelog topic. We're assuming that our application has no stateful tasks, so no restoring is needed. We discussed the restoring process in chapter 7 on stateful Kafka Streams.

Now that we've determined there's no restoring needed, it's time to process some records. Using figure D.16, let's discuss how Kafka Streams chooses the next record to process for each task.

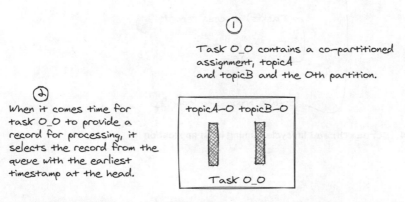

Figure D.16   Records for processing are chosen by selecting the queue with the smallest timestamp of the head record.

Kafka Streams processing proceeds by iterating over tasks. It selects a record from the internal queue with the smallest timestamp for each task. After all tasks have processed some records, some bookkeeping needs must be attended to.

> **NOTE**   Restoring and processing don't run until full completion in one step. During this process loop, each will make some progress and then return so that each step in this loop gets a chance to execute.

The first bookkeeping item on the list is to execute any necessary punctuations. Punctuation is an arbitrary action you can schedule with the Processor API. We covered punctuation in chapter 10 on the Processor API. After checking for punctuation, the next bookkeeping item is to determine whether it's time for a commit. Remember, committing is when a consumer stores the offset (+1) of the latest record successfully processed. Kafka Streams uses an embedded consumer, so it follows the same procedure.

In Kafka Streams, the default commit interval configuration is 30 seconds. The determination for committing is if at least 30 seconds (or whatever the configured commit interval is) has elapsed since the previous one. The time used for this calculation is the system time from the Kafka Streams application environment.

The last step at the bottom of the processing loop potentially does one of two things: either adjusting the number of records each task will process and/or checking whether it's time to break out of the processing loop and return to the consumer and poll for new records.

While it's important to process as many records as possible, ensuring that the processing loop terminates in time to make another poll call is equally important. Otherwise, the Kafka Streams application gets kicked out of the consumer group, causing a rebalance, only to trigger another rebalance when rejoining the group. Let's look at an illustration demonstrating the checks Kafka Streams makes to maximize processing while ensuring liveliness (figure D.17).

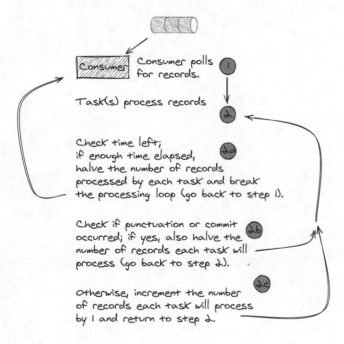

Figure D.17   Kafka Streams does some checks at the bottom of the processing loop.

Kafka Streams checks how much time has elapsed since the last poll, and if the time elapsed is more than half of the configured maximum poll interval, it reduces the number of records each task processes by one half and then breaks the processing loop. If there's enough time left to continue in the processing loop but there have been punctuations or a commit, the number of records processed by each task is halved at this point as well. Otherwise, the maximum number of records each task can process is incremented by 1.

Let's consider one final point before we wrap up this appendix. While Kafka Streams has measures to ensure liveliness, it's still the developer's responsibility to ensure processors are taking only a short time to complete, as the resulting rebalances from missing a poll call are a source of frustration and negatively affect throughput.

index

RELATED MANNING TITLES

Kafka in Action
by Dylan Scott, Viktor Gamov, and Dave Klein
Foreword by Jun Rao

ISBN 9781617295232
272 pages, $44.99
January 2022

Microservices Patterns
by Chris Richardson

ISBN 9781617294549
520 pages, $49.99
October 2018

Grokking Streaming Systems
by Josh Fischer and Ning Wang

ISBN 9781617297304
312 pages, $59.99
March 2022

Spring Microservices in Action,
Second Edition
by John Carnell and Illary Huaylupo Sánchez

ISBN 9781617296956
448 pages, $59.99
May 2021

For ordering information, go to www.manning.com

 MANNING

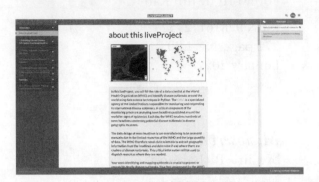

Hands-on projects for learning your way

liveProjects are an exciting way to develop your skills that's just like learning on the job.

In a Manning liveProject, you tackle a real-world IT challenge and work out your own solutions. To make sure you succeed, you'll get 90 days of full and unlimited access to a hand-picked list of Manning book and video resources.

Here's how liveProject works:

- **Achievable milestones.** Each project is broken down into steps and sections so you can keep track of your progress.

- **Collaboration and advice.** Work with other liveProject participants through chat, working groups, and peer project reviews.

- **Compare your results.** See how your work shapes up against an expert implementation by the liveProject's creator.

- **Everything you need to succeed.** Datasets and carefully selected learning resources come bundled with every liveProject.

- **Build your portfolio.** All liveProjects teach skills that are in demand from industry. When you're finished, you'll have the satisfaction that comes with success and a real project to add to your portfolio.

Explore dozens of data, development, and cloud engineering liveProjects at www.manning.com!